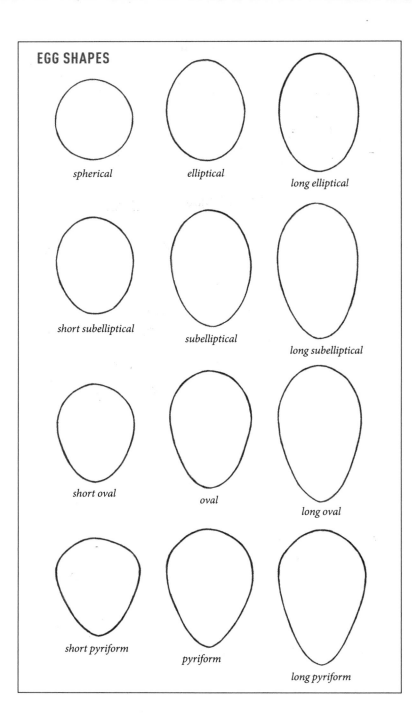

EGG SHAPES

spherical

elliptical

long elliptical

short subelliptical

subelliptical

long subelliptical

short oval

oval

long oval

short pyriform

pyriform

long pyriform

EGG MARKINGS

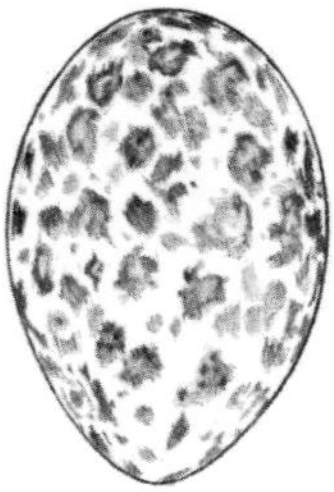

blotched

capped

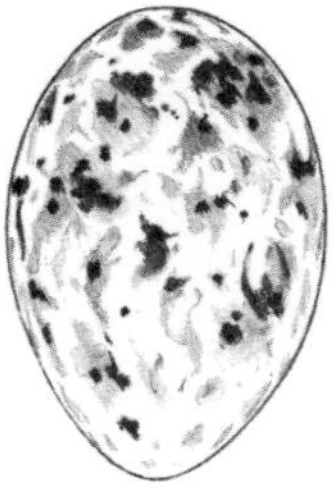

overlaid

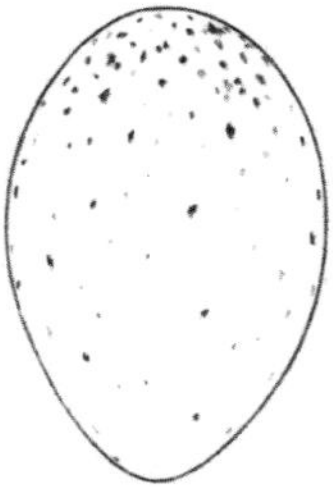

speckled

spotted

streaked

wreathed

marbled

PETERSON FIELD GUIDE TO
NORTH AMERICAN
BIRD NESTS

PETERSON FIELD GUIDE TO

NORTH AMERICAN BIRD NESTS

Casey McFarland, Matthew Monjello, and David Moskowitz

WITH CONTRIBUTIONS BY
Emily Gibson

HOUGHTON MIFFLIN HARCOURT
BOSTON   NEW YORK   2021

For information about permission to reproduce selections from this book,
write to Permissions, Houghton Mifflin Harcourt Publishing Company,
3 Park Avenue, 19th Floor, New York, New York 10016.

www.hmhbooks.com

PETERSON FIELD GUIDES and PETERSON FIELD GUIDE SERIES
are registered trademarks of
Houghton Mifflin Harcourt Publishing Company.

Library of Congress Cataloging-in-Publication Data is available.

ISBN 978-0-544-96338-2

Book design by Eugenie S. Delaney

Printed in China

SCP 10 9 8 7 6 5 4 3 2 1

ROGER TORY PETERSON INSTITUTE
OF NATURAL HISTORY
Jamestown, New York

Continuing the work of Roger Tory Peterson through Art, Education, and Conservation

In 1984, the Roger Tory Peterson Institute of Natural History (RTPI) was founded in Peterson's hometown of Jamestown, New York, as an educational institution charged by Peterson with preserving his lifetime body of work and making it available to the world for educational purposes.

RTPI is the only official institutional steward of Roger Tory Peterson's body of work and his enduring legacy. It is our mission to foster understanding, appreciation, and protection of the natural world. By providing people with opportunities to engage in nature-focused art, education, and conservation projects, we promote the study of natural history and its connections to human health and economic prosperity.

Art—Using Art to Inspire Appreciation of Nature
The RTPI Archives contains the largest collection of Peterson's art in the world—iconic images that continue to inspire an awareness of and appreciation for nature.

Education—Explaining the Importance of Studying Natural History
We need to study, firsthand, the workings of the natural world and its importance to human life. Local surroundings can provide an engaging context for the study of natural history and its relationship to other disciplines such as math, science, and language. Environmental literacy is everybody's responsibility—not just experts and special interests.

Conservation—Sustaining and Restoring the Natural World
RTPI works to inspire people to choose action over inaction, and engages in meaningful conservation research and actions that transcend political and other boundaries. Our goal is to increase awareness and understanding of the natural connections between species, habitats, and people—connections that are critical to effective conservation.

For more information, and to support RTPI, please visit rtpi.org.

CONTENTS

PREFACE

Our copies of Hal H. Harrison's Peterson Field Guides to birds' nests are tattered and soft from heavy use both in the field and at the desk. The first, for eastern species, was published in 1975, just a year or so before we were born. Hal's books were groundbreaking works on the subject and spoke to his skill, passion, and phenomenal perseverance. They were filled with excellent descriptions, detailed notes, and hundreds of beautiful photographs of nests brimming with full clutches. To accomplish what he did just in the field alone is profoundly inspiring, and having embarked on such a journey ourselves, still leaves us awestruck. In our own process of learning, searching, and thrashing about in woods and swamp and desert, Hal was there to guide us. For us, and many others, those first Peterson Field Guides to birds' nests are timeless and indispensable. And yet, when we proposed the idea for a new guide, the first was out of print and plans were underway to pull the second (for western species) from press. It was time to build on Hal's work and offer a fresh look at bird nests and how to identify them in the field.

Much has changed since that first printing in 1975. The human population has nearly doubled, habitat loss continues at breakneck speed, and the global climate crisis gains momentum with each passing year. Yet birds still carry out their annual rituals, many traveling cross-continent (or cross-hemisphere) from one particular kind of habitat to another and stopping along the way in many others. Theirs is an ecological knowledge of Earth not possessed by the majority of *Homo sapiens*, and one can only imagine the changes they witness. At its heart, a study of bird nests is a path that reveals and revels in the astonishing beauty and diversity of Earth's environments and the life forms they support. It simultaneously issues devastating perspective on how much is deeply threatened or already lost, and the impact that has and will have on birds, and ultimately on all of us. In 2016, the North American Bird Conservation Initiative found that 37 percent of all the species found in North America (including Mexico) are "at a risk of extinction without significant action."

A 2019 study published in *Science* estimated that about 3 billion birds in North America have been lost since 1970. Those numbers are broad and complicated, and require careful consideration to make sense of from a conservation standpoint. Regardless, bird populations on average are decreasing, not at all surprising given rapid human-driven landscape change globally, and far too many species are definitively in real peril. Importantly, though, populations of some species (notably waterfowl) have made impressive comebacks with focused management and restoration of healthy habitat, highlighting the fact that with concerted effort humankind can affect real change. We hope that this book offers another means to appreciate, connect with, and help conserve the amazing diversity of bird life with which we still share our planet today.

Learning to identify nests starts first with acknowledgment that, for birds, nesting is a difficult and taxing endeavor that often ends in failure. Our interest in this aspect of avian life can have real consequences for the birds we so admire—many eggs and nestlings never make it through the hurdles of this initial stage of life, even without human disturbance contributing to the odds. Many adult songbirds may face about a 50 percent annual mortality rate, and the stress and constant activity of raising young leave them particularly vulnerable. A practice of observing nests, then, must be done with the utmost care. Learning about nests should, by default, offer tools for better stewardship: increased knowledge facilitates greater awareness of nesting behavior and helps develop an eye for critical nesting habitat.

Well-intended curiosity about a nest can too easily lead to the demise of its contents. Aside from the host of difficulties nesting birds already face, major advancements in online media and cellphone cameras can pose additional challenges by encouraging our human desire to inspect, photograph, and share this exciting but vulnerable part of birds' lives. Ours should be a role to support birds in their nesting activities and help minimize and mitigate the challenges they face. At the simplest level, *this is mostly achieved by steering clear of active nests.* Those who contribute to research and conservation projects through nest monitoring adhere to strict guidelines for nest approach, and our hope is that this book can be used in tandem with these efforts and facilitate similar awareness and caution. If you are interested in contributing information about breeding activity in your area, the Cornell Lab of Ornithology's NestWatch program (nestwatch.org) provides excellent resources for certification to monitor nests and collect data. It also provides tips for recognizing bird behavior that indicates the presence of nests, and what to do if you've discovered one (so as not to draw the attention of watchful corvids, apply scent from hands that may alert keen noses, or create a direct path to vulnerable young, for example). NestWatch also allows you to explore data that have been previously collected, to get a feel for nesting birds in your area or across North America.

What's Inside

This book covers more than 650 species—the vast majority of breeding birds in North America from the United States–Mexico border north into the Arctic Circle—and is the first of its kind to do so while including nearly 750 photographs, a comprehensive key, useful species organization by nest design, and detailed group accounts and species accounts. We've combined the most useful features from many excellent texts on birds' nests, while also providing new tools that help categorize a tremendous array of nests into manageable sections. This book includes introductory chapters that cover nest evolution, breeding behavior and biology, nest design, and other topics, to offer more context about nests in general and perspective on how birds interact with their favored environments.

The purpose of this book is twofold. First, this is a field guide to bird nests, intended to help readers identify a particular nest. Second, it is a broad exploration of bird nests of North America, an ecological study of how hundreds of species have adapted nests and nesting behavior to environs that range from tropical deciduous forest to Arctic tundra. We believe the combination to be crucial. Nest identification is immensely enjoyable, and many nests are readily recognized and unforgettable once you learn them. In some ways, identifying nests is like birding, as there are many birds that are quickly (and satisfyingly) identified in the early stages of learning. But as with every skill, layers abound. Another level of nest identification is more akin to identifying gulls in various stages of development, or warblers by sight as they flit above in canopies. Unoccupied or out-of-season nests pose challenges that include sorting through many species whose nests are alike, and variation among nests of the same species. Identification often requires a naturalist's approach, one that studies not only a nest itself but the landscape that contains it. Knowing just a little about the nest of one species will help with the identification of another, and vice versa. It is a holistic process that builds on itself.

In truth, whole volumes could be dedicated to just one or a handful of species to allow for more thorough descriptions of nests and the behaviors that surround them. It was impossible to provide all the details we would have liked in one book, and some accounts are streamlined to their bare essentials. Where photographs or measurements are not included for a particular species, information can often be inferred from similar species with which it is grouped. This book aims to make the process of narrowing down a family, subfamily, or small group of unrelated species with look-alike nests quick and efficient—especially when identification to a single species is challenging. The information that follows provides the tools necessary to learn about what different birds do, whose nests are alike or dissimilar, useful features to look for, and how to go about learning to identify each species' nests in your home region.

Photographs

We've selected photographs that offer a range of perspectives to give a sense, as best as possible, of what nests actually look like in the field. Nests are not only photographed from above; many are photographed at angles, or directly from the side to better show the body of a nest. Nests are shown with nestlings, with full or partial clutches, or empty just prior to laying. Nests are frequently framed to show materials and design, but many photographs are taken from farther away to demonstrate the context in which the nest is commonly found, such as tucked beneath a shrub or hidden in emergent vegetation.

PETERSON FIELD GUIDE TO
NORTH AMERICAN
BIRD NESTS

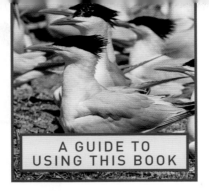

A GUIDE TO USING THIS BOOK

Identifying Birds, Songs, and Calls

There are some instances when the discovery of a nest can announce the presence of a species in a region. Mostly though, identifying a nest *requires birding basics*. The correct identification of bird nests often relies heavily on knowledge of species that occupy an area during the breeding season. There is simply too much overlap in nest design among many species for a nest itself to always belie its owner. A good deal of sleuthing is usually required, and that begins with a short list of potential species.

Learning to Identify Nests

Many who pick up this book likely know a great deal about birds, but learning about their nests may be an entirely new endeavor. With time in any particular region, it is more than possible to learn what to expect from the variety of species that breed there, and nest identification becomes considerably easier with experience. In addition to finding a nest and looking it up, take time to flip through this book leisurely, as you would a field guide to birds of another continent. Explore characteristics of nests in various families and study the photographs so that you begin to train your eye for where to look and what to look for. Make sure to peruse the species that nest in your area, perhaps starting first with those in the neighborhood or down by the local pond, and then moving outward to various habitats across your home state or region.

When you find a nest, be open to a range of possibilities, and look for interesting clues. Older nests built of finer materials lose shape rather quickly as they weather. Nests that have fledged young can be quite battered down, wider cupped, and have a significantly different appearance than when originally built. Some nests are torn apart by birds to construct another nest. Many birds will also start a series of nests and never finish, or build "dummy nests" that may serve other purposes such as mate attraction or, in the case of House Wrens and nest boxes, to discourage other nesters from settling in. See "dummy nests" on p. 22.

Range Maps and Geographic Explanation

"North America" in this work refers to the area north of the United States–Mexico border. Several species that breed in northern Mexico but only rarely in the United States are not included, and the number of nesting species in N. America listed in group accounts is based upon this geographic delineation. Most North American Arctic species are covered, although those that are particularly rare are not. Range maps are located in the species accounts, but note that some species with limited ranges do not have a map, and instead have their geographic location described in their species account.

Order of Species

Many species are organized within their family by what's usually most useful for nest identification: they are grouped by nest design, by placement (e.g., Cavity-nesting Ducks, p. 76), by region, or by some combination thereof (e.g., Wood-Warblers, p. 405). Within these groupings, species are listed in taxonomic order (according to the American Ornithological Society's *Checklist of North American Birds*, 7th edition through the 59th supplement). In other families, species are listed taxonomically throughout when their nests are similar across the entire group, or in large families, with slight variation due to a few species sharing a particular region or nest type.

Accounts

Each substantial family of birds contains two primary sources of information: one or more group accounts that give a broad overview of the species that follow, and individual species accounts.

READ THE GROUP ACCOUNTS FIRST. *When information is presented in the group account, it usually is not repeated in the species accounts that follow.* Group accounts provide general information, including nest structure and placement, breeding systems and behavior, incubation and nestling periods, and egg size and description. The number of species in the family that nests in North America is usually listed as well. Group accounts also serve to reduce redundancies and to capture information in one

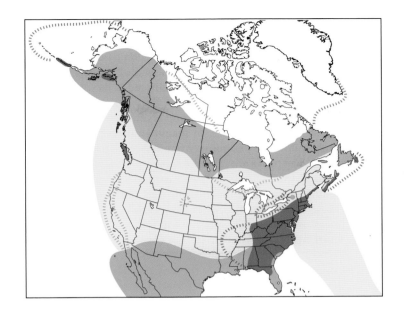

Legend to Range Maps

RED: summer range

BLUE: winter range

PURPLE: year-round range

YELLOW: traditional migration range

RED DASH LINE: approximate limits of summer range and/or postbreeding range

BLUE DASH LINE: approximate limits of irregular winter range

PURPLE DASH LINE: approximate limits of year-round range

YELLOW DASH LINE: approximate limits of migration range

place; reference them when particular details are not listed in the species accounts. Note also that *information has in many cases been generalized to provide an overarching view of what occurs within a family and may not always apply to every member.* When an individual species varies from the typical characteristics of its group, this is noted in its species account.

GROUP ACCOUNTS VARY. Sometimes a group account covers all the species in the family (e.g., Wrens, p. 307). In many families, however, separate group accounts follow the preliminary family account (e.g., Tyrant Fly-catchers, p. 246). Egg plates are sometimes present in group accounts and are intended to give an idea of the size or patterning across robust families with significant variation, or to show eggs that are less commonly seen (e.g., those of cavity nesters, or of hawks and eagles).

Structure of Species Accounts

Species accounts typically contain four categories: Habitat, Location and Structure, Eggs, and Behavior. Not all species accounts have each category heading, however. Where information has been covered in a preceding group account, the heading is often not present, and information within a category varies depending on the text above it.

HABITAT: Describes *typical breeding habitat.*

LOCATION AND STRUCTURE: Describes nest placement, shape, and features, and common materials used by the species. Keep in mind that details for nest height, placement, and other characteristics presented in species accounts often come from a particular region; variation can and does occur. Nest measurements represent a *typical or average size* and serve as a general guide; measurements can vary significantly from nest to nest and by region. (Placement often dictates size as well.) Measurements are not included for some species; infer approximate sizes from the nest descriptions of similar species. Most measurements are presented as follows: outside diameter (outer width of nest); height (outer height of nest); inside diameter (for distinct cups); depth (depth of cup). Some nest types have different shapes, and measurements may vary accordingly. (See Nest Types, p. 49.)

EGGS: The *typical clutch size* (often a typical range) is listed, not the full range that is possible. The numbers indicate how many eggs are generally laid in a single nest, and are not for identification purposes. Egg measurements are averaged for each species, given in millimeters and denoted as length x width. Egg shape and markings are usually described as well. "IP" indicates incubation period, the number of days that the eggs are incubated after a full clutch is laid (not age of clutch from first egg). "NP" indicates nestling period, the time that young birds remain in the nest before fledging. Both categories are usually followed by a notation about which sex incubates or feeds young: "(F, M)" indicates either that the female puts in more effort or that both sexes share the tasks fairly equally, while "(M, F)" indicates that the male is primarily responsible. "H" indicates helpers in colonially nesting species. In many cases, *IP or NP is not included in the species account* and instead is presented in the group account above. Sometimes NP or IP is absent because data are lacking for certain species.

BEHAVIOR: This information varies by species and typically includes breeding behavior and systems and which sex builds the nest. Occasionally there are also notes unrelated to nesting behavior, such as conservation status.

DIFFERENTIATING SPECIES: This appears when multiple species have been combined into a single account (and also in some group accounts), and offers tips to differentiate their nests by design, habitat, or other features.

Nest Key

The following nest categories (adapted from Baicich & Harrison 2005) will help narrow down nests by size, design, features, placement, or some combination thereof. Because there is so much variation across species, and even among the nests of a single species, this key is designed to facilitate a quick assessment and provide a short list of species to look into. Some species are found in multiple categories.

Pensile or Suspended Nests (hanging from nest rim, or rim and sides, often in horizontal fork or at end of drooping branch)
Some flycatchers (Acadian Flycatcher); Rose-throated Becard (hanging globular nest); Bushtit; kinglets (beneath conifer branches); vireos; orioles; some warblers (parulas, in clumps of hanging moss).

Globular or Domed/Roofed Nests (constructed overhead canopy)
Globular Nests: Some flycatchers; magpies; Verdin; some wrens (Marsh, Sedge, Pacific, Winter, Cactus Wrens); American Dipper (on

ledge or crevice near fast-moving water); House Sparrow; some warblers (Lucy's, in Verdin nest; parulas, in clumps of hanging moss).

Domed or Roofed Nests: Some rails; magpies; some wrens; Arctic Warbler (on ground); Sprague's Pipit (on ground); some sparrows (including Olive Sparrow, in shrub or tangle; Grasshopper, Vesper, Bachman's Sparrows, on ground); meadowlarks (on ground); Ovenbird (on ground).

Outer Layer Decorated with Lichen Flakes

Most hummingbirds; Eastern and Greater Wood-Pewees; Olive Warbler; most gnatcatchers; some vireos (pensile nests, often with spider egg sacs: Yellow-throated, Plumbeous, Cassin's Vireos); Bushtit (suspended and socklike).

Very Small Cup Nests (inside diameter ~2–5 cm), Usually Made of Fine Fibrous or Cottony Materials and Large Quantities of Spider or Caterpillar Silk in Shrubs and Trees

Hummingbirds; Least Flycatcher; gnatcatchers; Cerulean Warbler; Olive Warbler (larger outside diameter).

Small to Medium-Small Cup Nests on Ground

Yellow-bellied Flycatcher; longspurs; Horned Lark; wagtails and pipits; some sparrows (see Sparrow Chart, pp. 386–87) and towhees; juncos; many warblers; some icterids (meadowlarks, Bobolink, Brewer's Blackbird); some thrushes; Dickcissel; Common Ground Dove, occasionally Mourning Dove (but flattish, saucer-shaped).

Passerine Cup Nests Built in Emergent Vegetation or Elevated in Wetland or Marsh Grasses

Some sparrows (Seaside, Nelson's, Saltmarsh, Swamp Sparrows); some wrens (Marsh and Sedge Wrens); some icterids (Red-winged, Yellow-headed, Tricolored Blackbirds, grackles).

Small to Medium-Small Passerine Cup Nests in Shrubs, Trees, and Tangles, Built with Fine Materials (grasses, forb stems, plant or bark fibers or strips, fine twigs, rootlets, mosses, skeletonized leaves, or similar); Silk Possible

Many *Empidonax* flycatchers; some sparrows; Cedar Waxwing; many warblers; Morelet's Seedeater; House, Purple, Cassin's Finches; Goldfinches; buntings and Blue Grosbeak; Pyrrhuloxia; Wrentit; Phainopepla; Yellow-breasted Chat; some tanagers.

Small to Medium Passerine Cup Nests in Shrubs, Trees, and Tangles, Built with Coarser Materials (twigs, pine needles, rootlets, dry leaves, or similar); Mud or Leaf Mold Possible

Solitary Sandpiper (in nest of other bird, e.g., Rusty Blackbird, jays); most thrushes; waxwings; some towhees; some finches (Pine and Evening Grosbeaks, Common and Hoary Redpolls, Red and White Crossbills, Pine Siskin); Swainson's, Blackburnian, Yellow-rumped Warblers; some icterids (Brewer's and Rusty Blackbirds, grackles); some tanagers; Rose-breasted and Black-headed Grosbeaks; Northern Cardinal.

Commonly with Large Bulky Outer Structure or Base of Robust Twigs: Most kingbirds; mimids; shrikes; some corvids (jays, Clark's Nutcracker); Rusty and Brewer's Blackbirds possible.

Cup Nests Made Mostly of Mud, or Plant Material Glued Together with Saliva, Stuck on Ledge or Side of Building or Rock Faces or Cliffs

Swifts; Black and Eastern Phoebes (occasionally Say's Phoebe); Cliff, Cave, Barn Swallows.

Cup Nest Set on Rafters, Eaves, Ledges of Buildings, or Bridges

Rock Pigeon, Eurasian Collared-Dove, Spotted Dove; phoebes; Pacific-slope and Cordilleran Flycatchers; ravens (large nests); occasionally Barn Swallow; Canyon Wren, occasionally Rock Wren; Townsend's Solitaire; American Robin; juncos; rosy-finches.

Small to Medium Saucerlike Cup Nests or Platforms in Trees, Shrubs, and Tangles, Built of Twigs and Other Coarse Materials

Plain Chachalaca; most doves; cuckoos; small herons and egrets; Bonaparte's Gull; Brown Noddy; Olive-sided Flycatcher and tanagers (*inner cup* can be saucerlike but is built of finer materials).

Large Stick Nests in Shrubs and Trees, or Similar

Geese (on nests of other birds); Magnificent Frigatebird; herons and egrets; Limpkin; Bonaparte's and Mew Gulls; Brown Noddy; Wood Stork; cormorants; Anhinga; pelicans; ibises; Roseate Spoonbill; larger owls (on nests of other birds); Osprey; kites, eagles, accipiters, and hawks; Crested Caracara; some falcons (on nests of other birds); crows and ravens.

Large Nests on Rock or Cliff Ledges, Sea Cliffs, Rocky Islands

Northern Gannet; cormorants; some herons and egrets; Osprey; some hawks; eagles; Crested Caracara; some falcons (on nests of other birds); many gulls; kittiwakes; Brown Noddy; crows and ravens.

Nests in Earthen Burrows (excavated)

Some Storm-petrels; Barn, Great-horned, Spotted, Burrowing Owls; kingfishers; some swallows; Razorbill, guillemots, some murrelets and auklets, puffins; Manx Shearwater.

Nests in Cavities or Crevices in Ground, Rocks, Cliffs, or Buildings

Passerines: Cordilleran and Pacific-slope Flycatchers; some swallows; some wagtails; Townsend's Solitaire; rosy-finches; American Pipit; Northern Wheatear; Snow and McKay's Buntings; some sparrows (buildings); Canada Warbler; waterthrushes.

Non-Passerines: Swifts; occasionally Red-breasted Merganser; Dovekie; Razorbill, guillemots, some murrelets and auklets; some storm-petrels; Manx Shearwater; California Condor and vultures.

Nests in Tree Cavities (in woodpecker hole or natural tree cavity)

Woodpecker Hole (in tree or standing cactus): Some ducks; woodpeckers; some owls; Elegant Trogon; some falcons; some flycatchers; European Starling; House Sparrow; some warblers (Prothonotary, Lucy's); some wrens; some swallows; some thrushes (bluebirds); chickadees; titmice; nuthatches.

Natural Tree Cavity: Some ducks; some owls; Elegant Trogon; some flycatchers; some swallows; some swifts; European Starling; House and Eurasian Tree Sparrows; chickadees; titmice; nuthatches; Brown Creeper; some thrushes (bluebirds); some warblers (Prothonotary, Lucy's); some wrens; some icterids (grackles).

Small to Large Saucerlike Platform Nests with Shallow Cup, in Water, Built of Coarse Vegetation
(often a mound of reeds, cattail, rotted vegetation, or similar)

Some ducks, geese, and swans (down usually present); rails (often elevated in emergent vegetation); Limpkin; coots and gallinules; bitterns (often elevated in emergent vegetation); grebes (nest and eggs often wet); cranes; Little and Franklin's Gulls; Black and Forster's Terns; loons; sometimes ibises.

Mounded or Platformlike Nests on Ground with Shallow Cup Next to or Near Water, Built with Coarse Vegetation, Sticks and Twigs, Debris; Mud or Rotted Vegetation Possible

Some ducks, geese, and swans (usually down lining on an older mound); cranes; some gulls (Glaucous, Laughing, Mew, Herring Gulls); loons; cormorants; pelicans; Northern Harrier.

Eggs on Bare Ground (often in clear scrape) or with Little Nest Material, on Cliff or Sea Cliff Ledges, Rocky Islands
Some geese (e.g., on Aleutian Is.); Razorbill, murres, guillemots; Northern Fulmar; California Condor and vultures (in cavities, pot-holes); some falcons (on nests of other birds).

Eggs on Bare Ground (often in clear scrape) or with Little Nest Material, Next to or Near Water
Some ducks, geese (usually in down-lined bowl); loons; pelicans; oystercatchers; Black-necked Stilt; American Avocet; some plovers; sandpipers; Wilson's Snipe; phalaropes; jaegers; many gulls; most terns; Black Skimmer.

Eggs on Bare Ground (often in clear scrape) or with Little Nest Material, in Tundra, Woodlands, or Open Areas
Jaegers; ptarmigans, Chukar; most plovers; some sandpipers; some murrelets; some owls; some nightjars.

Eggs on Bare Ground (often in clear scrape) or with Little Nest Material, Beneath Grasses or other Vegetation, Bushes, or Trees
Some ducks (though bowl often lined with down); most gamebirds; some plovers; most sandpipers and phalaropes; Marbled Murrelet (northern range); Short-eared and Long-eared Owls; some nightjars.

No Nest, in Divot in Moss on Branch of Mature Conifer in Old-Growth Forest
Marbled Murrelet.

CHAPTER 1

NESTS, BREEDING BEHAVIOR, AND EVOLUTION

The diversity of bird nests is staggering. In every landscape imaginable, from frozen, windswept tundra to mist-shrouded rainforests, there are nests tucked low among the stones or herbaceous plants or placed high in the canopies of old-growth forests. They are hidden in rocky crevices and holes in living trees. They are fastened to branches, cliffs, and buildings, and placed deep in earthen burrows. They range in design from simple scrapes on bare ground to decorative, stretchy woven pouches that sway gently at the tips of twigs. In each of these designs lies a clue as to how a particular species solved an ecological challenge and created a tiny, climate-controlled world essential to producing and protecting offspring.

What can nests tell us about birds and their evolution? How did such a striking architectural variety come to be, and how is that diversity significant to the world of birds as we know it? Beyond the identity of the builder, bird nests offer a window into the ecology and evolution of birds.

Origins of Bird Nests

The evolution of bird nests remains a complex puzzle. Birds evolved from theropod dinosaurs and from fossil records we know that dinosaurs laid eggs and that at least some of them constructed nests. It is likely that early dinosaur nests were similar to the nests of two modern-day reptiles: those of sea turtles, which contain round eggs buried in sandy pits; and those of some crocodilians, which contain elongated eggs buried in mounded-up material.

A third design offers even more intriguing clues to the origins of modern avian nest construction. Fossilized scrape nests (shallow bowls dug

The fossilized nest of an oviraptor.
Mark Kang O'Higgins

into the ground) of an oviraptor, a fierce-looking, relatively birdlike theropod, were found with the builder atop its clutch in what appears to be a perfect brooding position. Whether the oviraptor was shielding the eggs or brooding them at the moment of death remains unclear. There are other recognizable characteristics of modern birds in these fossilized remains. Oviraptor eggs were found carefully oriented in the nest, suggesting they were turned and pushed about to ensure proper development, a behavior still common among living bird species. Embryos inside the fossilized eggs of a few dinosaur species—and certain dinosaur hatchlings in the fossil record—are also strikingly similar to those of typical, developing songbirds, and thus may have required similar care in order to survive.

All of this suggests that much of the avian nesting behavior we are familiar with today predates early bird life and reaches all the way back to the time of early dinosaurs. But how did nests change from a simple mound of earth made by a non-avian reptile 80 million years ago into the complex architecture of a pendulum nest of an oropendola or Bushtit today?

There are a handful of bird species that, like living species of non-avian reptiles, bury their eggs. They have thus clung to some of the most archaic (but clearly also effective) nesting strategies. Megapodes, taxonomically ancient, turkey-like birds of Australasia, employ the original mounding strategy of their terrestrial dinosaur ancestors. Some species create heaps of vegetation whose decomposition generates heat to incubate eggs, while others use geothermal features to find just the right spot for nests. Males are exclusively responsible for building and caring for the nest and continually manipulate its temperature by adding or removing material as needed. The eggs are massive compared with the size of the female, among the largest of any bird species, and the young that hatch exceed all other species in their ability to care for themselves almost immediately.

Megapodes represent the remnant of a nesting strategy that may once have been dominant in early avian life. Digging and burying eggs in mounded material—a non-avian reptilian habit that we still see in Megapodes—likely evolved into the making of the simple scrapes on the ground that left the eggs more exposed and required physical contact with an adult body; such scrapes made by early avian dinosaurs are still exhibited by numerous bird species, including grouse, quail, and terns. The strategy of depositing eggs in scrapes or natural cavities may have spurred the refinement of cavity renovations as well as the collection of crude nesting materials, which may have inspired intentional cavity excavation and constructed platforms. Platforms—structures built to provide sturdy surfaces in locations that otherwise lack a repository for eggs—could have lent a framework for what would one day become the cup nest, a simple but effective cradle that could be rapidly constructed and fitted into new environments. Today, birds use all of these designs, with each species adding its own unique adaptations. By and large, though, the most common nest type is that of the open cup, a design whose functionality opened up a world of possibility to birds. The most complex and diverse nests are those built by **passerines** (order Passeriformes), commonly known as perching birds or songbirds.

All forms of modern birds are likely to have had their origins in the late Cretaceous—about 95 million years ago—but a rapid proliferation in the diversification of bird species is noted primarily after the Cretaceous–Paleogene extinction event (about 66 million years ago), when the planet was pummeled by a massive asteroid and experienced extreme volcanism that wiped out the last of the non-avian dinosaurs and destroyed as much as three-quarters of all life on Earth. The development of the modern passerine is currently dated back roughly 50 million years, though the vast majority of passerine species living today arose within the last 16 million years.

Today passerines are the largest order of birds, comprising three-fifths of all living bird species. Over 5,000 species strong, they are widely considered to be the most diverse group of terrestrial vertebrates. The dramatic diversification of passerines and other modern birds coincided with, and was probably enhanced by, extraordinary global climatic changes. Wide swaths of forest gave way to grasslands, open woodlands, and deserts. Glacial and interglacial periods formed ecological barriers, isolating bird populations and forcing rapid speciation. Seasonal extremes increased across the planet as well, issuing new demands for novel adaptations, and the number and diversity of flowering plants and insects exploded. Passerines adapted exquisitely to it all.

Why did passerines do so well? It's suggested that among their advantages were their small bodies and large brains, diet of abundant insects, and rapid reproduction rates. But another factor that's widely acknowledged to account for their impressive success is their hard-won, tried-and-tested nest-building finesse. They developed the ability to construct a home for young in nearly any environment, no matter how challenging, allowing these small, adventurous creatures to choose sites at their own discretion, untethered to the otherwise limited offerings of naturally suitable spots for laying eggs. Instead they could venture into difficult territory and craft insulated microclimates to mitigate the ever-changing forces the world presents. Their clever innovations keep their young warm, dry, well hidden, and secure.

The Nest and Bird Biology

Climate, Predation, and Sexual Selection

In cold environments birds may choose protected nest sites under over-hanging rocks or branches, or in caves or cavities, or they may shelter their eggs beneath vegetation to minimize heat loss. Ducks pluck insula-tive down from their breasts to lay atop their eggs. For protection from the elements, many birds build their nests on the leeward side of trees and shrubs that naturally block the winds and rains of prevailing storms; meadowlarks have been observed to orient the entrances of their domed nests away from cooling breezes. In hot climates, species such as Verdins and Cactus Wrens build enclosed structures to provide protection from the sun. Both species will roost in nests year-round and may build mul-tiple nests with entrances either oriented away from wind during cooler parts of the season, or toward prevailing winds during the heat of sum-mer. Other birds place nests under shady vegetation or use their bodies to shield eggs and nestlings from the sun. In regions where little insulation is needed, some species spare themselves the trouble of building a nest altogether.

In addition to creating climate-controlled environs for eggs and nest-lings, parent birds often take great care to minimize predation of their young, and mitigating these risks is a strong driver of nesting strategies. The choice of nesting sites is often closely correlated with the types of predation threats they face, and locations must be assessed, monitored, and defended. Responses to predators may be immediate and fluctuate each season or may be ingrained patterns of behavior developed over millennia. Veeries respond to high levels of predation by moving their ground nest sites to areas with fewer egg-eating mice, while many sea-birds construct nests in sites that are relatively free of predation, such as offshore islands with no terrestrial predators, or on cliff faces inaccessible to most predators. Songbirds such as Ruby-crowned Kinglets hide their globular nests high in mature trees, and White-throated Swifts nest in

A Verdin nest's dense construction and feather lining insulate it against the summer sun.
David Moskowitz, CA

crevices in vertical rock walls. Other birds show flexibility in their choice of nest site that is carefully calibrated to a particular location. Some multiple-brood open-cup nesters, such as Song Sparrows, alter their nest's height and location for progressive broods throughout the breeding season in response to shifting changes in breeding habitat. In the spring, before ample leaf cover is available, nests are often found on or near the ground concealed in vegetation. Once leaves have emerged on shrubs, or trees and grasses have grown taller, subsequent nests are positioned higher above the ground; the new plant growth not only provides better nest concealment from above, but also minimizes access to terrestrial animals.

Other species will seek the "protection" of other animals, birds or otherwise, benefitting from some additional neighborhood muscle. Using the defensive aggression of another species increases the odds of reproductive success, and such associations can reduce nest predation as well as brood parasitism. Bullock's Orioles, for example, may build their nests near those of the predator-mobbing Yellow-billed Magpie. Many species may occupy nest sites near wasp or ant nests, limiting the approach of terrestrial reptiles and small mammals, and species comfortable among human habitation (e.g., Barn Swallows) effectively minimize the risks posed by wild predators. Similarly, some passerines such as House Sparrows will nest within the large nests of predatory birds such as eagles.

Beyond incubating eggs and rearing young, many species build nests solely to attract mates. Quintessential to this behavior are the bowerbirds of New Guinea and Australia. The male creates an elaborate nest site, meticulously crafted and decorated with colorful items, which the female carefully inspects to see if it meets her particular tastes. Male Marsh Wrens and Pacific Wrens build surplus "dummy nests"—as many as 12 in a breeding season—most of which aren't actually used for rearing young. Instead their purpose appears to be, at least in part, to attract females, which inspect the nests with the male in tow. Similarly, nests that are bi-parentally built may offer birds an opportunity to assess the level of their counterpart's building skills and help them determine the value of their reproductive investment in a particular individual. A nest built solely by a female may also signal her maternal qualities to the male. It appears that nests of many bird species likely serve as "extended phenotypic" signals; similar to the vibrantly colored plumage of many male birds, the design and quality of a nest itself may communicate particular traits about its builder.

Energetic Costs of Nest Construction

Building a nest is energetically expensive. Many birds will make well over 1,000 trips to gather the nesting material needed to complete a single nest. In roughly 13 days, a pair of Cliff Swallows may make more than 1,500 trips to build their nest, traveling a total of some 400 miles on collecting forays—the equivalent of flying across the entire state of Oregon.

One of the most important factors dictating how much energy birds put into nest construction is whether the nest will be used simply for incubation or whether it will serve as the home for nestlings for days or weeks after they hatch. Young birds that are able to leave the nest within minutes or hours after hatching are **precocial**. The chicks of sandpipers and plovers, for example, can follow their parents and feed on their own almost immediately after hatching, and the minimal scrape typical for these types of birds is not used at all for raising young. Passerine and some non-passerine families produce **altricial** young that are blind and often featherless when they emerge from the egg. For these species, the nest continues to be a vital part of rearing nestlings, so a great deal more energy may be put into its construction. There are varying degrees of dependence or independence among young in either category: some species are **semiprecocial**, for instance, and require more care after hatching than precocial species.

Research suggests that birds with access to plentiful food, or those in better body condition, invest more energy in construction and produce larger, better-built nests. Healthy males of species that build surplus nests

The altricial young of an Eastern Towhee, hatched blind and featherless.
Justin Sweitzer, PA

The precocial young of a Long-billed Curlew, just after hatching. John Pulliam, MT

often build higher numbers of them, while males of the same species affected by food scarcity or parasites build more conservatively.

However, spending too much time and energy on nest building may also negatively affect egg production. Birds show a wide range of behaviors that demonstrate their interest in cutting corners where possible.

Precocial Mallard chicks are able to follow their mother into the water shortly after hatching. David Moskowitz, WA

Some birds build new nests, while others of the same species simply refurbish an old one. They may apply techniques that simplify the building process but that at the same time increase the quality of their nest. Eastern Phoebes and Barn Swallows are reported to attach their nests to existing mud dauber wasp nests, increasing the ease with which the nest is built and its likelihood of staying secured in place. Stealing nesting material is also a common practice, particularly in colonial nesters such as Great-tailed Grackles. Solitary nesters, such as some flycatchers and hummingbirds, have also been reported to pilfer from their neighbors. Other species can be distinctly more aggressive: House Sparrows, for example, can take over active nests of Eastern Bluebirds and Cliff Swallows, sometimes tossing out the eggs or killing chicks or an incubating parent in the process.

What Starts Nest Building?

Birds must reproduce within windows of opportunity that provide their offspring with both favorable climate conditions and a plentiful food supply. The start of breeding behavior in birds is precipitated by a mix of internal physiological rhythms and changes in environmental conditions. The most influential trigger is the lengthening of daylight hours as winter wanes: the increase in the photoperiod, or amount of daily sunlight, stimulates hormonal changes in birds. In migratory species, this drives the impulse to begin their journeys toward summer habitats that provide the resources needed to breed, nest, and rear young.

Generally, species with large geographic ranges show significant differences in nesting timing. Members of a species breeding in warmer, southerly portions of the range might begin a month or more before their northern counterparts. Great Horned Owls begin breeding from as early as November to January in the South, versus early April farther north. Aside from differences in temperature and daylight from north to south, timing variation allows hatching to coincide with spikes in rodent populations in the spring.

Other factors specific to a particular ecosystem or species also influence the start of reproductive activities and nest building. For many birds that feed on insects and nest in temperate regions of North America, spring breeding coincides not only with increasing temperatures and daylight, but also with spikes in insect abundance. Some seedeaters, such as American Goldfinchs, which rely primarily on seeds of mature thistles to feed their young, delay nest building and reproduction until seeds become available. Crossbills, largely dependent on conifer seeds, may breed at any time of year: their reproduction is dictated by food supply rather than by photoperiod, and mirrors the fluctuating occurrence of their most important food source.

Nests and nest building may also serve to stimulate breeding and ovulation in some birds. Studies providing captive doves with nesting

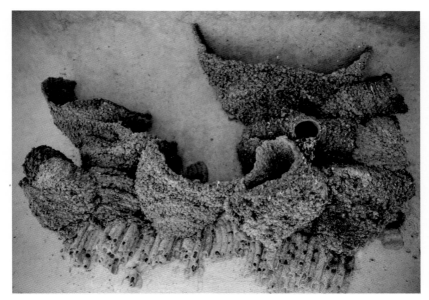

Cave Swallow nests anchored atop the nests of mud dauber wasps. Jason Kleinert, TX

materials induced breeding behavior, and in other experiments, prohibiting birds from building nests delayed egg laying.

Avian Mating Systems and Behavior

Some birds, such as the Common Loon and Bald Eagle, exhibit **monogamy** (or **genetic monogamy**), meaning a particular male and female breed only with each other. Partnerships, or pair bonds, may change each season, or last for many years or for the entire lifespan of one or both partners. For these species, nest construction and young rearing are typically a shared responsibility. True monogamy was once considered widespread among birds, but despite the appearance of mate fidelity among many species, actual mating behavior has proven complex. Far more common is **social monogamy**, seen in species such as tanagers, in which a particular male and female form a pair bond but one or both partners may engage in extra-pair copulations. This frequently results in a single clutch of eggs with multiple paternities.

Polygyny refers to a pattern in which males partner with more than one female at the same time. While relatively rare overall, it is most common in wetland- and grassland-nesting birds, including taxonomically unrelated species such as Red-winged Blackbirds, Marsh Wrens, many species of grouse, and a variety of shorebirds. Polygynous males often provide little to no assistance with nest building and rearing young. Several polygynous species are highly sexually dimorphic. For instance, female Boat-tailed Grackles are only half the size of males.

Less common than polygyny is **polyandry**, in which a female mates with multiple males. In North America this behavior is limited primarily to shorebirds, the Spotted Sandpiper being a prime example. A female lays eggs in separate nests, and each nest is tended by its respective male. Rarer still is **polygynandry**, in which both males and females form multiple mating pair relationships. This behavior is reported in just a handful of species, among them the Acorn Woodpecker, in which family groups work together throughout the year to stock, maintain, and defend a granary of acorns. Females often lay eggs in the same nest, and multiple males may mate with various females. **Cooperative breeding**, found in only about 3 percent of bird species (including some jays, some nuthatches, Acorn Woodpeckers, and Bushtits, is a system in which a breeding pair (or socially bonded pair) is supported by **helpers** —other adults or juveniles—to care for the young of a nest. **Promiscuity** refers to a breeding behavior in which pair bonds are not established between breeding males and females. This appears to be the case with many species of hummingbirds.

The role that each sex plays in nest building varies by family and species. In sexually dimorphic species in which the male is brightly colored and the female drab, the female is often the sole builder. When the color and plumage pattern is similar in both sexes, as in crows and loons, males and females often share the responsibility of nest building (as well as incubation and care of young). Certain tasks, however, such as collecting materials or working only on particular portions of the nest, may be divided.

In the case of Wilson's Phalarope, a polyandrous species, the responsibility of nest construction falls to the male. In many polygynous species, such

A singing Spotted Towhee.
David Moskowitz, WA

as grouse and grackles, the burden of nest building and caring for the young typically falls to the female. In promiscuous hummingbirds, it is females that build the nest, incubate, and rear young on their own.

Territory

A **territory** is typically defined as any area actively defended by a bird against conspecifics—that is, other individuals of the same species (though aggression toward other species is certainly possible). By defending a territory, birds effectively decrease competition that otherwise would interfere with their own breeding success. Breeding territories can be categorized into five types.

1. ALL-PURPOSE TERRITORIES, common to many North American songbirds, are areas that support all activities during the breeding season, such as mating, nesting, and feeding.

2. MATING AND NESTING TERRITORIES do not necessarily fulfill feeding requirements, and both sexes may pursue other foraging (and breeding) opportunities outside their established territory. This type of territory is typical among Red-winged Blackbirds.

3. GROUP TERRITORIES are areas in which two or more females may lay eggs in the same nest, and nonbreeding helpers, typically juveniles or mature birds that are the offspring of the nesting parents, assist with nest construction and parental duties. Some species may still aggressively defend territories from other individuals outside their extended family group. Perhaps the most elaborate example of this behavior in North America is that of the Acorn Woodpecker, previously mentioned. Bushtits can also be cooperative nesters, with multiple individuals assisting in building, incubation, and feeding nestlings.

4. COLONIAL NESTING TERRITORIES are areas where species nest close together and generally defend only the immediate area surrounding their nest. Roughly 13 percent of the world's bird species are colonial breeders. This strategy may have evolved in response to limited nesting sites and fluctuations in the abundance or location of food resources. The size of a colony can vary from dozens to hundreds of breeding pairs, to tens of thousands or more. While there are negative costs associated with nesting colonially, the benefits of this strategy generally offset the drawbacks. Large nesting colonies, for example, may increase the likelihood of attracting predators, but decrease the chance that any one individual or nest within the group will be preyed on. Additionally, the breeding behavior of many colonial species is synchronized, thus producing high numbers of eggs and offspring at the same time, which restricts predation to a shorter time span. Many gulls, terns, and other pelagic birds also occupy nesting sites that are difficult to access. Cliff Swallows build nests on sheer and overhanging rock faces, and grebes make floating platforms surrounded by water.

5. PAIRING AND MATING TERRITORIES, OR LEKS, are areas where several males gather solely to engage in competitive displays to attract the attention of females. These rituals are demonstrated by some of North America's polygynous species, such as Greater Sage-Grouse.

Birds defend territories by engaging in various behavioral displays. Singing is one of the most common means by which males establish boundaries and advertise for mates. Other behaviors include drumming (woodpeckers, grouse), visual displays and posturing, building of multiple nests, chasing, and in some instances physically attacking nearby competitors and destroying their eggs.

Generally, the quality of a territory is directly related to the reproductive success of the individual or pair that defends it. Birds that secure high-quality territories may attract a mate more quickly, or more mates, than those occupying poorer habitat. When a species' population density is low within rich habitat, all individuals may breed. When density is high, however, some individuals must settle for lower-quality areas still suitable for breeding. In many instances, high population levels fill both high- and low-quality habitats, and some birds become **floaters**, forced into habitats unsatisfactory for breeding. These individuals commonly move inconspicuously among territories, vying for an opportunity to mate or lay claim to a site should a vacancy occur.

Egg Laying, Incubation, and Nestlings

Eggs

There is much variation in the shape, size, and coloration of birds' eggs, reflecting the many habits and niches of the birds that lay them. The common shapes and markings used in egg descriptions throughout the species accounts follow.

Egg size obviously varies among bird groups, and also varies within a single species, and even among eggs in the same clutch. Larger birds generally produce larger eggs, but smaller birds may have proportionally larger eggs: the egg of a hummingbird may represent as much as 10 percent of a female's weight, compared with less than 2 percent in the case of an ostrich. In many species, a female's eggs increase in size as she ages. Birds that lay larger clutches generally lay smaller eggs. Egg size also varies between precocial species, whose chicks hatch in a more developed state, and altricial chicks, which hatch blind and often

The egg of a North American hummingbird, placed atop the egg of a now-extinct elephant bird. The larger egg would have weighed about 11 kg (24 lbs.), or the equivalent of roughly 2,700 live Ruby-throated Hummingbirds.

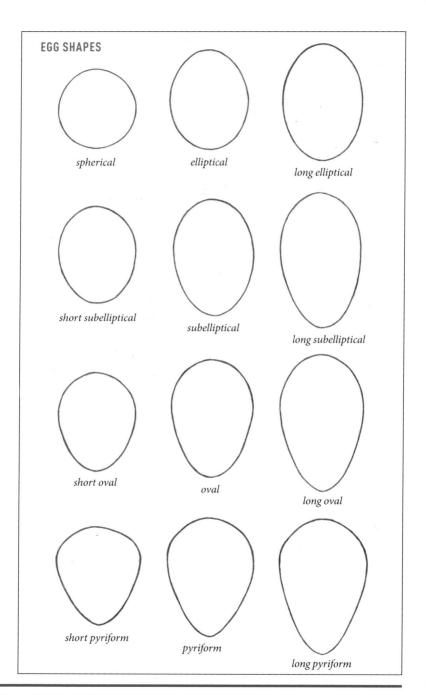

EGG SHAPES

spherical

elliptical

long elliptical

short subelliptical

subelliptical

long subelliptical

short oval

oval

long oval

short pyriform

pyriform

long pyriform

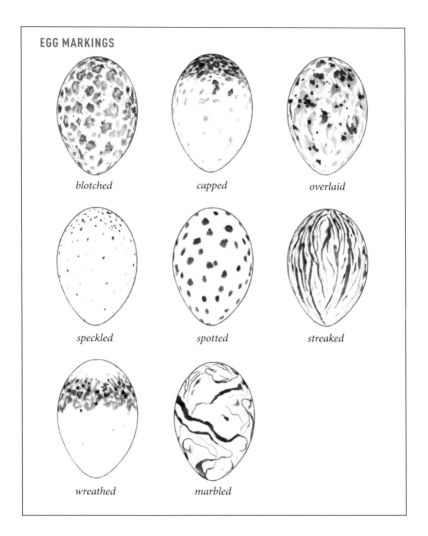

EGG MARKINGS

blotched

capped

overlaid

speckled

spotted

streaked

wreathed

marbled

naked. Precocial species require eggs to have comparatively larger energy stores so that the hatchling is less dependent on parental care, and thus they produce noticeably larger eggs.

Egg Color

The wide variation in egg color and markings is derived from two pigments, biliverdin and protoporphyrin. Biliverdin is responsible for the base color we see on eggs, is usually deposited first as the eggshell is forming and therefore when produced occurs throughout the shell. It ranges

A Common Nighthawk on her nest. David Moskowitz, ID

from blue to greenish blue in color. Protoporphyrin produces the markings and patterns we see on many eggshells and vary in color from yellow to brown. When these two pigments interact, they create an assortment of colors that range from green to purple.

Species that nest in the open and on the ground often lay eggs that are well camouflaged, helping decrease detection and predation. There are exceptions, however: species such as Common Poorwill lay stark white eggs that easily contrast with the forest floor, but the cryptic patterning of the incubating adult hides them. The white eggs of many cavity-nesting species are well hidden inside holes and excavations, and some ducks and grebes cover their eggs with feathers and vegetation when they leave the nest to forage and preen.

Some species, particularly those that nest in colonies, exhibit a wide variety of coloration and patterning of eggs. Such differences in appearance may be an adaptation that helps adults identify their own eggs among the many others.

Clutch Size

Clutch size is the number of eggs laid by a single bird during one breeding attempt. Most species are **indeterminate layers,** able to continue to lay eggs until a full clutch is reached, replacing as needed any eggs that are lost. Other species, such as Mourning Doves and crows, are **determinate layers:** the number of eggs they lay is predetermined, or fixed, and cannot be adjusted by the female when an egg is removed from the nest.

Clutch size tends to be similar among closely related species, but various factors determine clutch size within species, populations, and individuals, including regional climate, general body condition, and food supply. Nest type also affects clutch size. Birds that build open-cup nests are prone to a higher rate of brood failure than cavity nesters. To cushion potential loss, females of these species tend to lay comparatively smaller clutches.

Incubation

Incubation is the period when parent birds must keep their eggs at a suitable temperature to ensure the embryos' development. This responsibility may be assumed by both parents or by one adult (usually the female). Very rarely—in only about 5 percent of bird species—is incubation solely the responsibility of the male. Wilson's Phalarope is one example.

Most species incubate their eggs by sitting atop them, and some use the webs of their feet to cover the eggs. Birds that incubate using their

BROOD PARASITISM

Brood parasitism (also referred to as "egg dumping") is a strategy in which a female lays eggs in the nests of other birds, effectively increasing her reproductive potential by transferring the energetic costs associated with nest defense, incubation, and rearing to the host parent. There are two types of parasitism. **Intraspecific** brood parasitism occurs when eggs are deposited in the nest of a conspecific. This can be seen among a variety of colonial nesters and other species, including cavity-nesting ducks. **Interspecific** brood parasitism occurs when eggs are laid in the nest of a different species; see Ducks, Geese, and Swans, p. 58. Obligate interspecific brood parasites, such as the Brown-headed Cowbird, do not build nests and rely entirely on host species to raise their young. Some hosts have developed effective defense mechanisms against cowbird parasitism, and larger species are often less affected by raising cowbird young. Smaller, less adaptable species, such as the endangered Least Bell's Vireo (*Vireo bellii pusillus*), have seen alarming declines in population sizes as a result of cowbird parasitism and habitat loss. See Cowbirds, p. 400.

Two additional brood parasites, Yellow-billed and Black-billed Cuckoos, have also been observed taking advantage of the nests of other species, though both species frequently build nests of their own. Interestingly, the Yellow-billed Cuckoo employs both parasitic strategies, laying eggs in nests of other birds of the same and different species.

Wetted belly feathers help keep this Killdeer's clutch cool in the summer sun.
Bill Summerour, AL

breast and belly develop a **brood patch**, a bare and swollen patch of vascularized skin that is exposed in a particular pattern and provides a direct source of warmth to the eggs. Males of many species share in incubation duties, though some lack a brood patch. Their effectiveness as egg warmers is certainly less than that of the female, but they at least can maintain an adequate temperature while simultaneously serving to guard the nest. Generally, the larger the egg, the longer the incubation period. For example, incubation in many songbirds averages 10–14 days, in corvids 16–22 days, and in waterfowl 21–31 days. The California Condor, the largest land bird in North America, has an incubation period of 53–60 days.

Embryos are more likely to be killed by overheating than they are by cooler temperatures. In hot climates, species have developed a handful of ways to regulate egg temperature: they may lay eggs in shaded areas, soak their belly feathers in water before brooding, build roofed nests, stand above the eggs to shade them, or build nesting burrows.

The timing when birds initiate incubation determines whether or not a brood will see hatchlings of different ages. Species such as owls and hawks and eagles begin incubation immediately after laying their first egg, facilitating **asynchronous hatching**, which results in a disparity in age and size among the chicks. During times of food shortage this particular pattern of hatching may cause a **brood reduction** as a safeguard against complete brood failure. Larger nestlings have more advanced and aggressive begging abilities and are thus fed first and more often than younger chicks. Unable to compete with older siblings, the smaller nest-

The asynchronously hatched young of a Snowy Owl, showing the striking differences in development. Note the surrounding cache of dead rodents. Garrit Vyn, NU

Young Common Yellowthroats, hatched at nearly the same time. The shells have been removed from the nest. Matt Monjello, ME

The eggs of a Wild Turkey, opened by corvids. Matt Monjello, NH

lings often starve, or in many species may be killed and consumed by their larger nest mates.

Most species delay incubation until most if not all of their eggs are laid, resulting in **synchronous hatching**. This strategy, which produces a brood of the same age and size, is especially important in birds such as ducks, geese, and swans, whose precocial young are required to leave the nest as a group after hatching.

Ectoparasites

Both eggs and nestlings in the nest are threatened by a variety of parasites and pathogens. In addition to external parasites that feed on the birds themselves (mites, ticks, fleas, lice, leeches), nests also contain a

Sharp-tailed Grouse naturally hatched. Michael A. Schroeder, Washington Department of Fish and Wildlife, WA

Brown Pelican eggs in the process of hatching. The chicks in the foreground and background are beginning to open their shells; the third may be considerably smaller than its siblings by the time it hatches. Matt Monjello, NC

COMPETITION FOR NEST LOCATIONS

A long history of fierce competition over quality nesting locations is likely one of the primary drivers behind the wide variety of nest designs, placements, and behavioral strategies.

When resources required for breeding success are broadly distributed across a large area of desirable habitat, competition to secure a nest site may be reduced. In breeding habitats where nesting sites are in short supply, competition generally increases and may decrease the population density among subordinate species. For example, nonnative European Starlings may drive small species of woodpeckers from areas that have quality nesting cavities.

variety of pathogens, such as viruses, fungi, and bacteria. Behaviors such as preening, dust bathing, and sunning aid birds in combating parasites that feed on their blood, skin, and feathers. The molting of feathers also helps thwart parasites. Numerous species keep nests clean by removing the fecal sacs their young produce, and they may construct new nests for successive broods; many woodpeckers excavate new cavities, thereby ensuring a clean, parasite-free environment. Raptors can have multiple nests in their territory and may use different nests during different breeding seasons. Some species line their nests with fresh, green plant material, sometimes replenishing them daily. The volatile chemicals in some of these plants may act as inhibitors or as an insecticide and thus limit the effects of parasites.

CHAPTER 2

AVIAN ARCHITECTURE: NEST DESIGN, MATERIALS, AND CONSTRUCTION

It's easy to think of a nest as rudimentary, albeit beautiful, architecture—a simple pile of twigs and perhaps lined with a bit of soft material. And sometimes, this is true. But many nests, particularly the highly diverse, elaborate nests of passerines, are far from haphazard. On any given landscape, the diversity of species is generally mirrored by the variety of design, placement, and specific way in which commonly available materials are used.

Birds must assemble a variety of materials in such a way that they effectively bind together to retain a nest shape that endures the development of young and the constant activity required to rear them. The nest also needs to stay solidly fixed in place, and its particular structure serves in varying degrees to insulate its contents from the outside world—even if only to keep the growing chicks from tumbling to the ground. Many nests often have multiple components, or layers, each of which provides a distinctive purpose. Twigs may be used to create an initial platform and the structural outer wall, weed stems and grasses may form a padded inner cup, and plant down and feathers may constitute the lining. Examining nests for the presence or absence of particular layers is highly useful for nest identification, and also helps develop an understanding of nest design and structure. For instance, some ground-nesting warblers use wet, dead leaves to form the outer wall of their nests. The outer leaves dry in place and retain their shape, creating a delicate shell that can persist for many months. The appearance of this layer is not only distinctive, but also lends insight into other, larger species' nests that undergo a similar construction process and strategy.

A close-up view of the cross section of an American Robin nest, showing the nest layers. From left to right: the bulky outer cup, an inner cup of mud, and a lining of fine grasses.
Casey McFarland, WA

Nest Features and Composition

By understanding the general patterns and features typical of many types of nests, we begin to develop an eye for the distinctive ways that birds use certain materials. The following descriptions of nest components and layers offer insight into the essential structure of nests and provide guidelines for making detailed observations about nests found in the field. There are, of course, nests of numerous species that exhibit many similar features: a bunting nest may look like that of a small flycatcher, for example. Careful attention should be given to the often subtle differences that occur in the nests of various groups and species, and their particular locations.

As described by Hansell (2005), the layers of a nest can be divided into four basic categories, and each layer is usually fairly distinct. Not all layers are always present, however, and are at times difficult to distinguish from one another. For instance, the soft "lining" of a hummingbird's nest may not actually be an added lining at all, but may instead be the soft inner cup that is also the structural wall.

Many nests have portions of their outer wall that are clearly bound in some fashion to a supportive structure. This attachment—perhaps more appropriately referred to as a nest component than a specific layer—is any

Spider silk binding an old, weathered nest securely in place. Casey McFarland, AL

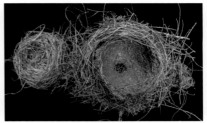

The grass lining of an old American Robin nest, easily removed to reveal the inner cup of mud. Casey McFarland, WA

LEARNING TO BUILD

Despite birds' clever adaptations and impressive building skills, it is important to note that not all nests are built well. At times, birds may build suboptimal, poorly insulated nests, perhaps resulting in the stunted development of young or an energetically costly increase in brooding time. Indeed, the fact that individual birds may build increasingly higher quality nests over the course of their life, as has been documented in many species, suggests that nest building is a mix of instinctual behavior and developed skill. Nests that are oddly or precariously placed, or built of materials that hold together poorly, may be the work of a first-time nester.

part of the nest architecture that is purposed to hold it securely in place. Some nest designs, such as a simple platform of sticks assembled on a cliff ledge, or a cup of twigs wedged in the upright fork of a branch, require no specific points of attachment and are instead held in place by gravity. Others are carefully secured with strands of spider silk or wrappings of plant fiber or glued in place with mud or saliva. Ensuring that their nest will not sag or fall from its place is of obvious importance, and many species devote considerable energy and expertise to securing it against wind, precipitation, and the weight and activity of developing chicks.

The **outer cup,** or **structural layer,** forms the overall body of the nest, providing the integrity, strength, and general shape. It may consist of many types of materials: twigs, sticks, reeds, cattails, and other vegetation. Even plant down, combined with spider silk or lichens, can be formed into a tough outer wall. Many nests have multiple structural layers. The American Robin builds an outer cup of grasses, fine twigs, mud, and other materials, and then applies another inner layer of mud that substantially increases the nest's structural integrity. Robin nests are among the longest-lasting songbird nests in North America, and one of the most commonly observed. The final lining of fine plant stems or grasses can be easily removed to inspect the hardened clay bowl beneath.

The **inner cup**, or inner layer, when present, is usually distinctive from both the outer structural layer and the final lining (see below). It is often a well-defined, padded cup of soft or fine materials. Shrikes build a bulky, messy, and loose structural cup of twigs, to which they add an impressive, thick inner cup of grasses, weed stems, sage leaves, plant fibers, and other similar materials.

A bird may add a layer of finer insulative materials to the inner cup to create the nest **lining**. Fine grasses, mosses, horsehair fungus, feathers, fur, plant down, and human-made materials are among the materials commonly used. The lining helps maintain the microclimate essential for the development of young and can vary in mass depending on climate and latitude, as can the structural wall. The addition of a separate, softer lining isn't an essential feature for many birds, which instead simply lay eggs directly on the often rough-hewn structural layer.

The **outer**, or **decorative**, **layer** lends to the appearance (or disappearance) of the nest but doesn't necessarily increase its overall strength. The outer layer may help camouflage the nest, or it may act to repel water and retain heat. The nests of Blue-gray Gnatcatchers and of many hummingbirds are heavily flecked with bits of lichen, dramatically altering their appearance. Strips of plastic, shed snakeskin, fine plant leaves and fibers, catkins, and other materials are present on the nests of many other species as well. This layer is most common among birds that use large amounts of spider silk in construction.

A Loggerhead Shrike nest, showing the distinctive inner and outer layers, or cups. Casey McFarland, TX

A Blue-gray Gnatcatcher nest, thoroughly covered with lichen flakes. Matt Monjello, NC

Thistle down. This plant material and others like it are frequently used in both the inner and outer cups of many passerine species. Casey McFarland, AL

Materials

While countless different materials are used to build bird nests, they are sourced from only a few broad categories: plant life, animals, insects, and mud or other inorganic materials. Despite this immense diversity of materials, most nests—even those that are quite elaborate—are built with fairly few. Often, only three or four primary building blocks are used to construct and finish a nest, and they may vary each season depending on where a species builds and what is available in any given year. Birds apparently choose a material mostly for the purpose it serves, and they may substitute an array of items for another as long as they fulfill a desired function. Frayed bailing twine may take the place of fine grasses and plant fibers, fishing line may replace fine rootlets, and stuffing from an old discarded pillow makes a sufficient substitute for plant down. Other materials, however, serve very specific

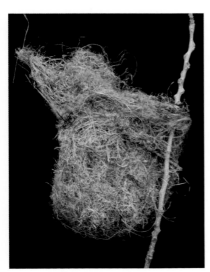

A Bullock's Oriole nest made almost entirely of bailing twine, a handy replacement for long plant strands. Casey McFarland, NM

A female Purple Martin collects the inner bark of an aspen tree to line her nearby nest cavity. David Moskowitz, CO

purposes and are difficult if not impossible to substitute. Bushtits and other species that build nests of a Velcro-like "fabric" rely on particular types of lichens with stiff, barblike hairs that neatly entangle with the spider silk with which they are combined.

The size of a bird influences what is used for construction, and larger birds are obviously able to gather and manipulate larger items. Smaller birds generally use the finest materials, and typically create more intricate and often more "compact" or tightly built final products. Careful inspection of an old nest will reveal the primary items a bird chose to build with, and provide clues as to where on the landscape the bird may have spent time gathering.

PLANT LIFE: From mosses to tree branches, plant life is the primary building block of most nests worldwide. As both birds and humans have discovered, cellulose, the main constituent in the cell walls of plants, has characteristics perfectly suited for building a home. It is light, wonderfully tough but flexible, and does not dissolve readily in water. Plants—alive, dead, or decaying—are widely used in each layer of a nest. Thin fibers are bound to stems to fasten a nest in place, twigs are used for structural walls, seaweed as adhesive, leaves for padding, and rootlets, fine bark fibers, and plant down for lining. Green plant materials or flowers may be added for their antiparasitic qualities and to signal that a nest is in use. Beside plants, certain fungi are also used. Horsehair fungus, whose long, dark, hairlike strands are frequently mistaken for actual hair, is considered to be one of the most widely used lining materials in bird nests across the planet.

SILK THREAD: Silk is commonly used to bind nest materials together, and for fixing a nest in place. Birds frequently gather the sheetlike webbing from funnel web spiders, tent caterpillars, and spider egg cocoons. Silk can be used as a scaffolding of sorts to create a foundational platform between twigs on which to build (hummingbirds), and when combined with certain plant materials, can produce a flexible, feltlike structural material. Note that when silk is used in nest construction, it can be difficult to see; birds apply it to varying degrees and sometimes it is highly visible, sometimes not.

FEATHERS AND FUR: Feathers provide excellent insulation and are used in the nests of many species, primarily as lining material, though they may also be incorporated into the structural wall. Female ducks, geese, and swans pluck feathers from their breast to line their nests and to cover their eggs. Smaller passerines such as Barn and Tree Swallows use them frequently as well, collecting molted feathers or those from the carcasses of dead birds. Fur is also commonly used, though to a lesser extent. It is most common in the nests of some cavity-nesting species, such as nuthatches, chickadees, and titmice.

MUD: Only about 5 percent of the world's bird species use mud for nest construction, and only a small number of these species, such as the Cliff Swallow, use *only* mud. Mud is used primarily as a cementing agent that reinforces structural layers. When mixed with plant materials, it forms an adobe of sorts that strengthens the inner cup, as commonly seen in the nests of grackles and Steller's Jays. A layer of pure mud may be added to form an inner structural cup, referred to as "plastering" (see "structural layer" above). In nests built with large amounts of mud, such as those of Cliff Swallows and phoebes, distinct color striations may be visible, showing that the mud was collected from different sources in a landscape.

Cliff Swallows collecting mud for nests. David Moskowitz, WA

An American Robin nest, freshly plastered with mud. The mud is allowed to harden before a lining of fine dry grasses is added. Casey McFarland, CT

Building Techniques

As described by Hansell (2005), nests can be placed in two categories of construction techniques containing several building methods. Each nest found in the field can be roughly evaluated to ascertain the method by which it was built, which provides a starting point for understanding typical categories of nest architecture. **Sculpting** is the removal of material to form a desired receptacle, which can range from a subtle scrape on the ground to a deep cavity diligently chipped into the living or dead wood of a tree. **Assembling** is the collection and joining together of materials to make the nest. Assembly techniques employed by birds fall into additional categories. **Molding** is

The intricately twined attachment point of a Hooded Oriole nest to a palm leaf. Casey McFarland

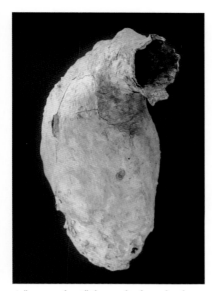

the shaping of mud and clay, typified by Cliff Swallows. **Sticking together** refers to combining and manipulating mud (or dung, or saliva in some species) with vegetation and woody materials to hold them together, as discussed above (see "Mud"). Chimney Swifts, for instance, create a nest completely of small twigs glued together with a salivary mucus, the entire nest neatly fixed to a vertical surface.

A "saguaro boot," the result of woodpeckers excavating, or "sculpting" away, the pulpy innards of a saguaro cactus. The wound heals in a year or so, creating an internal, hardened shell that can be used again and again by woodpeckers and other species alike. Casey McFarland

INTERLOCKING is widely used in passerine nests and includes an impressive range of techniques. Three subcategories have been described: "Entanglement" is commonly used to manipulate finer materials by means of wrapping, poking, and tucking items into a structural wall. Fine materials such as grass and stems may be entangled to create a matted nest of sorts, like that of the Common Yellowthroat and Yellow Warbler. "Stitching and pop-riveting" materials together help fasten nests in place, or fold materials together to create a pocket. The Hooded Oriole provides an example of an uncommon tactic in North America—it often "sews" its pendulous nest to the underside of palm leaves by poking strands of fibrous material through one side and pulling them out the other. "Velcro" refers to the use of fine materials in combination with silk, as demonstrated among Bushtits, hummingbirds, and many other species of birds. The resulting tacky "fabric" is stretchy but strong, and parts of the nest may be pulled free and reattached as the builder sees fit.

PILING UP describes a straightforward building process of collecting twigs and branches and simply dropping them into place. It mostly pertains to platforms built in trees but also occurs in nests built on the ground or in water. Unlike in interlocking nests, where materials are intensely manipulated, piling up generally requires considerably less effort to work items into the nest in ways that lock them together. There is, however, variation

A close-up of a Bushtit nest. It contained finely bristled leaves, stringy lichens, fine twigs, plentiful silk, and other items. Note the spider egg sacs. Casey McFarland

in the degree to which materials are manipulated; some birds may only pile materials atop each other, and others spend more time poking an item into a proper place. Regardless, the appearance of these nests generally is somewhat messy or simplistic. This technique is broadly used by many species, including members of the grebe, heron, dove, and hawk families. Gravity generally holds the nest together and in place, and a recognizable cup or bowl is usually present. Nests created in marshes and wetlands, like those of some waterfowl, rails, and coots, are built in similar fashion, but by bending and pulling down aquatic vegetation to form a mat or mound.

WEAVING is an elaborate technique wherein birds skillfully and meticulously intertwine fibers by means of warp and weft, loops, knots, and hitches. Weaving is often used to construct suspended nests and is performed by North American orioles as well as oropendolas of Central and South America.

A close-up view of an Orchard Oriole nest, showing the impressive intertwinement of material. Casey McFarland

The meticulous over–under weave of nesting material to upright supports. Red-winged Blackbird. Matt Monjello, NC

Interestingly, the knots used by these species as well as by the true weavers (Ploceidae) of Africa are strikingly similar. They all use long, fine materials and have developed parallel techniques for manipulating them.

Nest Types

Bird nests can be placed into several categories defined primarily by their structure and shape. Some taxonomic groups have a high degree of fidelity to one nest type, such as gulls, most of which use a bare or sparsely lined scrape or depression on the ground. Alternatively, some families of birds construct a broad variety of nest structures. The New World flycatchers create multiple types of open-cup nests, some build globular nests, and others use tree cavities.

SCRAPE: Many species nest in a simple depression in the ground; this may be situated within a shallow natural divot or hollow, or the bird may create it by scraping. The nest may lack any lining whatsoever, or be lined with vegetation, feathers, leaves, or just a few pebbles or other items. Occasionally, birds lay eggs on bare rock that has no noticeable depression. Many species use scrapes, including terns, plovers, and gamebirds.

PLATFORM NEST: This nest is generally created by means of piling up (p. 47), wherein materials are stacked somewhat crudely together to form a mound or platform, generally with a slight depression on top. This design ranges greatly in size and shows a wide array of adaptations and

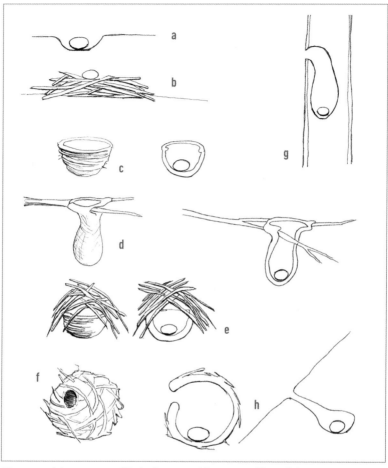

Nest types: (a) scrape nest; (b) platform nest; (c) cup nest; (d) pendulous or pensile cup nest; (e) domed nest; (f) globular nest; (g) cavity nest; (h) burrow nest.

applications, and frequently has a bowl or saucerlike upper surface. Platform nests may be constructed of mud and plant debris along a wetland edge, like those of loons; built of aquatic vegetation to form a floating mat on water, like those of grebes and coots; or placed atop a cliff ledge as in the nest of the Golden Eagle. Most commonly they consist of twigs and sticks placed among the branches of a tree or shrub (e.g., in herons). A Mourning Dove's nest is just a thin, meshlike dish of small twigs, while a decades-old Bald Eagle nest is a massive, heavy, and sturdy heap of large sticks and other items.

CUP NEST: This is the most common nest design and is typical among passerines. Generally this structure is hemispherical in shape with a distinctive cup. Depth varies from about half as deep as wide (appearing saucerlike or like a shallow bowl) to deep enough to conceal the entire body of the parent bird. Common cup nesters include thrushes and sparrows. Size, placement, and building materials vary tremendously; cup nests are built from almost any imaginable material and can be fitted into hollows in stumps, banks, burrows, and cliffs, but most commonly are constructed in a tree or shrub. Cup nests are typically supported from the bottom: wedged into an upright crotch or fork of twigs, saddled atop a horizontal branch, or placed on the ground or a small ledge.

PENDULOUS OR PENSILE CUP NESTS: Some cup nests, such as those of vireos, orioles, and bushtits are suspended primarily from the nest rim, and range from neatly woven open cups to long, enclosed pendulous structures.

DOMED NEST: This is generally a cup nest with a constructed roof that helps conceal the nest, maintain heat, and protect the nest from wind, rain, and sun. North American species that construct domed nests include meadowlarks and Ovenbird.

GLOBULAR NEST: This unique, readily identifiable nest type is roughly spherical in shape and completely enclosed except for a small entrance, usually on the side. Globular nests are frequently visible in shrubs or trees but are also placed on ledges or tucked in natural cavities, hidden from sight. While characteristic of several passerine species in North America, globular nests are the common architecture of the wren family: Cactus Wrens construct grassy oblong spheres in the spiny arms of cholla cactus and scrubby trees, Pacific Wrens make a ball of moss and wispy twigs in the temperate forests of the Pacific Northwest, and Marsh Wrens weave a tidy hutlike nest of cattail leaves in emergent wetland vegetation. Other globular nest builders include American Dipper, Great Kiskadee, and Verdin. If the spherical structure has a distinctive, igloo-like entrance tunnel, it is referred to as a **retort nest**.

CAVITY NEST: This is an empty space within a larger substrate, such as naturally occurring, rotted hollows in a tree; rock crevices; or holes excavated by woodpeckers. This is one of the most important nest types from an evolutionary perspective; about half of all orders of birds include one or more species that nest in holes. Cavity nests tend to be safer, with overall fledgling rates higher than those of open-cup nests. Some species excavate their own cavities in live or rotted wood, including woodpeckers, nuthatches, and chickadees (the latter two are equipped only for opening softer, more punky wood). "Cavity adopters" are species that use a range of holes—natural crevices in snags, stumps, rocks, cliffs, and similar substrates. "Obligatory secondary cavity nesters" depend specifically on the holes excavated by other species to hatch and rear their young, and include a wide variety of birds, from passerines to numerous species of small owls. Naturally occurring cavities such as hollows in trees attract cavity-nesting ducks such as Hooded Merganser and Common Goldeneye. Many species readily use human-constructed nest boxes that mimic natural or woodpecker cavities; these have been important for conservation efforts.

BURROW NEST: A burrow nest is defined as a hole or cavity in the ground that is generally longer than it is high, and is usually accessed by a sloping or horizontal entrance tunnel. This is perhaps the most overlooked nest type: earthen holes and burrows are common but are frequently associated with mammal activity, and the fact that these burrows are used or even excavated by a bird can be surprising. Quintessentially, puffins, shearwaters, and auklets use burrow nests, wriggling in and out of their earthen tunnel and typically using the same hole year after year. Belted Kingfishers excavate burrows in the eroded banks of rivers and streams, and Burrowing Owls typically use holes made by mammals.

Nest Placement

Most of the nest types above can be placed in a variety of locations. Throughout the group and species accounts, particular placements are mentioned frequently, including variations of "in upright fork or crotch"; "saddled atop a horizontal branch"; "suspended from the rim between a horizontal fork"; "attached to vertical surface"; "on floating mat of vegetation"; "tucked beneath a grass tussock"; and others. The following illustrations show common sites where North American bird species choose to build their nests on a broad variety of landscape features, and provide examples of descriptions found in the text.

Nest placement in trees, shrubs, and other settings: (a) scrape in ground; (b) cup nest built at base of shrub; (c) cup nest built in crotch or upright forks of bush or shrub or tree; (d) small to medium-sized platform nest in tree; (e) large platform nest in tree; (f) pensile nests supported from horizontal forks or drooping twigs; (g) cup nest saddled atop horizontal branch; (h) natural tree cavity or beneath loose bark; (i) excavated tree cavity; (j) cup nest built in upright stems of grasses or other vegetation; (k) cup nest built at base of grass clump or tuft. Drawn after Suzuki, 2010.

Nest placement on cliffs and similar settings: (a) eggs laid on bare rock on ledge; (b) cup nest built atop ledge; (c) nest attached to vertical surface; (d) eggs laid on bare ground in cliff recess or cave; (e) platform nest built on ledge; (f) nest or eggs placed in natural crevice; (g) burrow nest. Drawn after Suzuki, 2010.

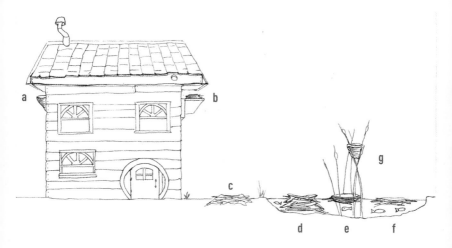

Nest placement on human structures and in wetlands: (a) attached to vertical surface; (b) on ledge beneath overhang; (c) platform near water's edge; (d) platform built in water from bottom up; (e) platform built in emergent vegetation at water level; (f) floating platform; (g) cup nest built in upright stalks of emergent vegetation.
Drawn after Suzuki, 2010.

Placement of two types of platform nests found in or on water: built from the pond-bottom to above water, or floating, either attached by roots or tendrils below or fastened to emergent vegetation.

SPECIES
ACCOUNTS

DUCKS, GEESE, and SWANS (Anatidae)

More than 40 species of ducks, geese, and swans nest in N. America. Nests can be divided into three distinct structure types: *shallow, lined depressions on the ground*; *lined platforms constructed in emergent vegetation over water*; and *tree cavities*. They are often in close proximity to water, but there is variation both within and among species. Most species use down plucked by the female from her own breast as part or all of the inner lining of the nest, which may be pulled over the eggs when the incubating female leaves the nest. Goose nests are similar to those of ground-nesting ducks, but larger. Swan nests are raised mounds, significantly larger than those of ducks.

Breeding behavior varies: some species find new mates each year (most ducks), while others primarily form long-term or lifelong pair bonds (most geese and swans). Among dabbling ducks, the male is typically intolerant of other males entering the territory during early incubation, though the pair bond generally dissolves by the second week of incubation.

All species have elliptical, solid-colored eggs ranging in color from cream to drab gray, creamy yellowish, greenish, or olive. Young are precocial and leave the nest shortly after hatching, often tended only by the female; the young of some species, including geese and swans, stay with

From left to right:
Cinnamon Teal,
Wood Duck,
Redhead.

Two smaller
Northern Pintail
eggs deposited in a
Ruddy Duck nest.
Mark Nyhof, BC

both parents until the following spring, and whistling-ducks stay with both parents through the fall or perhaps longer. Nest parasitism, or egg dumping, is common among ducks, both within and among species and across families. Some "dump nests" have been found to contain an impressive 80 eggs. Females are the sole incubators in most species (both male and female whistling-ducks incubate). All species are primarily single-brooded.

GEESE

Geese build ground nests, typically with a design similar to that of ground-nesting ducks but proportionally larger. Usually the female scrapes away a hollow depression, layers it with vegetation, and then lines it with finer plant materials or down. Canada Geese are ubiquitous across much of N. America, and their nests are commonly discovered along waterways and rivers. Geese are monogamous, usually maintaining lifelong pair bonds. Eggs are whitish to creamy white.

ROSS'S GOOSE *Anser rossii*

EMPEROR GOOSE *Anser canagicus*

SNOW GOOSE *Anser caerulescens*

HABITAT: Open tundra along coast or by body of fresh water. **LOCATION AND STRUCTURE:** Similar to that of Canada Goose and its kin; a large version of the classic ground duck nest. A substantial lining of vegetation forms a rim around the depression. Geese often add a substantial amount of down to nests after laying eggs. Ross's Goose: Outside diameter 47 cm; inside diameter 16 cm; depth 7.8 cm. Emperor Goose: Slightly smaller overall but with relatively larger cup and depth. Snow Goose: Variable; occasionally overlaps with swan nests in size. **EGGS:** Ross's: 3–4; 70 x 47 mm; IP 21–23 days. Emperor: 5–6; 78 x 52 mm; IP 24–25 days. Snow: 4–7; 79 x 52 mm; IP 22–25 days. **BEHAVIOR:** All *Anser* geese nest colonially in Arctic. Ross's and Snow Geese hybridize in parts of their breeding range. **DIFFERENTIATING SPECIES:**

Emperor Goose.
Gerrit Vyn

Snow Goose. Susan N. Felage, ME

Emperor Goose. Tom Bowman, AK

Emperor Goose nests in coastal w. AK. Ross's and Snow Geese nest along Arctic coast and islands in Canadian high Arctic, often found together on breeding grounds.

CANADA GOOSE *Branta canadensis*

CACKLING GOOSE *Branta hutchinsii*

BRANT *Branta bernicla*

HABITAT: Canada and Cackling Geese: Wide variety of wetland habitats across N. America, tundra near water, boreal forest, meadows and grasslands, rocky shores in Aleutian Is. Brant: In low Arctic salt marshes and estuaries and in high Arctic near river deltas or inland freshwater lakes. **LOCATION AND STRUCTURE:** Raised location near water. Nest is built into an excavated depression in the ground or with a collection of coarse vegetation lined with dry grass and down feathers. Canada Geese also occasionally nest on raised platforms, including large raptor nests (the young sometimes plunging from great heights to join adults below). *Nest dimensions are similar.* Canada Goose: Outside diameter 41–64 cm; inside diameter 16–25 cm; depth 10 cm. **EGGS:** Cackling and Canada: 5–6; size varies among subspecies, 72–87 x 48–58 mm; IP 25–30 days. Brant: 3–5; 71 x 47 mm; IP 22–26 days. **DIFFERENTIATING SPECIES:** Eggs and nests are indistinguishable.

Canada Goose. Matt Monjello, ME

Canada Goose. David Moskowitz, WA

SWANS

Swans are the largest anatids. Their nests are similar to but larger than those of their smaller relatives: they construct large vegetative mounds, but with very little down incorporated into the lining. They generally nest as isolated pairs. Like geese, swans are monogamous, typically maintaining lifelong pair bonds. Swans don't breed until 3–5 years of age. Eggs are generally creamy white to yellowish, those of Mute Swan pale gray blue. Clutch size is usually ~5 but can vary by latitude and time of year.

TRUMPETER SWAN *Cygnus buccinators*

TUNDRA SWAN *Cygnus columbianus*

MUTE SWAN *Cygnus olor*

HABITAT: Coastal tundra near edges of pools, wetlands, and wet meadows. Less frequently inland. **LOCATION AND STRUCTURE:** A large mound of plant material with a distinct central hollow on the edge of fresh water or surrounded by water on an island. Tundra Swans occasionally construct nests that float in water. Often nest on beaver or muskrat lodge. Trumpeter Swan: Outside diameter 1.2–3.6 m; height 0.4 m; inside diameter 20–40 cm; depth 10–20 cm. Tundra Swan: Outside diameter 1.2–1.8 m; height 0.6 m; depth 10–20 cm. Mute Swan: Similar to Trumpeter. **EGGS:** 5–6; off-white; subelliptical. Trumpeter: 111 x 72 mm; IP 33 days (F, M). Tundra: 107 x 68 mm; IP 32 days (F, M). Mute: 80 x 54 mm; IP 34–38 days (F, M). **BEHAVIOR:** Though female does majority of incubation, male participates. Eggs are often covered

Trumpeter Swan, WY. National Park Service

Trumpeter Swan. Carroll L. Henderson, AK

Tundra Swan. Gerrit Vyn, AK

Tundra Swan. Dan Fontaine, AK

with vegetation. Young stay with adults until following spring. Mute Swans, introduced to N. America from Eurasia, occasionally nest in proximity to Trumpeter Swans.

DUCKS

GROUND-NESTING DUCKS

Nests of these species are shallow, excavated depressions, sometimes with an outer lining of vegetation and often with an inner lining of down plucked from the female's breast. Construction often happens concurrently with egg laying rather than preceding it. Nests are frequently near water and islands in ponds, and ducks often use wetlands when available, though some species may nest 480 m or more (up to 1.6 km in the case of pintails) from water.

A great deal of nest appearance overlap in this group makes identification difficult, and the nests of some species may vary considerably across their range or in different habitats. The overall size of the nest, its location, and the height and type of surrounding vegetation can help narrow down who the maker may be; the coloration and patterning of breast feathers in the nest are also a reliable way to distinguish among species if adult birds are not present. See Down Plates , pp. 63–65.

BLUE-WINGED TEAL *Spatula discors*

HABITAT: Prairie habitat near shallow, often seasonal, wetlands. **LOCATION AND STRUCTURE:** In low, dense, homogenous grasses or herbaceous cover, rarely brushy areas, within 150 m of water. On dry ground, though may be in areas where water and emergent vegetation are intermixed. Well concealed in vegetation, often with nearby vegetation arched over entire nest. Inner lining of grass and down. Down is drab brown with large white center. Outside diameter 19 cm; inside diameter 13.5 cm; depth 5.5 cm. **EGGS:** 10; 47 x 33 mm; IP 19–29 days. **BEHAVIOR:** Multiple pairs may nest in proximity to one another in good location. Sometimes nests in large numbers outside usual range.

CINNAMON TEAL *Spatula cyanoptera*

HABITAT: Edges of shallow bodies of fresh or brackish water, favoring areas with bulrushes. **LOCATION AND STRUCTURE:** Well concealed beneath mat of dead stems in vegetation at water's edge, or beneath bush with tunnel through emergent vegetation for access. Less frequently beneath shrubs. Occasionally over water. Usually within 1 m of water, almost always within 10 m. In dry location, a simple hollow with slight vegetation lining; in wetter location may be more substantial collection of vegetation to raise nest above water. Abundant down is drab brown with large white center, *similar to that of Blue-winged Teal*. Outside diameter 19 cm; inside diameter 12.5 cm; depth 5 cm. **EGGS:** 9–10; 47 x 34 mm; IP 21–25 days. **BEHAVIOR:** Seasonally monogamous.

GEESE

Canada Goose

Cackling Goose

Brant

Emperor Goose

Greater White-fronted Goose

Snow Goose

SEA DUCKS

White-winged Scoter

Black Scoter

Common Eider

Spectacled Eider

SEA DUCKS (CONTINUED)

King Eider

Steller's Eider

GROUND-NESTING DUCKS

Mallard

Northern Pintail

Northern Shoveler

Mottled Duck

American Green-winged Teal

Harlequin Duck

Greater Scaup

American Wigeon

Long-tailed Duck

IN-WATER DUCK NESTS

Canvasback

Redhead

TREE-CAVITY DUCK NESTS

Barrow's Goldeneye

Common Goldeneye

Bufflehead

Hooded Merganser

Red-breasted Merganser

NORTHERN SHOVELER *Spatula clypeata*

HABITAT: Open habitats and treed parklands near shallow wetlands. **LOCATION AND STRUCTURE:** In low vegetation on a dry site at edge of wetlands and slow-moving streams. Typically 2–50 m from water, less commonly up to 100 m. Down is brown with light center, *like that of Northern Pintail*. Outside diameter 20 cm; inside diameter 14 cm; depth 9 cm. **EGGS:** 10; 52 x 37 mm; IP 25 days. **BEHAVIOR:** Most territorial of N. American dabbling ducks.

Northern Shoveler nest. The larger eggs were deposited by a Common Eider.
Susan N. Felage, MB

GADWALL *Mareca strepera*

HABITAT: Prairie, taiga, parkland, and brushy habitats near wetlands. **LOCATION AND STRUCTURE:** Usually under shrubs, forbs, or in dense grassland vegetation. Usually less than 150 m from water but up to 2.4 km away. On islands in freshwater wetlands when available and relatively close to water (some within 2 m). Down is brown, with pale center and tip. Outside diameter 20–30

Thomas Erdman, WI

cm; inside diameter 18–20 cm. **EGGS:** 7–12; 55 x 40 mm; IP 26 days. **BEHAVIOR:** May nest in close proximity to other pairs, especially on islands.

AMERICAN WIGEON *Mareca americana*

HABITAT: Tundra and boreal forest with brush and grass near fresh water. **LOCATION AND STRUCTURE:** Concealed in tall vegetation, typically shrubs but also dense grassland. Sometimes in leaf litter by tree or under overhanging shrub. Average distance to water varies across its range, from 20–320 m, exceptionally up to 800 m in northern locations. May be lined with leaves, grass, and other vegetation. Down is dark with indistinct pale center and light tips, *smaller and lighter than that of Gadwall*. Outside diameter 20 cm. **EGGS:** 7–9; 53 x 37 mm; IP 23–28 days.

MALLARD *Anas platyrhynchos*

HABITAT: Grasslands, thickets, and forests associated with wetlands and ponds. **LOCATION AND STRUCTURE:** Often within 150 m of water, occasionally much farther away or set in wetland or riparian zone. Occasionally builds overwater nests in wetlands, or in other bird nests, in agricultural fields, or in buildings. Commonly found in clumps of tall bunch grass. Typically with overhanging vegetation or cover of some sort. Nest hollow in vegetation is lined with plant debris and down. Down tufts are brown with white center and tips. Outside diameter 26–29 cm; inside diameter 14–22 cm; depth 2.5–14 cm. **EGGS:** 6–11; 58 x 42 mm; IP 28 days.

Matt Monjello, NH

AMERICAN BLACK DUCK *Anas rubripes*

HABITAT: Mixed hardwood and boreal forests near freshwater and saltwater marshes, streams, ponds, and bogs. Also coastal islands, cultivated areas. **LOCATION AND STRUCTURE:** Nest site is highly variable; in stumps, brush piles, hay roles, artificial structures, rock crevices, trees, abandoned hawk nests; under shrubs or logs. Nest bowl with sparse covering of vegetation and down added during incubation. Down has pale center, but less *obvious than in Mallard down*. Outside diameter 30 cm; height 8 cm. **EGGS:** 9–10; 59 x 43 mm; IP 26 days.

Christian Artuso, MB

MOTTLED DUCK *Anas fulvigula*

HABITAT: Pastures, pine flatwoods, oak scrub, citrus groves, and scrub oak near freshwater and brackish marshes, irrigation ditches, rice fields, and coastal waters. **LOCATION AND STRUCTURE:** Typically within 200 m of water in dense grassland or cultivated fields, often at base of shrubs, or concealed from above by vegetation. Occasionally an overwater nest in marshes. Depression is lined with grass and down. Down feathers are uniformly brown. Outside diameter 27 cm; inside diameter 15 cm; depth 4–8 cm. **EGGS:** 8–12; 57 x 42 mm.

Bill Summerour, AL

NORTHERN PINTAIL *Anas acuta*

HABITAT: Grassland and tundra with shallow wetlands. Also tidal areas, meadows in boreal forests, and tilled cropland. In relatively open locations compared with other upland-nesting ducks. **LOCATION AND STRUCTURE:** In open with minimal cover, in short grasses, occasionally in shrubs. Typically well away from water (2–3 km), rarely close to water's edge. Female constructs nest scrape, *top of nest flush with or below ground level.* If in emergent vegetation, usually built on platform of vegetation. Lining is a mix of vegetation, down, and feathers. Down is relatively long, brown, with light center. Outside diameter 19–26 cm; height 0–18 cm; inside diameter 13–19 cm; depth 6–10 cm. **EGGS:** 8; 54 x 38 mm; IP 24 days. Color varies from yellowish to blue. **BEHAVIOR:** Nests of different pairs are often relatively close to each other.

Tim Bowman, AK

Gerrit Vyn, AK

GREEN-WINGED TEAL *Anas crecca*

HABITAT: Deciduous parkland, prairies, boreal forest near ponds and marshes. Near saltwater shorelines in parts of range. Occasionally near streams. **LOCATION AND STRUCTURE:** Usually within 200 m of water. Well concealed in heavy vegetation: often sedges, high grass, or shrubs. Nest hollow is lined with dry plant material and down. Down tufts are small and very dark with white center. Outside diameter 15–18 cm; height 7–15 cm; inside diameter 14 cm; depth 9 cm. **EGGS:** 9; 47 x 34; IP 20–23 days.

Tim Bowman, AK

GREATER SCAUP *Aythya marila*

HABITAT: Treeless wetlands with shallow lakes and ponds. **LOCATION AND STRUCTURE:** On open ground with little to no cover, often less than 1 m from fresh water, sometimes up to 40 m. Occasionally on floating vegetation. In tidal areas may be built up to almost 0.5 m to keep above high tide. Location is selected for presence of last year's growth of grasses or sedges. Down is dark sooty brown with light center; large quantity amassed by the end of incubation. Outside diameter 23–31 cm; inside diameter 18–20 cm; depth 8–13 cm. **EGGS:** 8–9; 63 x 44 mm *(larger than those of Lesser Scaup)*; IP 23–27 days. **BEHAVIOR:** Often many pairs nest in proximity to one another.

Kate Fremlin, NU

LESSER SCAUP *Aythya affinis*

HABITAT: Boreal forests, prairie potholes, and parklands associated with lakes and wetlands of various sizes. **LOCATION AND STRUCTURE:** In bushes or high grass, often with overhanging vegetation. Uses islands when available. Usually within 1 m of open fresh water. Simple scrape is lined with grass and some down. In some locations nests ar constructed over water; these nests have significant vegetation added to keep nest above water. *Down is similar to that of Greater Scaup.* **EGGS:** 8–10; 57 x 40 mm; IP 21–27 days. **BEHAVIOR:** Often multiple pairs nest relatively close to one another.

STELLER'S EIDER *Polysticta stelleri*

HABITAT: Limited range in N. America. Tundra with abundant freshwater ponds, usually within 30 km of coast, sometimes farther. **LOCATION AND STRUCTURE:** In open or among small shrubs, often within 2–3 m of water. Less commonly 50–300 m from water. Down is dark brown with occasional white tufts. Outside diameter 23–51 cm; height 14–25 cm; inside diameter 14–15 cm; depth 9–10 cm. **EGGS:** 5–8; 61 x 42 mm; IP 26–27 days.

SPECTACLED EIDER *Somateria fischeri*

HABITAT: Very limited range. Coastal estuaries and salt marshes, tidal rivers. **LOCATION AND STRUCTURE:** Close to water, usually within 2 m, often on island or peninsula in lake. Nest bowl is filled with a layer of vegetation. *Down is similar to that of Common Eider.* Outside diameter 24 cm; inside diameter 14 cm; depth 5 cm. **EGGS:** 4–5; 68 x 45 mm; IP 24 days.

Tim Bowman, AK

Gerrit Vyn, AK

KING EIDER *Somateria spectabilis*

HABITAT: Arctic tundra near fresh water close to coast, *unlike other eiders, which more commonly nest adjacent to salt water.* **LOCATION AND STRUCTURE:** In fairly open areas, sometimes concealed with willows. May be adjacent to or more than 300 m from water. Nests are solitary. *Down is darker than that of Common Eider.* Outside diameter 25 cm. **EGGS:** 4–5; 67 x 45 mm; IP 22–24 days.

COMMON EIDER *Somateria mollissima*

HABITAT: Most frequently marine islands and inaccessible coastal locations, occasionally on islands in lakes or rivers close to coast. *Extremely oriented to marine habitat.* **LOCATION AND STRUCTURE:** Often in south-facing and leeward sites, typically with some vegetation providing cover for nest. *Often next to prominent landscape feature* such as a log, rock, or structure. Scrapes from previous years are often reused. Collection of down can be substantial. Down is light gray-brown with lighter center and tip. Outside diameter 26 cm; inside diameter 21 cm; depth 7 cm. **EGGS:** 3–5; 77 x 51 mm; IP 24–25 days. **BEHAVIOR:** Often nests in large colonies, sometimes thousands of pairs. Young often form creches. Female may not leave nest at all after first week of incubation unless disturbed.

HARLEQUIN DUCK *Histrionicus histrionicus*

HABITAT: Along mountain streams and rivers. Also along coast and on offshore islands. **LOCATION AND STRUCTURE:** On mid-stream islands or brushy locations along inland rivers. Typically well concealed within 5–10 m from water. On ground, stumps, tree cavity, upturned rootwad of fallen trees, cliff ledges. Shallow scrape is lined with vegetation and eventually down. Down is light brown with pale center. Outside diameter 20–28 cm. **EGGS:** 5–6; 56 x 41 mm; IP 28 days. **BEHAVIOR:** Multiple pairs nest in proximity at times. Sometimes reuses nest sites in subsequent years.

SURF SCOTER *Melanitta perspicillata*

HABITAT: By ponds, lakes, rivers in boreal forest. **LOCATION AND STRUCTURE:** Well concealed and sheltered by low conifer branches, deadfall, or other vegetation. Average 19 m from water (reported range 3–49 m). Down is dark brown with light center. Outside diameter 31 cm; inside diameter 17 cm; depth 6 cm. **EGGS:** 6–9; 62 x 42 mm.

WHITE-WINGED SCOTER *Melanitta deglandi*

HABITAT: Large boreal-forest lakes, wetlands. **LOCATION AND STRUCTURE:** Well concealed in shrubs, often thorny. Rarely open tundra with some sort of concealment. Often well away from water; average 96 m, up to 800 m. Down is dark brown with light center; down feathers are larger than those of Black Scoter. Often reuses old scrapes, including those excavated by other species. Outside diameter 20 cm; depth 9 cm. **EGGS:** 8–10; 67 x 46 mm; IP 25–30 days. **BEHAVIOR:** Nests in association with gulls and terns in some locations.

BLACK SCOTER *Melanitta americana*

HABITAT: Near small and relatively shallow freshwater ponds, lakes, and coastal marshes in boreal forest and tundra. **LOCATION AND STRUCTURE:** In grass clumps in tundra. Typically within 30 m of water. Down is dark brown with light center. Outside diameter 22–28 cm; inside diameter 15–20 cm; depth 10–15 cm. **EGGS:** 8–9; 64 x 44 mm.

LONG-TAILED DUCK *Clangula hyemalis*

HABITAT: Subarctic and Arctic wetlands, including on offshore islands, high-elevation tundra. **LOCATION AND STRUCTURE:** On islands or peninsulas of lakes or ponds, 10 m or less from water. Occasionally in uplands up to 143 m from water's edge. Typically some concealment from at least one side, rarely with overhead cover. Shallow depression is lined with leaves or grass and eventually down during incubation. Outside diameter 19 cm; inside diameter 13 cm; depth 8 cm. **EGGS:** 7–8; 53 x 37 mm; IP 25–26 days. **BEHAVIOR:** Pairs may nest very close to one another (7.5 cm), especially on islands. Avoids nesting in proximity to nesting Herring Gulls, Pacific Loons, Common Eiders.

Long-tailed Duck.
USFWS, AK

RED-BREASTED MERGANSER *Mergus serrator*

HABITAT: Boreal and tundra freshwater lakes or rivers close to coast. Also along coast directly or on offshore islands. **LOCATION AND STRUCTURE:** Hidden under thickets, among rocks, tree roots, or rarely in a burrow. Typically less than 30 m from water, not more than 70. Down is brownish with light tips and center. Outside diameter 21–35 cm; inside diameter 17–25 cm; depth 5–10 cm. **EGGS:** 10 (5–25); 63 x 45 mm; IP 30–31 days. **BEHAVIOR:** On some islands multiple pairs may nest close together.

Thomas Erdman, WI

OVERWATER-NESTING DUCKS

These ducks typically nest in freshwater wetlands among emergent vegetation such as cattails and bulrushes. Nests are built of last year's vegetation, woven into a platform within the emergent stems of living plants. Several species do not include down in the nest. In addition to the species below, Mallard and some other species that typically build onshore will occasionally construct overwater nests.

FULVOUS WHISTLING-DUCK *Dendrocygna bicolor*

HABITAT: Limited range in N. America, primarily in cultivated areas such as flooded rice fields. Also freshwater marshes. **LOCATION AND STRUCTURE:** In flooded fields or wetlands in dense vegetation, floating or emergent. Less commonly on dry ground along edges of wetland. Nest begins as flimsy platform, developing more solid structure and bowl shape during egg laying. Often sheltered by a canopy of vegetation. Lined with grass and other vegetation. *No down.* Outside diameter 30–44 cm; inside diameter 17–27 cm; depth 7–14 cm; top of nest 0–50 cm above water. **EGGS:** 9–13. 53 x 41 mm; IP 24–25 days. **BEHAVIOR:** Pair bonds are believed to be continuous over multiple breeding seasons, or potentially the lives of individuals. Nests in colonies.

CANVASBACK *Aythya valisineria*

HABITAT: Wide variety of wetlands, most commonly shallow wetlands with emergent vegetation. **LOCATION AND STRUCTURE:** Floating nest amid emergent vegetation such as cattails or bulrush. Large, bulky foundation of coarse plant material. Shallow depression is lined with finer plant material and down. *Down is darker gray than that of Redhead.* Often with overhanging vegetation that hides nest, and one or more ramps of vegetation leading from nest to water. Rarely on dry ground. Size varies, as females continue to add to nest with changing water levels. Outside diameter 30–89 cm; inside diameter 13–26 cm; depth 5–17 cm; top of nest 2–69 cm above water. **EGGS:** 6–8; may be blue-green; 63 x 44 mm; IP 25 days. **BEHAVIOR:** Females frequently dump eggs in nests of other Canvasbacks and are also primary hosts for parasitic Redheads and Ruddy Ducks.

Fred Greenslade/Delta Waterfowl, SK   Fred Greenslade/Delta Waterfowl, SK

REDHEAD *Aythya americana*

HABITAT: Wide variety of wetlands. **LOCATION AND STRUCTURE:** Well concealed in emergent bulrush and cattail, often with overhanging vegetation above nest. Sometimes on top of muskrat lodges. Rarely in dry locations. Well-crafted, deep cup is constructed of old plant stems woven into emergent vegetation. Usually a foundation of plant material extends down into water. Inner lining often includes both cattail fluff and duck down. Down is pale gray-white. Outside diameter 23–58 cm; inside diameter

R. Wayne Campbell, BC

Redhead nests are typically concealed in tall emergent vegetation. Delta Waterfowl, MB

13–25 cm; depth 7.5 cm; top of nest 18–25 cm above water. **EGGS:** 7–8; 60 x 43 mm; IP 24–25 days. **BEHAVIOR:** Will dump eggs in nests of other Redheads. Both a victim and perpetrator of nest parasitism with a wide variety of other ducks, as well as bitterns, coots, harriers, pheasants, and terns.

RING-NECKED DUCK *Aythya collaris*

HABITAT: Shallow freshwater wetlands with emergent vegetation, most commonly *Carex* sedges but also cattails, bulrushes. **LOCATION AND STRUCTURE:** Over water among dense emergent vegetation, often sedges. At first a flimsy platform of vegetation that is enhanced into a minimal bowl-shaped structure after egg laying begins, often with a ramp of vegetation constructed to access nest. Lined with down. Outside diameter 16.5–28 cm; inside diameter 13–18 cm; depth 5–10 cm. **EGGS:** 7–9; 57 x 40 mm; IP 26 days.

Ring-necked Duck eggs close up.
Delta Waterfowl, MB

Delta Waterfowl, MB

RUDDY DUCK *Oxyura jamaicensis*

HABITAT: Ponds and wetlands with emergent vegetation in prairie and other western landscapes. **LOCATION AND STRUCTURE:** Partially floating, attached to emergent vegetation. Relatively flat platform of local plant materials, usually previous year's growth, but sometimes green material. Vegetation arches over top of nest, concealing it. Built up above water level, 2–30 m from open water. Typically no down used in inner lining; if present, it is white. Outside diameter 30 cm; inside diameter 10–30 cm; depth 5–9 cm; top of nest 10–19 cm above water. **EGGS:** 7–10; 62 x 46 mm; IP 24 days.

Mark Nyhof, BC

CAVITY-NESTING DUCKS

Several ducks use natural tree cavities and woodpecker cavities (made by large species such as Pileated Woodpecker and Northern Flicker for nests, typically close to water. They will all use human structures and constructed nest boxes when provided for them.

Common Goldeneye nesting in a natural cavity in a dead ponderosa pine tree. David Moskowitz, WA

BLACK-BELLIED WHISTLING-DUCK *Dendrocygna autumnalis*

HABITAT: Along shallow wetlands with mesquite savanna, live oak, and Bermuda grass surroundings. **LOCATION AND STRUCTURE:** Usually in a natural tree cavity, favoring ebony, mesquite, and cactus thickets. Less commonly a ground nest with cover of prickly pear or shrubs. May also use nest boxes. *No down in lining.* In cavity, uses only wood chips. Average cavity entrance is 18 cm wide by 32 cm tall. Smallest cavity entrances recorded are 10 x 12 cm. Up to 6 m high. **EGGS:** 13; 52 x 38 cm; IP 28 days. **BEHAVIOR:** Frequent brood parasite.

Bill Summerour, AL

WOOD DUCK *Aix sponsa*

HABITAT: By freshwater streams, ponds, and wetlands with extensive emergent shrubs and trees that tolerate flooding. Riparian and upland forests up to 2 km from water. **LOCATION AND STRUCTURE:** Natural cavity or rarely a large woodpecker cavity in a large tree, often very near or over water. Will use nest box,

Bill Summerour, AL

rarely other human structures. White down lining is added to cavity. Averages 7.3 m high, with higher cavities preferred. Prefers entrance diameter of 9–13 cm but will use much larger holes. **EGGS:** 10–12; 50 x 38 mm; IP 30 days. **BEHAVIOR:** Frequent nest parasite (egg dumping), mostly among other Wood Ducks but also Black-bellied Whistling-Duck.

BUFFLEHEAD *Bucephala albeola*

HABITAT: Near freshwater ponds and lakes in forested areas, especially with aspen and poplar. **LOCATION AND STRUCTURE:** Often uses unenlarged Northern Flicker holes, which are too small for any other cavity-nesting duck. Occasionally uses Pileated Woodpecker hole, natural tree cavity, or nest box. Distance to water from nest tree varies across range; in some regions within 25 m, in other areas up to 425 m. Pale gray-brown down with light center is added to interior, but no other materials. Uses less down than Goldeneye. Nest hole is 0.6–14 m off ground, rarely much higher. Entrance diameter as small as 6 cm. **EGGS:** 9; 51 x 36 mm; IP 30 days. **BEHAVIOR:** Mixed clutches with Goldeneyes occur, probably because of overlapping nest site selection.

COMMON GOLDENEYE *Bucephala clangula*

HABITAT: Wetlands, lakes, and rivers in areas of mature coniferous or deciduous forests. **LOCATION AND STRUCTURE:** Natural cavities or Pileated Woodpecker holes in any species of tree. Readily uses nest boxes. Sometimes other locations, such as crevices in rock. Nest hole is typically 1.3–13 m high, but sometimes at ground level. Down is whitish. **EGGS:** 7–10; 59 x 43 mm; IP 28–32 days.

BARROW'S GOLDENEYE *Bucephala islandica*

HABITAT: Boreal and montane forests by alkaline or freshwater lakes, beaver ponds, and sloughs. **LOCATION AND STRUCTURE:** Natural cavities or, less commonly, Pileated Woodpecker or enlarged flicker holes. Usually in aspen or Douglas-fir. Typically high off ground, but recorded 2–15 m high. Also rarely recorded in a variety of other settings, including crow nests, rock ledges, and marmot burrow. Gray down is added to interior. **EGGS:** 7–10; greenish hue; 62 x 44 mm; IP 28–32 days.

HOODED MERGANSER *Lophodytes cucullatus*

HABITAT: Usually in wide variety of forested wetlands, but occasionally in nonforested wetland areas. **LOCATION AND STRUCTURE:** Pileated Woodpecker hole or natural tree cavity. Highly variable in terms of height, dimensions, and tree species, with most important trait apparently being proximity to water. Will readily use nest boxes if close to water. Down is the only material added to cavity. **EGGS:** 9–11; *whitish*; 54 x 44 mm; IP 29–32 days.

Jack Bartholmai, WI

COMMON MERGANSER *Mergus merganser*

HABITAT: Mature coniferous or mixed forests adjacent to lakes and rivers. **LOCATION AND STRUCTURE:** Pileated Woodpecker hole, natural cavity, or rotten top of a tree. Readily uses nest boxes. Nest is lined with down and other materials. Reported up to 30 m high in tree and more than 0.5 km from water. Occasionally builds nest on ground under overhanging rocks, in hollow logs, or other cover; these nests include a bulky collection of vegetation and *may be difficult to distinguish from that of Red-breasted Merganser, which always nests on ground.* Down is light gray-white. **EGGS:** 9–12; 67 x 46 mm; IP 32 days. **BEHAVIOR:** May use same nest site year after year.

Dana Visalli, WA

UPLAND GAME BIRDS
(Cracidae, Odontophoridae, Phasianidae)

Twenty-three species of game birds, almost exclusively ground nesters, breed in N. America. They include 3 distinct taxonomic families: chachalacas (Cracidae), New World quail (Odontophoridae), and pheasants, grouse, and allies (Phasianidae). For most species, the nest is a *simple oval or round scrape with a small amount of vegetation lining*, collected from the immediate vicinity. Nests are often located in brush, under logs, or near similar structures that help obscure the nest. Breeding behavior varies among species from monogamous to polygamous or polygynous. In most species only the female incubates, and her cryptic color helps camouflage the nest. Eggs are short subelliptical to oval and either a solid color or lightly marked. Young are precocial and often leave the nest within a day of hatching; they can feed on their own with guidance from an adult almost immediately. In many instances broods stick together until the following breeding season.

CHACHALACAS

PLAIN CHACHALACA *Ortalis vetula*
HABITAT: Lower Rio Grand Valley thorny scrub and brushland, forest edges. **LOCATION AND STRUCTURE:** Arboreal, in tree forks, horizontal limbs, brush tangles. Nest is flimsy, flattish, and refurbished from or built atop that of another bird (cuckoo, thrasher, etc.). Often appears too small for the size of bird. Outside diameter ~21 cm. **EGGS:** 3; buff white; ovate to long ovate; 59 x 42 mm.

NEW WORLD QUAIL

Four of these 6 species construct a concealed scrape nest common among game birds. Exceptions are Montezuma Quail and Northern Bobwhite, both of which form an arched covering of vegetation over the nest. Breeding structure varies. New World quail are apparently socially monogamous for individual broods, though partners may change between broods. In many species, the male assists with nest construction and in some cases with brooding young.

MOUNTAIN QUAIL *Oreortyx pictus*

HABITAT: Forest edges and desert scrub. Strongly associated with shrubby habitat with dense cover. **LOCATION AND STRUCTURE:** Often on or near a steep slope under a shrub or other cover. Basic scrape is lined with vegetation. Outside diameter 15 cm; depth 8 cm. **EGGS:** 7–10; completely unmarked; 35 x 27 mm; IP 24–25 days. **BEHAVIOR:** Two females may lay eggs in the same nest, leading to larger numbers. Male may help make the nest scrape, which female lines.

Robert Gundy, KY

NORTHERN BOBWHITE *Colinus virginianus*

HABITAT: Various scrublands, grasslands, and woodlands. **LOCATION AND STRUCTURE:** Often located on edge of or within 20 m of open area. Commonly among bunch grasses but will also nest in pricky pear cactus. Vegetation-lined scrape oftentimes has adjacent vegetation pulled over top of nest, concealing it. Outside diameter 15 cm; depth 5 cm. **EGGS:** 12–16; unmarked; 30 x 25 mm; IP 23–24 days (M, F). **BEHAVIOR:** When conditions are optimal, may successfully raise 2 or more broods. Male occasionally assists with incubation.

SCALED QUAIL *Callipepla squamata*

HABITAT: Arid grassland and desert. **LOCATION AND STRUCTURE:** In dense vegetation with visual obstruction. Scrape is lined with grass stems, *sometimes with a canopy*. Outside diameter 23 cm; depth 8 cm. **EGGS:** 12–14; finely marked; 33 x 25 mm; IP 22–23 days. **BEHAVIOR:** In dry seasons, breeding may not occur; in favorable conditions can be double-brooded. Two females may lay in same nest.

CALIFORNIA QUAIL *Callipepla californica*

HABITAT: Wide variety, from forest to desert. **LOCATION AND STRUCTURE:** Typically a simple lined scrape on ground, concealed in brush. Occasionally raised off ground on stump or among branches, or in an old nest of another bird. Outside diameter 13–18 cm; depth 2–4 cm. **EGGS:** 12–17; marked with brownish spots, blotches; 31 x 24 mm; IP 21–23 days (F). **BEHAVIOR:** Occasionally double-brooded. Male assists with rearing; female may leave first brood with male to start second brood with a different mate.

Mark Nyhof, BC

GAMBEL'S QUAIL *Callipepla gambelii*

HABITAT: Desert scrub. **LOCATION AND STRUCTURE:** Sheltered under shrub, fallen branches, or tall vegetation, occasionally off ground (up to 10 m) if a platform is available. Scrape is bordered with twigs and sparsely lined. Outside diameter 13–18 cm; depth

John Alstrup, AZ

4 cm. **EGGS:** 10–12; *more heavily marked than those of California Quail*; 32 x 24 mm; IP 21–23 days. **BEHAVIOR:** Sometimes double-brooded. Nests with more than 15 eggs may be from more than 1 female laying in same nest.

MONTEZUMA QUAIL *Cyrtonyx montezumae*

HABITAT: Montane pine and oak forest, often with steep hillsides and tall grass or other cover. **LOCATION AND STRUCTURE:** In brush, often tucked into clump of grass or other vegetation. Lined with grass and leaves. *Grass stems are pulled together around and over nest to form an enclosed nest cavity.* Outside diameter 13–15 cm; height 10–13 cm. **EGGS:** 6–14; unmarked; 32 x 25 mm; IP 25–26 days (M, F). **BEHAVIOR:** Both male and female incubate eggs and care for young. When conditions are optimal, female may lay multiple clutches in succession, which male tends.

Note the constructed roof over this Montezuma Quail nest and the recently hatched young in the foreground.
Rick Bowers, Bowersphoto.com, AZ

PHEASANTS, GROUSE, and ALLIES

The breeding behavior of this family is generally very conspicuous and notable. The strut of Wild Turkey males and the large gatherings of male sage-grouse, known as leks, where males perform for females to attract a mate, are examples. As with other game birds, nests are a shallow scrape with a variable amount of lining added to it. Specific locations for nests vary with habitat and species.

RUFFED GROUSE *Bonasa umbellus*

SPRUCE GROUSE *Falcipennis canadensis*

DUSKY GROUSE *Dendragapus obscurus*

SOOTY GROUSE *Dendragapus fuliginosus*

HABITAT: Various forest habitats, depending on species. Dusky Grouse also nests in shrub-steppe. **LOCATION AND STRUCTURE:** A bowl lined with feathers and vegetation from surrounding area. Dusky, Sooty, and Spruce Grouse nests are typically in a well-concealed location under brush or other cover in forest, almost always with significant overhead cover. Ruffed Grouse prefers deciduous forest, either near cover or in the open, often offering female a clear view of her surroundings. Inside diameter 16–22 cm; depth 4–8 cm. **EGGS:** Ruffed: 9–12; unmarked or finely speckled; 40 x 30 mm. Spruce: 7–8; marked with irregular blotches; 44 x 32 mm. Dusky and Sooty: 6–8; *lighter base color with fewer and lighter marking than in Spruce*; 50 x 35 mm. **BEHAVIOR:** Male is polygamous and carries out distinctive mating displays in spring to attract female. Female may breed with more than 1 male. Nesting and young rearing by female only.

Ruffed Grouse.
Thomas Schultz, WI

Ruffed Grouse. Brian McConnell, WA

Sooty Grouse. Michael A. Schroeder, WA

Spruce Grouse. Bill Summerour, AB

Spruce Grouse. Mark Peck, ON

GREATER SAGE-GROUSE *Centrocercus urophasianus*

GUNNISON SAGE-GROUSE *Centrocercus minimus*

HABITAT: Obligates of sagebrush habitat; populations of these species have declined precipitously because of the conversion of many landscapes for agriculture, mining, and other uses. **LOCATION AND STRUCTURE:** Under sagebrush or similar structure. Nest is a bowl lined with sagebrush, other vegetation, and feathers. Outside diameter 17–22 cm; depth 4–9 cm. **EGGS:** 7–8; 56 x 38 mm; IP 25–27 days. **BEHAVIOR:** Polygamous; males gather to strut in groups at predictable location (lek). Females gather to watch, and breeding occurs there. Males take no part in nesting and rearing young.

Bottom left image caption: *Greater Sage-Grouse.* Stephanie Beh, NV

Greater Sage-Grouse. Joe Smith, MT

WILLOW PTARMIGAN *Lagopus lagopus*

ROCK PTARMIGAN *Lagopus muta*

WHITE-TAILED PTARMIGAN *Lagopus leucura*

HABITAT: Alpine or Arctic tundra. **LOCATION AND STRUCTURE:** A scrape, with vegetation pulled in from around nest by female as she incubates. She adds some body feathers. Willow Ptarmigan nests are often near boggy locations and typically partially obscured by vegetation. Rock Ptarmigan nests are often in open in rocky locations and may be very shallow. White-tailed Ptarmigan nests are in alpine areas, in rocky areas, or often in a clump of stunted trees and shrubs on snow-free slopes. Inside diameter 15–20 cm; depth 4–16 cm. **EGGS:** Willow: 6–11; 44 x 32 mm; IP 20–26 days; Rock: 5–10; 43 x 31 mm; IP 24–26 days. White-tailed: 4–7; 43 x 29 mm; IP 22–23 days. *Eggs of Willow and Rock are heavily marked*, those of *White-tailed often less so. Markings and background color of Rock eggs show higher contrast than those of Willow.* **BEHAVIOR:** Typically monogamous. Occasionally polygynous.

Willow Ptarmigan. Chris Smith, AK

Rock Ptarmigan. Kate Fremlin, Amaroq Wildlife Services, NU

Rock Ptarmigan. Cameron Carroll, Alaska Department of Fish and Game, AK

GREATER PRAIRIE-CHICKEN *Tympanuchus cupido*

LESSER PRAIRIE-CHICKEN *Tympanuchus pallidicinctus*

SHARP-TAILED GROUSE *Tympanuchus phasianellus*

HABITAT: Grassland and arid scrub and woodland. Sharp-tailed Grouse is also in canyons and some arid woodland. Prairie-chicken ranges do not overlap. There is some overlap in range between Sharp-tailed Grouse and Greater Prairie-chicken. **LOCATION AND STRUCTURE:** Typical game bird nest. Sharp-tailed nest is usually within cover of shrubs, well concealed by vegetation. Nest bowl is often slightly oval rather than round. Outside diameter 18 x 20 cm; depth 7 cm. Prairie-chicken nests are often in more open location, though surrounding grass may be used to obscure nest. Outside diameter 18 cm; depth 7 cm. All typically have some overhead cover. **EGGS:** 12–14; spotted or unmarked; 42–45 x 32–34 mm; IP 23–26 days. **BEHAVIOR:** Polygamous; males form leks, females nest alone. Lesser Prairie-chicken populations are in decline.

Sharp-tailed Grouse. Megan Milligan, MT

Greater Prairie-Chicken.

Sharp-tailed Grouse. Megan Milligan, MT

WILD TURKEY *Meleagris gallopavo*

HABITAT: Various habitats; open mature timber, patchwork landscapes, swamps, and grassy savanna are examples. **LOCATION AND STRUCTURE:** Often against a tree or under a log or overhead vegetation. A simple depression with minimal lining consisting of surrounding vegetation. Often oval rather than round. Inside diameter 20–34 cm. **EGGS:** 8–12; heavily marked; 63 x 47 mm; IP

Wild Turkey.
Matt Monjello, NH

Wild Turkey.
Bill Summerour, AL

28 days (F). **BEHAVIOR:** Males are polygamous. Females nest alone, though sometimes multiple females share a nest. Nest may be covered with leaves to protect eggs when female is absent.

CHUKAR *Alectoris chukar*

HABITAT: Introduced. Sagebrush-grassland. **LOCATION AND STRUCTURE:** Typically on a slope, often under sagebrush, saltbush, or similar, sometimes under rocks. Shallow depression is sparsely lined with grass and feathers. Larger than most quail nests. Outside diameter 20 cm; depth 5 cm. **EGGS:** 8–15; lightly speckled; 43 x 32 mm; IP 22–23 days. **BEHAVIOR:** Typically monogamous. Female incubates eggs alone, but male may incubate a separate clutch.

GRAY PARTRIDGE *Perdix perdix*

HABITAT: Introduced. Grassland, agricultural areas. **LOCATION AND STRUCTURE:** On ground in fields and pastures, shelterbelts, and fencerows. Scrape has minimal lining of vegetation and feathers. Inside diameter 14–29 cm; depth 4–11 cm. **EGGS:** 9–20;

36 x 27 mm; IP 23–25 days. **BEHAVIOR:** Rarely 2 females may lay eggs in same nest. Monogamous; male attends to female while incubating and may help brood young.

RING-NECKED PHEASANT *Phasianus colchicus*
HABITAT: Introduced. Agricultural edges. **LOCATION AND STRUC-TURE:** Generally on ground within high vegetation or brush. Rarely on straw bales or in old squirrel nests. Shallow depression is sparsely lined with vegetation that is within reach of nest, or unlined. Inside diameter 18 cm; depth 7 cm. **EGGS:** 7–15; solid olive brown; 46 x 36 mm; IP 23–27 days. **BEHAVIOR:** Male is polygamous; a single male travels with and mates with a group of females.

David Pierce, IA

GREBES (Podicipedidae)

Seven species of grebes nest in N. America. While several winter in salt water, most nest on fresh water. A few will nest in tidal wetlands and sheltered coastal bays. *Some are solitary while others are colonial, a good starting point for differentiating species.* Grebes sometimes nest in association with other wetland birds, such as ducks, coots, and rails. All species are monogamous, and both parents share in incubation.

Nests are constructed in water and consist of a *pile of wet vegetation*. This may be floating, anchored to emergent vegetation, or built from the bottom up in shallow water. Coarser plant material is mixed with sodden, decomposing matter, the latter of which may add heat to the eggs and assist with incubation. *Rarely, the nest is constructed on dry ground or atop a structure sticking out of the water*, such as a rock. The amount of material in the nest, and thus its height above water, varies among species. Grebes also sometimes construct platforms for copulation but not for nesting.

A Red-necked Grebe parent next to their nest attached to emergent vegetation. Les Dewar, BC

Surprisingly, eggs in the nest are typically wet or sometimes partially submerged. Eggs are often covered with vegetation when adults are absent. Eggs of all species are long elliptical; most are light blue initially but fade to white and often become stained brown by wet vegetation in the nest. Some are white when laid. Young are precocial and leave the nest soon after hatching, often transported on the back of a parent. Grebes sometimes use nests to roost and brood for a week or more after chicks hatch.

COLONIAL NESTERS

Mixed colony of Western and Clark's Grebes. Gae Henry, CA

Two Eared Grebes (left) and one Western Grebe (right) in a mixed colony in SK. Paule Hjertaas, SK

EARED GREBE *Podiceps nigricollis*

HABITAT: Associated with shallow, often alkaline bodies of water with aquatic vegetation, usually with limited or absent trees around perimeter. **LOCATION AND STRUCTURE:** Along shoreline away from trees, among sparse emergent vegetation with easy escape by diving and swimming. Nests may be tightly spaced (0.5–3 m apart), constructed with bent-over emergent plants as foundation, with plant material from lake bottom piled on top. Outside diameter 20–34 cm; inside diameter 13–23 cm; top of

Eared Grebe nest with eggs lightly covered.
Paule Hjertaas, SK

nest 8–15 cm above water. **EGGS:** 3–4; 44 x 30 mm; IP 20–24 days. **BEHAVIOR:** Males and females tend to young. Often in dense colonies of up to hundreds of pairs.

WESTERN GREBE *Aechmophorus occidentalis*

CLARK'S GREBE *Aechmophorus clarkia*

HABITAT: Large freshwater lakes with open water and emergent vegetation around edges. Occasionally tidal marshes. **LOCATION AND STRUCTURE:** Built into emergent vegetation in water, often more than 25 cm deep. Usually close to edge of open water, either floating anchored to vegetation or built up from bottom in shallower water. Often well concealed. If water recedes after start of nesting season, nest may be entirely out of water by hatching. Outside diameter 54–120 cm; height 9–15 cm. **EGGS:** 3–4; 58 x 39 mm; IP 23 days. **BEHAVIOR:** Typically colonial; can be 100 pairs or more. Occasionally a single pair nests alone.

SOLITARY NESTERS

LEAST GREBE *Tachybaptus dominicus*

HABITAT: Still or slow-moving brackish and fresh water. **LOCATION AND STRUCTURE:** Floating platform in water up to 1 m deep, among emergent vegetation or in open but anchored to vegetation. Outside diameter 20–30 cm; inside diameter 7 cm; depth 2 cm; top of nest 5–6 cm above water. **EGGS:** 4–6; whitish to pale blue; 34 x 23 mm; IP 21 days.

PIED-BILLED GREBE *Podilymbus podiceps*

HABITAT: Often in small ponds and wetlands, occasionally brackish water. **LOCATION AND STRUCTURE:** Floating platform anchored to either tall emergent or underwater vegetation. Sometimes in open water. If in emergent vegetation, nest will be close to open water. Nest cup is relatively deeper than in other species. Outside diameter 19–25 cm; depth 3 cm. **EGGS:** 2–10; 44 x 30 mm; IP

Pie-billed Grebe. David Moskowitz, WA   *Pie-billed Grebe.* Eric C. Soehren, AL

23 days. **BEHAVIOR:** Occasionally double-brooded, with pair reusing same nest. May build mating and brooding platforms in vicinity of nest.

HORNED GREBE *Podiceps auritus*

HABITAT: Small lakes, ponds, and marshes in a variety of ecosystems. Occasionally larger lakes. **LOCATION AND STRUCTURE:** Nest is often built among tall emergent vegetation for cover but within 3 m of open water. Sometimes built on rock, anchored to bottom of pond, or rarely built on shore. A mass of vegetation and mud. Outside diameter 30–40 cm; top of nest 0.5–14 cm above water. **EGGS:** 4; white when laid, staining brownish; 44 x 30 mm; IP 22–25 days. **BEHAVIOR:** Typically nests in isolated pairs (1 pair per small pond), with breeding pair defending territory from all others. On larger lakes with plentiful food, small loose colonies of up to 20 pairs can be found.

Paule Hjertaas, AB

RED-NECKED GREBE *Podiceps grisegena*

HABITAT: Northern freshwater ponds and lakes with some emergent vegetation. Also sloughs, slow-moving rivers, bogs, and large irrigation ditches. **LOCATION AND STRUCTURE:** Often in slightly brackish or fresh water, in water greater than 20 cm deep. Typically in emergent vegetation but also in open water

Red-necked Grebe.
Mark Nyhof, BC

with submerged mats of vegetation. Size and shape highly variable. Outside diameter 33–275 cm; inside diameter 18–28 cm; depth 4 cm; top of nest 7 cm above water. **EGGS:** 4–5; 55 x 35 mm; IP 22–25 days. **BEHAVIOR:** Typically a single nest by a pair on a small body of water, sometimes multiple separated pairs on a larger body of water.

PIGEONS and DOVES (Columbidae)

These small to medium-sized birds feed largely on seeds, fruits, and grains. Eleven species breed in N. America in a wide variety of open and semiopen habitats, including woodlands, forest edges, riparian corridors, deserts, prairies, and areas near human habitation. Nests are typically *thin, shallow, saucerlike platforms of twigs with little to no lining*, and are located in trees, shrubs, cacti, rocks, ledges, buildings, and on the ground. Nests may be reused, and some species will use nests of other birds as a foundation on which to build. The male generally selects the nest site and presents nest material to the female, sometimes landing on her back when delivering items. Both sexes may build, but construction is typically done mostly by the female. Nests usually take 2–4 days to complete, sometimes up to 10.

Columbids typically are monogamous, and pairs may have multiple broods per year, with clutch sizes averaging 1–2. Eggs are white, unmarked, and vary in shape from subelliptical to long subelliptical. Both parents incubate: males generally during the day, females at night. IP is typically 12–14 days but may be as long as 22 days in larger species. Both parents begin to produce "pigeon milk" a few days before eggs hatch. They feed this nutrient-rich substance to the squabs (chicks) for about a week before the chicks begin to transition to a seed and fruit diet. Depending on the species, young hatch synchronously or asynchronously and fledge at anywhere from 11 to 45 days.

ROCK PIGEON *Columba livia*

HABITAT: Urban, suburban areas, farms; occasionally in caves, remote rocky cliffs. **LOCATION AND STRUCTURE:** On ledge beneath an overhang in caves and crevices; in narrow spaces, ledges of human-made structures. Typically 3–10 m high; may be as high as 30 m; occasionally on ground. A platform of small twigs, roots, pine needles, grass. Becomes solidified with droppings of young. Can grow quite large over years of use. Outside diameter up to 50 cm; height up to 20 cm. **EGGS:** 2; 39 x 29 mm; IP 18 days (F, M); NP 25–42 days (summer), up to 45 days (winter) (F, M). **BEHAVIOR:** 1–6 broods per year. May reuse nest repeatedly, building new nests atop old ones.

George Peck, ON

WHITE-CROWNED PIGEON *Patagioenas leucocephala*

HABITAT: FL Keys and southern tip of mainland FL: mangrove islands; rarely in woodlands and mangroves on coastal mainland. **LOCATION AND STRUCTURE:** Usually in dense thickets and well concealed by surrounding vegetation, in mangroves, shrubs, occasionally low in herbage, 0.1–7 m up. Nest is a weak platform of small twigs, roots, plant stems, often loose enough to see through. **EGGS:** 2; 37 x 27 mm; IP 13–14 days (F, M); NP 16–20 days (F, M). **BEHAVIOR:** Double- or treble-brooded. Usually nests in colonies. Numbers are declining due to habitat loss and hunting.

RED-BILLED PIGEON *Patagioenas flavirostris*

HABITAT: Rio Grande Valley. Floodplain forests and adjacent areas with tall trees. Populations are now more localized to islands where native habitat remains relatively undisturbed. **LOCATION AND STRUCTURE:** On horizontal branches or forks of trees, shrubs, saplings, also vegetative tangles, crown of spiny palm. Usually 1.2–7.6 m aboveground, but up to 24 m. Nest is a thin, flimsy platform of twigs, sometimes lined with grasses, stems. Outside diameter 20–25 cm; height 6.5 cm; depth 4–5 cm. **EGGS:** 1; 39 x 27 mm. **BEHAVIOR:** Several broods. Populations have declined due to habitat loss.

BAND-TAILED PIGEON *Columba fasciata*

HABITAT: Variety of mixed forest types; wooded urban and suburban areas; occasionally orchards. **LOCATION AND STRUCTURE:** High on a horizontal branch or fork, in both coniferous and deciduous trees. Average 9 m up but can be much lower or higher. *Nest is similar to that of Mourning Dove.* Sometimes contains pine needles, moss, and feathers. Outside diameter 20–24 cm; height 8.6–10 cm; inside diameter 13 cm; depth 3 cm. **EGGS:** 1; 40 x 28 mm; IP 16–22 days (F, M); NP 15–29 days (F, M). **BEHAVIOR:** Single- to treble-brooded. May be nomadic, depending on food availability. Colonial, often in small groups.

EURASIAN COLLARED-DOVE *Streptopelia decaocto*

HABITAT: Introduced; range expanding. Strong association with human development. Avoids large tracts of forest, also heavily farmed areas if there are no spots to nest and roost. **LOCATION AND STRUCTURE:** Nests in both coniferous and deciduous trees, shrubs; also on ledges of human-made structures. Usually 3–12 m up. Nest is a thin platform of twigs, stems, roots, grasses, occasionally feathers and human material. One nest (multiple use) in CO: outside diameter ~21 cm; height ~14 cm. **EGGS:** 2; 30 x 23 mm. **BEHAVIOR:** 3–6 broods. May reuse nests from previous seasons.

Casey McFarland, CO

SPOTTED DOVE *Streptopelia chinensis*

HABITAT: Introduced, limited to s. CA. Wooded urban and suburban habitats with dependable water source and access to open spaces; also tall tree groves around farms. Almost entirely in nonnative habitat. **LOCATION AND STRUCTURE:** In a fork or on a horizontal branch of a tree or shrub, sometimes on a building ledge. Nest is a platform of loosely constructed twigs, occasionally lined with rootlets or feathers. Outside diameter ~18 cm. **EGGS:** 2; 26 x 21 mm; IP 14–16 days (F, M); NP 10–15 days (F, M). **BEHAVIOR:** Double- or treble-brooded.

INCA DOVE *Columbina inca*

HABITAT: Typically in arid to semiarid areas (also humid areas in TX) near towns, farms, parks, campgrounds, and cities; usually not in riparian areas. **LOCATION AND STRUCTURE:** Often close to residences. In shrubs or trees, vines, cacti, human-made structures. Usually unshaded from sun and 3–3.6 m aboveground, sometimes 0.2–16 m. Nest is a small, compact platform of thin twigs, stems, leaves, rootlets, sometimes hair and bark strips, usually unlined. Often reused year to year, becoming larger and reinforced with excrement. Outside diameter 5 cm; height 2.9 cm; depth 1.25 cm. **EGGS:** 2; 22 x 17 mm; IP 13–15 days (F, M); NP 12–16 days (F, M). **BEHAVIOR:** 2–5 broods.

Casey McFarland, TX

COMMON GROUND-DOVE *Columbina passerina*

HABITAT: Common in semiopen, dry habitats undergoing early successional growth: woodland edges, pine woods, beach dunes, lake margins, brushy fields, scrublands, desert riparian thickets, dry washes, flatlands. Also around human habitation, including citrus orchards, irrigated areas, suburbs. **LOCATION AND STRUC-TURE:** Nest site varies, may be in relatively open spot or well con-

Matt Monjello, NC

cealed. On ground, horizontal branches of shrubs or trees, stumps, palmetto fronds, corn stalks, posts, cacti. Usually low, up to 3 m aboveground, sometimes higher. Ground nest is a small depression lined with scant plant material. Elevated nest is a thin, flat, weak platform of fine twigs or pine needles. Lined with grass, rootlets. Outside diameter 6.4–7.6 cm; height 3–5 cm. **EGGS:** 2; 22 x 16 mm; IP 12–14 days (F, M); NP 11–14 days (F, M). **BEHAVIOR:** Single- to treble-brooded. Will reuse nests of other species, including Mourning Dove, Northern Cardinal, Abert's Towhee.

WHITE-TIPPED DOVE *Leptoptila verreauxi*

HABITAT: Native brushlands and woodlands along Lower Rio Grande Valley, citrus orchards, suburbs. **LOCATION AND STRUCTURE:** Often in dense interior portions of habitat, nesting in substrates taller than 3 m with good canopy cover: tree or shrub or vine tangles, usually 2–3 m aboveground, though may be higher, especially in woodlands. Nest is a shallow bowl-shaped platform of thick twigs, grasses, or weed stems. *More substantial than nests of Mourning and White-winged Doves.* **EGGS:** 2, but subject to interspecific egg dumping; 21 x 33 mm; IP 12–14 days (F, M); NP 12–15 days (F, M). **BEHAVIOR:** Single- or double-brooded.

WHITE-WINGED DOVE *Zenaida asiatica*

HABITAT: Variety of humid to semiarid habitats, including thorny brushlands, tamarisk thickets, mesquite woodlands, oak-juniper forests, cactus and palo verde deserts, riparian zones; residential shade trees, orchards. Favors woodland interiors over edges. **LOCATION AND STRUCTURE:** On horizontal branch or crotch, shaded and close to trunk, 1.5–12.2 m high. Nest is a shallow platform of small twigs; occasionally weeds, grasses, moss. Rarely lined with feathers, other material. Outside diameter 11–15 cm. **EGGS:** 2; 31 x 23 mm; IP 14–20 days (F, M); NP 13–18 days (F, M). **BEHAVIOR:** Often nests in colonies in denser habitat. Usually double-brooded, or single-brooded in deserts. Range is expanding northward.

Bill Summerour, AL

MOURNING DOVE *Zenaida macroura*

HABITAT: Widely variable, including residential areas, windbreaks, orchards, open woodlands, grasslands, roadsides, deserts. Typically absent from dense or contiguous forests, common along habitat edges. **LOCATION AND STRUCTURE:** Nest substrate is highly variable, including coniferous and deciduous trees and shrubs, human-made structures, and on ground. Up to 80 m high. Nest is a small, flimsy platform of small twigs, pine needles, grass stems, often just large enough to hold eggs. Outside diameter ~7–13 cm. **EGGS:** 2; 28 x 22 mm; IP 14 days (F, M); NP 13–15 days (F, M). **BEHAVIOR:** 1–6 broods.

Casey McFarland, AZ

Mark Nyhof, BC

CUCKOOS, ROADRUNNERS, and ANIS (Cuculidae)

The 6 species of medium-sized birds that breed in N. America build fairly sizeable, somewhat messy nests of sticks and twigs loosely placed to form *a shallow bowl or platform* that is *typically lined with leaves and other debris.* Breeding behavior among many species is peculiar and interesting: some species are communal nesters, some have remarkably fast breeding cycles, and others have elaborate pair-bonding rituals. Males of many species attract females with cooing calls and offerings of food. Both sexes build, incubate (males at night), and care for young. Some cuculids (*Coccyzus spp.*) may lay eggs in the nest of other cuckoos, as well as those of other birds, particularly if food sources are abundant. Young generally leave the nest before they are capable of flying.

Eggs range from light blue or light greenish blue (cuckoos and anis) to white (roadrunner), and in some species have a chalky coating. Eggs hatch asynchronously, and if food is in short demand, incubation may be halted and chicks may die; on occasion parents kill some of the young to feed to others. With the exception of the roadrunner, these birds tend to be secretive, shy, and inconspicuous, particularly the cuckoos.

YELLOW-BILLED CUCKOO *Coccyzus americanus*

HABITAT: Open woodlands, moist thickets, streams and marshes, clearings with low, dense scrub, overgrown orchards, gardens, desert riparian woodlands. **LOCATION AND STRUCTURE:** In trees, bushes, vines, generally within dense foliage, usually not higher than 11 m, usually 0.6–6 m. In trees usually on horizontal branch, in fork, or in crotch, away from trunk and closer toward limb end. Nest is a smallish, loose, flat platform, frail and oblong, of dry twigs. Lightly lined with dry leaves, bark, moss, pine needles, weeds, catkins. Outside diameter 12.7–20.3 cm; depth 3.8 cm. **EGGS:** 3–4; pale bluish or bluish green; elliptical to subelliptical;

Mark Peck, ON

30 x 23 mm; IP 9–11 days (F, M); NP 7–8 days (F, M). **BEHAVIOR:** May be double-brooded in some regions, and breeding seems to coincide with food abundance. Breeding cycle happens incredibly quickly: 17 days from egg laying to fledging. Populations are declining rapidly throughout range.

BLACK-BILLED CUCKOO *Coccyzus erythropthalmus*

HABITAT: More northern range than Yellow-billed Cuckoo, but in similar habitats, though Black-billed tends toward more extensive, densely wooded habitat. Often associated with water. *Also in conifers, unlike Yellow-billed.* **LOCATION AND STRUCTURE:** *Similar to that of Yellow-billed*, but nest is often more compactly made and with more lining, sometimes including materials such as bracken ferns, spider web, etc. **EGGS:** 2–3; *like Yellow-billed's but smaller.*

Jeffrey A. Stratford,
PA

MANGROVE CUCKOO *Coccyzus minor*

Extremely secretive, and in U.S. only on southern tip and western coast of FL; also in FL Keys. Nest is like that of Yellow-billed Cuckoo, usually on horizontal branches in mangroves. Eggs are like those of other cuckoos.

GREATER ROADRUNNER *Geococcyx californianus*

HABITAT: Open, arid country with low trees, thickets, chaparral, cactus, dense shrubs. In eastern range, in pines, hardwood uplands. **LOCATION AND STRUCTURE:** Usually low in small tree, bush, or cactus, 0.9–2.7 m, rarely higher. Commonly in isolated

Greater Roadrunner.
Rick Bowers, Bowersphoto.com, AZ

patches of thickets or small trees, in crotch or on horizontal branch. Shallow, compact, platform-like cup nest is made of sticks, lined with leaves, grass, roots, feathers, snakeskin, and bits of manure or debris. Sometimes bailing twine, similar. Largest sticks tend to be under 1 cm in diameter, but many are much thinner. Outside diameter ~25–30 cm but may be more than 45 cm; height 15–20 cm; depth 5–10 cm. **EGGS:** 3–6; white, often with chalky, sometimes yellowish coating; short oval to subelliptical, elliptical; 39 x 30 mm; IP 20 days (F, M, male at night); NP ~11 days (F, M). **BEHAVIOR:** Female primarily builds, male brings materials. Double-brooded. Copulate at nest site. Monogamous, with long-term pair bonds; both defend territory. May start and abandon multiple nests.

GROOVE-BILLED ANI *Crotophaga sulcirostris*
HABITAT: Lower Rio Grande Valley. Open to semiopen country, orchards, scattered trees, brushy habitats of thick or tangled growth, thorny trees and shrubs, bushy pastures. **LOCATION AND STRUCTURE:** Often well hidden in thick foliage of trees or bushes, in forks or on twigs and branches, commonly below 3 m (range 0.6–7.6 m), sometimes higher. Nest is a large, bulky, shallow bowl of sizeable twigs, lined with fresh green leaves; leaves are added until eggs hatch, creating a thick layer. Outside diameter ~30 cm; inside diameter ~10–15 cm. **EGGS:** 3–20, with highest number known to hatch being 13; pale blue, but so covered in chalky coating that they appear white; 31 x 24 mm; IP ~13 days; NP 6–11 days. **BEHAVIOR:** Communal nesters; 1–5 pairs may defend a group territory and contribute to nest building, and females will each lay eggs in same nest. The whole group incubates (males at night) and tends to clutch. Each time a bird takes a turn at incubation, it adds a leaf to nest.

SMOOTH-BILLED ANI *Crotophaga ani*
Nest and nesting behavior are similar to that of Groove-billed Ani, though Smooth-billed may create layers of eggs and vegetation in nests, as each female that lays covers those laid prior with twigs and leaves.

NIGHTHAWKS and POORWILLS
(Caprimulgidae)

The 8 species that breed in open landscapes across N. America all lay their eggs *directly on the ground without any nest structure.* The highly cryptic feather patterns of adults are an important part of protecting the eggs, which are laid in open areas without cover, except for the occasional shade of a bush or small bit of vegetation. The parent will usually sit tight on the eggs when danger approaches, possibly resorting to an intimidation and distraction display. Eggs are typically cryptically colored (though those of Common Poorwill are white) and during the IP may be moved within the vicinity of where they were laid. Two-egg clutches are typical, and 2 broods in a season are possible for some species. In some species the female is the sole incubator, in others both adults participate. Birds may reuse an egg-laying location from season to season.

Egg shape and size, along with geographic location, can help differentiate among species, although

Chuck-will's-widow. Bill Summerour, AL

eggs are rarely found without an incubating adult during the day. IP is typically 18–21 days. Young are altricial, and it is 3–4 weeks before significant flight ability develops. The downy feathers of nestlings are cryptic and help hide them on the ground.

LESSER NIGHTHAWK *Chordeiles acutipennis*

HABITAT: Desert and arid landscapes, often in proximity to water sources. Where range overlaps with Common Nighthawk, *Lesser is typically found in valleys or basins, Common at higher elevations.* **LOCATION AND STRUCTURE:** Eggs are laid on gravel. **EGGS:** Cream-colored with heavy markings; oval to elliptical oval; 27 x 20 mm.

Bill Summerour, NM

COMMON NIGHTHAWK *Chordeiles minor*

HABITAT: Large variety of open landscapes, from grasslands to logged forests to human rooftops. **LOCATION AND STRUCTURE:** Eggs are laid on bare ground, occasionally in shelter of small grass clump or shrub. **EGGS:** Cream-colored with heavy markings; long elliptical to elliptical; 30 x 22 mm.

Nate Bacon, WA

Well-camouflaged Common Nighthawk nestlings. Aaron Van Geem, NV

COMMON PAURAQUE *Nyctidromus albicollis*

HABITAT: Brush or thornbrush habitat: mesquite, oak, similar. **LOCATION AND STRUCTURE:** Nests in open locations on bare ground or leaf litter. **EGGS:** Creamy buff to pinkish with markings; long elliptical to subelliptical; 30 x 22 mm. **BEHAVIOR:** Male and female incubate.

David Pierce, TX

COMMON POORWILL *Phalaenoptilus nuttallii*

HABITAT: Open or shrubby arid landscapes. *Avoids urban landscapes used by Common Nighthawk, and wooded landscapes used by whip-poor-wills.* **LOCATION AND STRUCTURE:** Eggs are often near shelter of a shrub. **EGGS:** White; oval to elliptical oval; 26 x 20 mm. **BEHAVIOR:** Both sexes incubate.

CHUCK-WILL'S-WIDOW *Antrostomus carolinensis*

HABITAT: Woodlands with open understory and adjacent openings. **LOCATION AND STRUCTURE:** Often lays eggs under dense cover by an opening. **EGGS:** White-gray with variable markings; subelliptical; 36 x 25 mm.

Bill Summerour, AL

BUFF-COLLARED NIGHTJAR *Antrostomus ridgwayi*

HABITAT: Dense brush in arid landscapes, often in canyons and ravines. **EGGS:** Cream-colored with purplish brown splotches; elliptical to subelliptical; 21 x 19 mm.

EASTERN WHIP-POOR-WILL *Antrostomus vociferous*

HABITAT: Woodlands with open understory and adjacent openings. *Where it overlaps with Chuck-will's-widow, Eastern Whip-poor-will is more common at higher elevations and in more heavily*

Bill Summerour, AL

forested areas. **LOCATION AND STRUCTURE:** Eggs are laid on leaf litter, often beside a small herbaceous plant. **EGGS:** Cream-colored with darker markings, fading as they age to match leaves; elliptical to short subelliptical; 29 x 21 mm.

MEXICAN WHIP-POOR-WILL *Antrostomus arizonae*
HABITAT: Oak-pine and other woodlands. **EGGS:** White with very indistinct markings; elliptical to subelliptical; 29 x 21 mm.

SWIFTS (Apodidae)

Four species of these aerial specialists and small insectivores nest in N. America and exhibit differences in breeding and nesting strategies. They occupy urban areas, coastlines, canyons, lowlands, and montane forests. Nests may be a small cup of moss located behind a waterfall, a half-cup of dry twigs glued with saliva to the inside of a tree hollow or chimney, or a cup of feathers and plant material built into crevices of vertical cliff walls.

Swifts are monogamous. Pairs are often single-brooded (sometimes double) with clutch size varying from 1 to 7. Eggs are long oval to long sub-elliptical, white, and unmarked. Both sexes incubate (16–27 days) and feed young. NP varies widely by species, 14–51 days. Cooperative breeding behavior has been observed in both Chimney and Vaux's Swifts, with pairs having 1 or more helpers in raising the young. Except in Vaux's Swift, species in N. America may reuse nests from season to season.

BLACK SWIFT *Cypseloides niger*
HABITAT: In areas with steep, vertical cliffs, along coasts and mountain canyons, from sea level to 2,600 m. **LOCATION AND STRUCTURE:** In damp, dark settings on a ledge or in a crevice; often in sea caves, near or behind waterfalls in montane areas. Nest structure and material vary based on habitat. Mountain nest: Small round cup primarily of moss, sometimes fern tips

Sue Hirshman, CO

and lined with pine needles. Coastal nest: Small mud depression with or without seaweed and moss. *Saliva is not used as a binding for either nest type.* **EGGS:** 29 x 19 mm; IP 23–26 days (F, M); NP 45–51 days (F, M). **BEHAVIOR:** May nest colonially, depending on site. Will reuse nests, adding small amounts of material each year.

CHIMNEY SWIFT *Chaetura pelagica*

VAUX'S SWIFT *Chaetura vauxi*

These closely related species build unique, shallow half-saucer-shaped nests of short dead twigs glued together and attached to a wall with saliva. Unlike Chimney Swift, which uses no lining material, *Vaux's may include pine needles and weed stems.* **HABITAT:** Chimney: Mostly in a variety of urban and suburban areas, in dark, well-sheltered locations such as chimneys, silos, air shafts, and other similar structures. Occasionally will nest in tree hollows, cavities, caves. Vaux's: Mostly in old-growth coniferous and mixed deciduous-coniferous forests inside a hollowed tree, accessed through broken-off top, or Pileated Woodpecker cavity, and usually placed on an underlying support. Will use nest boxes placed 10–15 m aboveground. **EGGS:** 4–5. Chimney: 20 x 13 mm; IP 16–21 days (F, M); NP 14–19 days (F, M). Vaux's: 18 x 12 mm; IP 18–19 days (F, M, H); NP 28 days (F, M, H). **BEHAVIOR:** Both sexes build in 18–30 days (Chimney on longer end). Nest construction continues during incubation. May reuse nests that are built on top of underlying support.

Chimney Swift. George Peck, ON

Chimney Swift with the peculiar long oval/long subelliptical egg shape common to swifts. Jack Bartholmai, WI

WHITE-THROATED SWIFT *Aeronautus saxatalis*

HABITAT: Arid mountains and canyons near open areas or dense forests, on coast along sea cliffs, also offshore on islands and isolated rocks. Will use a broad variety of human-made structures. **LOCATION AND STRUCTURE:** Often deep in narrow cracks and crevices in cliffs, canyons, and rocks, as well as in human-made structures. From 3 to 50 m high. Nest is shallow and saucer-shaped. Made from various materials such as feathers,

grasses, plant down, bark, moss, cotton. Glued together and to wall or ledge with saliva. **EGGS:** 4–5; 21 x 14 mm; IP 20–27 days, probably by both sexes. **BEHAVIOR:** Breeds in small colonies.

A White-throated Swift flying into their nest in a crack on a cliff face.
David Moskowitz, WA

HUMMINGBIRDS (Trochilidae)

Hummingbirds are the smallest of all birds. Roughly 14 species breed in N. America, several of them only on or near the U.S.–Mexico border. All polygynous, forming no pair bonds, and males and females keep separate territories; males hold areas of good feeding and access to females, and females choose areas with good nesting sites. Much commotion and squabbling occur during the breeding season, with highly aggressive territory defense, spectacular flight displays, and vocalizations. Males may be considerably depleted by the end of the breeding season, potentially losing a fifth of their weight. Hummingbirds can breed in their first spring, though males from later clutches may not have sufficient breeding plumage or vocalizations and must wait a year. Hybridization occurs among similar species and across genera. Most species are migratory, and many travel incredibly long distances.

Nests are generally very easy to recognize. Differentiating among species can be difficult, however, particularly where many species overlap. Familiarize yourself with which species are in the area, preferred habitat types, and subtle differences in nests as best as possible. The *typical nest is tiny and usually fastened to small horizontal or slanted twigs or branches, or attached atop small projections on walls or similar substrates.* It typically is composed of fine materials such as plant down and fine bark shreds, bound entirely in spider silk (as well as cocoon and caterpillar webbing, etc.) and has a soft, cottony inner cup. The outer wall is *often decorated in flakes of lichen,* small bits of leaves, bark scraps, and similar materials, likely helping camouflage the nest. The cup itself is often slightly hemispherical. Nests may be refurbished, or built on top of old ones (sometimes after several years of nonuse); the subsequent layering is distinct. Hummers will also reuse materials from old nests, especially in colder climes, thus saving energy essential for egg production. Stealing materials from the nests of other females is not uncommon. Because the silk content in nests is high, they are "stretchy," and niftily expand to accommodate the growing, oversized chicks. Nests are generally built in 5–14 days, but may be completed in a few days or over a month, and may be worked on simultaneously with laying and incubation. Eggs across genera and species are quite similar and impressively large relative to the size of the bird. They are white, elongated, and of fairly similar dimensions. For two-thirds of the species that follow, egg size is about 13 x 8–9 mm. Eggs of larger species (Rivoli's and Blue-throated Hummingbirds) are 15 x 10 mm; those of Costa's and Calliope Hummingbirds are the smallest: 12 x 8 mm. Most species are frequently double-brooded, with occasional third broods among some, while others are single-brooded with an occasional second (noted below). Females are the sole builder and incubator, and care for young entirely on their own. The nesting cycle takes ~42–44 days; IP varies across species, 14–18 days; NP is 19–30 days.

RUBY-THROATED HUMMINGBIRD *Archilochus colubris*

HABITAT: The only eastern breeding species, with the largest breeding range of any N. American hummer. Mixed woodlands, deciduous forest; parks, gardens, orchards, shade trees. In Southeast in pines, mixed pine, hardwood forests. **LOCATION AND STRUCTURE:** In a tree or shrub, usually close to tip or along a downward-sloping twig or branch, open to ground and sheltered from above by foliage. Height is 1.8–15 m, usually 3–6 m. Typical nest, the bulk of which is white plant down and bud scales, may use pine resin. Often *heavily decorated* with lichen. Bottom of nest may be stiff, wider than top. Outside diameter 4.5–5 cm (at base), height 4 cm; inside diameter 2.8 cm; depth 2.5–2.9 cm. **BEHAVIOR:** Unlike most other hummers, travels over open water, potentially traversing Gulf of Mexico.

Cyndi Shepherd, OH

BLACK-CHINNED HUMMINGBIRD *Archilochus alexandri*

HABITAT: Abundant, occupying diverse habitats. Semiarid to xeric landscapes, commonly in riparian floodplain and mesic canyon bottoms. Sycamore, cottonwood, ponderosa pine, pinyon-juniper, cypress, etc. Also in urban, suburban areas in shade trees, parks, gardens. *Usually in drier areas than Ruby-throated Hummingbird where they overlap.* **LOCATION AND STRUCTURE:** In a tree or shrub on a small horizontal branch or sloping to vertical hanging twigs; 1.2–2.4 m high, occasionally higher. Typical nest, but *may lack decoration, using uniform downy material for outer wall.* May include leaf fragments, seed pods and heads, bark scraps, and small feathers (which may be added to inner cup as well). Outside diameter 4 cm (range 3.6–4.5 cm); inside diameter 1.9–2.5 cm; depth 2.5 cm. **BEHAVIOR:** Nests are often within 100 m of one another along riparian areas.

Casey McFarland, NM

Casey McFarland, NM

ANNA'S HUMMINGBIRD *Calypte anna*

HABITAT: Diverse. In chaparral, especially partly wooded live oak canyon bottoms, coastal scrub, riparian woodland, savanna. Common in urban and suburban environs in gardens, parks, etc. **LOCATION AND STRUCTURE:** Broad variety of sites, including

Paul Suchanek, AZ

Anna's Hummingbird. Mark Nyhof, BC   *Anna's Hummingbird.* Matt Monjello, AZ

human-made items (e.g., windchimes). Some sites are similar to those of Black-chinned Hummingbird. Nest is 0.5–6 m up, sometimes to 15 m. Variable appearance. Walls contain plant down and other small vegetation, and cup is loosely lined with cottony material, feathers, hair. Bound with silks, also plant fibers or small mammal hair. Lichen decorations are common, also dead leaves, mosses, sometimes bark, algae. Cottony inner cup. Outside diameter 3.8–4.4 cm; height 3.2–4 cm; inside diameter 2.5 cm; depth 1.6–1.9 cm. **BEHAVIOR:** In BC, commonly treble-brooded, possibly a fourth. Early nester (Jan.).

COSTA'S HUMMINGBIRD *Calypte costae*

HABITAT: Harsh, xeric desert: habitats of ocotillo, yucca, creosote, cacti (saguaro, etc.); dry canyons and washes; coastal scrub, chaparral, sage scrub. Often far from water but will use riparian areas to some extent, particularly in regions such as Mojave Desert; usually in drier areas than other species, but may overlap with Black-chinned Hummingbird, with which females are nearly indiscernible. **LOCATION AND STRUCTURE:** On a branch, twig, stem in shrub or tree, near tips; also may build on leaves. Often 1–2 m high. Typical nest; often distinct in that it is more

Postfledging; note the rim of scats.
Scott Olmstead, AZ

shallow, relatively thin-walled, and smaller than other species' nests. Built of fine plant materials (tiny stems, bark shreds, lichen, tiny flower pieces, etc.) and feathers (appearing less cottony), and often has a gray appearance. Outside diameter ~4 cm (range 3.2–5 cm); height 2.5–3.8 cm; inside diameter averages 2.2 cm; depth averages 1.3 cm. **BEHAVIOR:** Often subordinate to other hummingbird species.

BROAD-TAILED HUMMINGBIRD *Selasphorus platycercus*

HABITAT: High mountains (3,230 m) and high desert. Pinyon-juniper, pine-oak, ponderosa pine, spruce-fir, aspen; riparian willow thickets. **LOCATION AND STRUCTURE:** Often on horizontal twig of a tree or shrub (alder, aspen, cottonwood, juniper, oak, conifer, etc.) with a protective overhead limb or bent trunk; usually 1–9 m aboveground but may be closer to ground, particularly in conifers (where nests are commonly over water). Typical nest; lichen exterior is common, also bark, moss. Interior of early nests is mostly of spider silk. Outside diameter 4.5 cm; inside diameter 1.9 cm. **BEHAVIOR:** Probably single-brooded. Builds while incubating.

RUFOUS HUMMINGBIRD *Selasphorus rufus*

HABITAT: The northernmost breeding hummingbird. Habitats diverse: second-growth and mature coniferous forests (fir, cedar, maple forests). In BC as many as 95 percent of nests are located in drooping limbs of western red cedar. Also in riparian areas, meadow and pasture edges, swamps, and areas of human habitation (farmlands, gardens and parks in towns, cities). Usually in older forests than Calliope's Hummingbird. **LOCATION AND STRUCTURE:** Often in an old tree, generally hidden on drooping lower branches or twigs, or in a shrub or bush, commonly in a horizontal fork. Also in vines, upturned tree roots. Height is 1–9 m, often low. Typical nest; variable. Similar to that of Allen's Hummingbird. Outside diameter 4.6 cm; height 3.2 cm; inside diameter 2.2 cm. **BEHAVIOR:** Double-brooded (BC), with second nest close to successful first. High site fidelity when successful; first nest of season is often built atop last year's. Makes the longest known avian migration relative to body length. May nest near conspecifics, with as many as 20 or more in close proximity. Very aggressive.

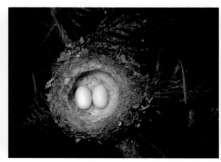

David Moskowitz, BC

Mark Nyhof, BC

ALLEN'S HUMMINGBIRD *Selasphorus sasin*

HABITAT: CA, in damp, coastal fog belt; mixed and riparian woodlands, coniferous woodlands; coastal scrub with scattered trees. Usually in denser vegetation than Anna's Hummingbird, and with some tree cover. **LOCATION AND STRUCTURE:** In a tree often saddling a small fork, or in a shrub (or blackberry tangle, fern, similar), from close to ground to 15 m, but often low, sometimes high up in tall trees. Typical nest; walls of downy material, bits of leaves, grass, and similar, decorated with lichens, mosses, sometimes bark. Cottony interior. *More commonly builds nest walls into surrounding twigs than Anna's, and nest is usually taller, bulkier.* Inside diameter 3.2–3.8 cm. **BEHAVIOR:** May have as many as 5 broods in prolonged breeding season, likely correlated with nectar availability. Flight-dive display of male is one of the most impressive and complex. Arrives at breeding grounds early, relative to most migratory hummers, often returning in Jan. Several nests may be clustered in close proximity, and male territories may contain 3 active nests.

CALLIOPE HUMMINGBIRD *Selasphorus calliope*

HABITAT: Breeds at wide range of elevations, 185–3,400 m (in Sierras), but most concentrated in mountainous areas, *often choosing younger (earlier) successional forest than Rufous Hummingbird.* Coniferous mountain slopes, meadows, willow thickets in drainages; mixed forests (i.e., red cedar, alder); birch, maple (BC). **LOCATION AND STRUCTURE:** Typically beneath an overhead branch for protection from weather, heat loss. Commonly built on surface where old pine cones have dropped from tree, *thus looking conelike.* Typical nest, decorated with lichens, moss, bark, *often giving non-cottony, rough-hewn appearance.* Outside diameter 3.8–4.5 cm; height 3 cm; inside diameter 2 cm; depth 1.6 cm. **BEHAVIOR:** Probably single-brooded. Smallest N. American hummingbird, and the tiniest long-distance avian migrant.

HUMMINGBIRDS THAT BREED ONLY NEAR U.S.–MEXICO BORDER

RIVOLI'S HUMMINGBIRD *Eugenes fulgens*

HABITAT: Mountainous country, occasionally north to CO. Canyons, drainages, to high-elevation forested slopes, usually in or near riparian habitat. Pine-oak, oak, pinyon-juniper, mixed conifers (fir), aspen. **LOCATION AND STRUCTURE:** Nest is usually high in tree (6 m or higher), saddled on a horizontal branch or fork well out from trunk, often above water. Typical nest, though more stout and squat, slightly larger than those of most other species, heavily covered in lichens, often with feathers. See Least Flycatcher, pewee. Outside diameter 5.7 cm; height 3.8–5.8 cm; inside diameter 2.9–3.8 cm; depth 1.9–3.2 cm. **BEHAVIOR:** The second-largest hummingbird in N. America (after Blue-throated Hummingbird).

BLUE-THROATED HUMMINGBIRD *Lampornis clemenciae*

HABITAT: Mountainous regions, generally in moist, shady zones; along streams in canyons in riparian woodlands; mixed and ever-green woodlands. **LOCATION AND STRUCTURE:** Often protected from above: beneath a rock overhang, undercut creek bank (on roots, leaves, stems, etc.), and commonly over water. Also on protected ledges, cave entrances, sometimes in old Black Phoebe nest; commonly around human habitation on exposed wires, nails, eaves, beams, wasp nests, etc. Typical nest, but large and often heavily swathed in silk. May appear "lumpy," with extra material added to bottom or when built on old nest. May incorporate cocoons and spider egg sacs, decorated with green mosses (though not in drier areas), mosslike lichen *(flakelike lichens common to other species are* not *used)*. Outside diameter 5–6 cm; inside diameter 3–4.5 cm; depth 1.5–3.2 cm. **BEHAVIOR:** The largest hummingbird in N. America (roughly 3 times the mass of many other species), with elaborate vocalizations.

Doug Backlund, AZ

LUCIFER HUMMINGBIRD *Calothorax lucifer*

HABITAT: Mostly in small portions of w. TX (Chisos Mts.). Arid deserts, talus slopes, dry washes, rocky hillsides. Along dry slickrock arroyos. Overlaps with Black-chinned Hummingbird. **LOCATION AND STRUCTURE:** On a shrub, cactus (live or dead cholla), ocotillo if leafed out, lechuguilla, agave. *Not found in riparian oak growth, unlike Black-chinned.* Usually low, 0.5–3m. Typical nest, of fibers, plant down, small stems, etc. *May be fairly thick-walled, decorated with small leaves.* Dimensions like that of Black-chinned and other similar-sized hummers.

BROAD-BILLED HUMMINGBIRD *Cynanthus latirostris*

HABITAT: Arid country in riparian canyons, creek bottoms. **LOCATION AND STRUCTURE:** Commonly at canyon edge, often on north-facing side. In thickets or small trees and shrubs, often on a thin, drooping twig or branch, at 1.5–3 m, often 2 m, sometimes lower. Typical nest, often with fine materials, mostly fine bark shreds, also *including grasses*; decorated with bark, dead leaf bits, etc., *rarely lichen.* Often lined with white plant down. Height ~2.5 cm; inside diameter 1.9 cm.

Casey McFarland, AZ

BUFF-BELLIED HUMMINGBIRD *Amazilia yucatanensis*

HABITAT: Arid country shrubland of Lower Rio Grande Valley to semihumid Gulf Coast woodland. **LOCATION AND STRUCTURE:** Small shrub or tree, usually in a *horizontal fork* (*Black-chinned Hummingbird nest is often situated differently*, but nests may be difficult to differentiate); 1–3 m aboveground. Typical nest; dimensions similar to those of Black-chinned.

VIOLET-CROWNED HUMMINGBIRD *Amazilia violiceps*

HABITAT: Rare, local. In deciduous riparian woodlands (sycamore, cottonwood, etc.). **LOCATION AND STRUCTURE:** Horizontal branches at or near tip, often shaded by a large leaf overhead; 7–12 m aboveground, sometimes lower. Typical nest. Cottony materials may give *whitish appearance*, decorated mostly with lichen, interior white plant cotton. *Tendrils of cotton or cobweb wrapped around exterior may be distinctive from other species, though not always present.* Outside diameter 3.8–4 cm; inside diameter 1.2–2.9 cm; depth 2–2.7 cm.

RAILS, GALLINULES, and COOTS (Rallidae)

All species nest in wetlands, next to or over water. Their nests are tight, sometimes bulky, cups of vegetation. They may be on the ground by water, attached to vegetation in water, suspended over water, or rarely in thick shrubs close to water. Nests are typically well concealed, with some hidden under a "roof" of natural vegetation. Monogamy is common, but behaviors may vary: polygyny, polyandry, promiscuity, cooperative breeding, and intraspecific brood parasitism possible. Males and females often share in incubation and rearing.

The egg of a Yellow Rail (right) and Sora (left). Note the unique distinctive patterning of the Yellow Rail egg, different from that of other rails (Black Rail can be similar).

RAILS

Nests of the 7 species that breed in N. America are all very similar in structure and location. They are typically in freshwater wetlands, often in cattail and sedge marshes, and are cup or basket nests of grasses, rushes, and sedges. Nests may be hidden on the ground in grasses or sedges or suspended over water in emergent vegetation, and are sometimes slightly submerged. They are typically well hidden; the birds often pull and weave surrounding vegetation over the nest to conceal it. Many species construct a ramp of vegetation leading into the nest.

Rails will defend their territory from other rails of the same or different species. They sometimes construct dummy nests for feeding and resting or in case of flooding of the primary nest. Eggs are subelliptical, but there is variation in size, color, and markings among species; Yellow Rail eggs are particularly distinctive. Chicks are precocial and leave the nest soon after hatching. Geographic range, habitat, and eggs can help differentiate among species.

SMALL RAILS

YELLOW RAIL *Coturnicops noveboracensis*

HABITAT: Fresh and brackish wetlands, wet sedge meadows, lightly brackish estuaries. Occasionally coastal salt marshes. Rarely associated with cattail. **LOCATION AND STRUCTURE:** In drier edges of wetlands, located on ground in a natural hollow of bent-over dead vegetation. Cup nest is made of fine grasses or sedges. Outside diameter 16 cm; inside diameter 7–10 cm; depth 3–8 cm. **EGGS:** 8–10; *off-white or buff, usually with speckling concentrated on larger end*; 29 x 21 mm; IP 1–18 days (F, M). **BEHAVIOR:** Second nest is constructed for brooding young, often without a covering canopy.

BLACK RAIL *Laterallus jamaicensis*

HABITAT: Limited breeding range in N. America. Coastal salt and brackish marshes, occasionally fresh water close to coast. **LOCATION AND STRUCTURE:** Built over moist soil or shallow water, often in a clump of vegetation. Relatively tight cup of grass or sedge with vegetation woven over nest. A ramp of vegetation is often constructed to access nest. Outside diameter 13 cm; inside diameter 7 cm; depth 4 cm. **EGGS:** 6–10; cream to pinkish, evenly spotted with fine dots or concentrated at larger end, can be somewhat similar to those of Yellow Rail; 26 x 20 mm; IP 17–20 (F, M).

VIRGINIA RAIL *Rallus limicola*

HABITAT: Freshwater wetlands, occasionally brackish water, rarely coastal salt marshes. **LOCATION AND STRUCTURE:** In areas of very shallow standing water or mud and plenty of vegetation. Nest is in emergent vegetation, away from open water, often along edge between vegetation types. Similar to Sora nest though typically in slightly drier areas. Loose cup of coarse plant

Virginia Rail.
David Moskowitz, WA

materials, often with canopy. Outside diameter 17 cm; inside diameter 12 cm; depth 3 cm. **EGGS:** 7–12; *paler and with smaller spots than Sora eggs*; 32 x 24 mm; IP 20 days (F, M). **BEHAVIOR:** Both sexes construct nest, including sometimes multiple dummy nests in vicinity of actual nest.

SORA *Porzana carolina*

HABITAT: Freshwater (occasionally brackish or saltwater) marshes with plenty of emergent vegetation. **LOCATION AND STRUCTURE:** Nest is built in a dense clump of vegetation near water, usually cattail or sedges, or suspended above water 6–12 cm. Loose cup of coarse local wetland plants with finer lining. Sometimes adjacent vegetation is pulled over nest. Often has runway of nesting material leading from nest to water. In areas of overlap with Virgina Rail, Sora typically selects wetter loca-

Dana Vissali, WA

tions; nests are very similar. Outside diameter 12–20 cm; inside diameter 9–13 cm; depth 6 cm. **EGGS:** 8–12; glossy cream to olive brown, irregularly spotted *(spotting larger than on Virginia Rail eggs)*; 32 x 23 mm; IP 18–20 (F, M). **BEHAVIOR:** Male brings material to nest site, and female constructs nest.

LARGE RAILS

CLAPPER RAIL *Rallus crepitans*

HABITAT: Saltwater marshes and mangrove swamps, often associated with ditches or tidal creeks. **LOCATION AND STRUCTURE:** Located just above high-tide line, usually within 15 m of open water. Just above ground to 1.5 m above water. Tall, bulky platform with varying amounts of overhead cover and variably with a ramp. Outside diameter 26 cm; inside diameter 15 cm. **EGGS:** 4–11; cream to variably colored with variable spots concentrated at larger end; 42 x 30 mm; IP 20 days (F, M). **BEHAVIOR:** Constructs additional brooding platforms.

Matt Monjello, NC

This Clapper Rail nest is well hidden within a bowerlike structure.
Matt Monjello, NC

Clapper Rail eggs in a nest. Liz Tymkiw, VA

KING RAIL *Rallus elegans*

HABITAT: Freshwater or tidal wetlands, rice fields. **LOCATION AND STRUCTURE:** In shallow water, in or on grass or sedge-type vegetation, rarely on dry ground away from water. Base of decaying vegetation with loose platform or cup of dry vegetation from surrounding area, usually with a canopy that may be domed or cone-shaped. Height above water varies. Outside diameter 28 cm; depth 2 cm. **EGGS:** 8–11; 41 x 30 mm; IP 21–24 days (F, M). **BEHAVIOR:** Male does most of nest construction. May build additional nests for brooding young, without canopies, in vicinity.

Bri Benvenuti,
USFWS, ME

RIDGWAY'S RAIL *Rallus obsoletus*
This rail (comprising 3 subspecies) is threatened and limited to narrow corridors in extreme sw. U.S. and along portions of Pacific Coast of cen. and s. CA. Nest is similar to that of other large rails.

GALLINULES and COOTS

Nests of these 3 species are found in a variety of still or slow-moving freshwater or tidal wetlands with a mix of open water and emergent vegetation. They are typically bulky floating nests constructed of a coarse outer layer of wetland plants, with a finer lining that is built after incubation starts. Sometimes the nest is suspended over water in emergent vegetation or is built on the ground adjacent to water. An entrance ramp is often constructed to access floating nests. All species often construct additional floating platforms for feeding, displaying, or brooding young. Eggs are sub-elliptical (to oval in Purple Gallinule), and their coloration and markings can help differentiate nests where species overlap. Young stay in the nest from 1 to several days.

PURPLE GALLINULE *Porphyrio martinica*
HABITAT: Freshwater wetlands with a mix of emergent and floating vegetation and open water; rice fields. **LOCATION AND STRUCTURE:** Either on floating mats of vegetation or suspended in emergent vegetation from water level to 60 cm above water. Local grass and vegetation are often bent into outer layer of nest cup without being severed from roots. Finer layer of vegetation is added after incubation begins. Surrounding vegetation may be pulled over to form partial covering. Ramp often constructed to access nest. Outside diameter 22 cm; inside diameter 11 cm. **EGGS:** 6; white to pink, irregularly spotted; 39 x 28 mm; IP 25 days (M, F).

Bill Summerour, AL

COMMON GALLINULE *Gallinula galeata*

HABITAT: Wide variety of freshwater and brackish wetlands, typically with a mix of tall emergent vegetation and open water. **LOCATION AND STRUCTURE:** Often on floating mats or attached to emergent wetland vegetation. Sometimes on water's edge and occasionally off ground in shrubs or in abandoned nest of another bird species. Nest platform consists of coarse parts of surrounding vegetation, with leaves and finer material added to create cup, occasionally with overhead vegetation pulled over to conceal nest and ramp for access. Outside diameter 24–31 cm; inside diameter 13 cm; depth 7 cm. **EGGS:** 5–11; gray with scattered markings; 44 x 31 mm; IP 19–22 days (M, F).

David Pierce, IL

AMERICAN COOT *Fulica americana*

HABITAT: Freshwater lakes, wetlands, and slow-moving water with emergent vegetation. **LOCATION AND STRUCTURE:** In areas of dense emergent vegetation, at least some of which is in relatively deep water. Nest is a floating platform with inner cup attached to emergent vegetation with overhead cover. Generally in water .1–1 m deep, often with a ramp of vegetation to access nest. Out-

Casey McFarland, WA

Casey McFarland, WA

side diameter 30–45 cm; inside diameter 18 cm; top of nest 20 cm above water. **EGGS:** 6–9; buff to pinkish with heavy, even speckling; 48 x 34 mm; IP 21–24 days (M, F). **BEHAVIOR:** Sometimes double-brooded. Socially monogamous. Populations often include floater females that lay eggs in nests of mated pairs. Coots construct multiple platforms prior to egg laying; those not chosen for egg nest may be used for roosting.

LIMPKIN (Aramidae)

LIMPKIN *Aramus guarauna*

HABITAT: Freshwater marshes, lakes, ponds, swamps. Range is closely connected to distribution of primary food source: the apple snail, an unusually large mollusk. **LOCATION AND STRUCTURE:** Often in thick emergent or floating vegetation, but also in shrubs and up to 12 m. high in trees. May be exposed or well hidden. A flimsy to sturdy platform of sticks and other vegetation from immediate vicinity, with a finer inner layer. Outside diameter 52 cm. **EGGS:** 3–8; light gray, with olive brown or purple-gray streaks; 60 x 43 mm; IP 27 days (F, M). **BEHAVIOR:** Often constructs a brooding platform over water, usually at a considerable distance from nest.

Tree nest of a Limpkin.
Jonathan Mays, FL

Ground nest of a Limpkin.
Linda H. Godwin, FL

CRANES (Gruidae)

SANDHILL CRANE *Antigone canadensis*

WHOOPING CRANE *Grus americana*

HABITAT: Freshwater wetlands and slow-moving waterways surrounded by grasslands, open forests, tundra, cropland, and savanna. Whooping Crane breeding areas are limited to 4 small distinct locations, in FL, LA, WI, and n. AB and s. NT. **LOCATION AND STRUCTURE:** Generally in standing water, a floating mound constructed of dominant local vegetation such as cattails, bulrush, or grass collected from immediately around nest. Around Sandhill Crane nests, often a distinctive zone denuded of vegetation. Sandhill Crane creates a lined cup of finer material built by female into nest mound. Whooping Crane builds a flat mound, sometimes with a slight depression for eggs. In some parts of its range, Sandhill nests on dry ground; these nests are typically much smaller with a less distinctive cup. Overall nest shape is oval rather than round. Sandhill Crane: Outside diameter 80–125 cm; height 10–15 cm. Whooping Crane: Outside diameter 60–160 cm; height 20–46 cm. **EGGS:** Sandhill: 1–3; pale brown-green and irregularly marked; ovate; 93 x 59 mm; IP 30 days (F, M). Whooping: 1–3; 98 x 62 mm; IP 29–31days (F, M). **BEHAVIOR:** Socially monogamous, typically pairing for life (up to 20 years). Young are mobile within 8 hours of hatching and follow parents for ~9 months. Sandhill Crane may not start breeding until 7 years old.

Sandhill Crane. Eric C. Soehren, IA

Sandhill Crane. Gerrit Vyn

Whooping Crane.
USFWS, WI

SHOREBIRDS

This large group includes 4 distinct families: stilts and avococets (Recurvirostridae), oystercatchers (Haematopodidae), plovers (Charadriidae), and sandpipers and phalaropes (Scolopacidae). With the notable exception of Solitary Sandpiper, which reuses the tree nests of perching birds, most shorebirds create simple scrapes on the ground (lined or unlined), lay eggs on bare ground or in burrows or crevices, or build simple mounds of debris. Breeding systems range widely from true monogamy to polygyny to polyandry (phalaropes and some sandpipers). Many species are solitary nesters, but others nest in massive colonies. A single brood per year is common among most species. Nestlings are semiprecocial or precocial.

A few contextual clues may help narrow down the family or species. For instance, plovers tend to nest in the open, while sandpipers tend to place nests where they are more obscured by vegetation. Sandpipers typically lay 4 eggs per clutch, while some of the plovers and other species lay 3. Distinguishing species by the nest structure itself is in many cases impossible; note context (geographic location combined with microhabitat features) and the number, shape, and size of eggs if present.

STILTS and AVOCETS

Two species nest in N. America, both on the edges of shallow freshwater, brackish, and alkaline wetlands, or ponds or lakes with shoreline areas of little to no vegetation. The nest is often completely exposed with no attempt to conceal it; instead, these birds rely on inaccessible locations (e.g., small isolated islands, mud bars, patches of emergent vegetation) for security. Both species often form loose colonies, with monogamous pairs defining a territory within the colony. Both parents incubate and tend to the young, which are precocial and leave the nest shortly after hatching. Single-brooded. **DIFFERENTIATING SPECIES:** Stilts tend to use areas with more emergent vegetation than avocets. Stilt eggs are smaller and more heavily marked than avocet eggs.

Black-necked Stilt.
David Moskowitz, WA

BLACK-NECKED STILT *Himantopus mexicanus*

LOCATION AND STRUCTURE: On small islands, on clumps of emergent or floating vegetation, on bare ground near shoreline. Nest may be a scrape, lined or unlined, or a platform built up of dead vegetation and other materials. Outside diameter 13 cm; depth 2 cm. **EGGS:** 3–4; subelliptical to oval; 44 x 31 mm; IP 25–26 days (M, F).

Bill Summerour, AL

AMERICAN AVOCET *Recurvirostra americana*

LOCATION AND STRUCTURE: Often near water's edge on bare ground with little to no vegetation cover. Often on islands where available. Shallow scrape, often lined with local vegetation and small objects such as pebbles and bones; other times unlined. Lining is added over the course of incubation. Outside diameter

Janet Bauer, WA

15 cm; depth 4 cm. **EGGS:** 4; subelliptical to oval; 50 x 35 mm; IP 22–24 days (F, M). **BEHAVIOR:** Intraspecific brood parasitism occurs rarely. Black-necked Stilts laying eggs in avocet nests has also been documented.

OYSTERCATCHERS

AMERICAN OYSTERCATCHER *Haematopus palliatus*

BLACK OYSTERCATCHER *Haematopus bachmani*

HABITAT: Coastal. American Oystercatcher: Sandy shores and tidal marshes. Black Oystercatcher: Rocky coastline. **LOCATION AND STRUCTURE:** Both species nest on ground in sand or other fine substrate. American Oystercatcher nest sites include dunes, marsh islands, and dredge spoil with no to moderate vegetative cover. Black Oystercatcher nests in a variety of locations along open, rocky coastlines, on headlands or islands, usually in areas with no vegetation and sometimes close to high-tide line. Nest is variable, from a bare scrape to a structure created and lined with pebbles, shells, tide wrack, and other debris. Nest dimensions variable. American Oystercatcher: Outside diameter 20 cm; depth 5 cm. Black Oystercatcher: Outside diameter 20 cm; depth 3 cm. **EGGS:** 3 (Black, 2–3); gray with darker spotting; oval to pyriform; 56 x 39 mm; IP 24–28 days (F, M); NP 12–18 days (M, F). **BEHAVIOR:** Single-brooded. Typically monogamous.

American Oystercatcher Matt Monjello, NC

Black Oystercatcher Peter Hodum, WA

Black Oystercatcher R. Adam Martin, WA

PLOVERS

Plovers create simple, shallow scrapes on open ground with little to no short vegetation. The scrape may be lightly lined with local available material, including vegetation and objects. Nine species commonly nest in N. America, and geographic location can help identify or narrow down species options for nests encountered. Microhabitat, as well as size, shape, and markings of eggs, can also help, but unless adults are seen it is often impossible to distinguish the nest of one species from another where there is range overlap.

N. American plovers are primarily seasonally monogamous and, depending on the species, typically lay 3 or 4 eggs in a single brood. Many species are single-brooded, but some can replace failed clutches, and temperate nesters can produce as many as 3 broods. Eggs show the classic pyriform shape common to most wading birds. They have varying amounts of markings, often concentrated on the larger end. Young are precocial, can run as soon as their down dries, and often leave the nest soon after hatching. They begin foraging on their own immediately, following guidance from one or both parents.

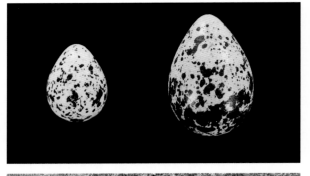

Semipalmated Plover (left) and American Golden-Plover (right), showing both the variety of size and similarity in markings across species.

Cryptic American Golden-Plover eggs disappear into the tundra.
Dan Fontaine, AK

KILLDEER *Charadrius vociferous*

HABITAT: Ubiquitous across N. America, overlapping with numerous other species of plovers and sandpipers with similar nests. Wide variety of habitats, including human-created and disturbed environments: pastures, sandbars, cultivated fields, gravel rooftops, golf courses, airports, etc. **LOCATION AND STRUCTURE:** Nest is in open, often on an area slightly raised relative to surroundings and near a landmark object such as a large stick, rock, manure pile, etc. Shallow scrape is sometimes lightly lined with bits of local material, particularly light-colored objects such as bones, shell fragments, cigarette filters, etc. Outside diameter 8–9 cm; depth 2.5–4 cm. **EGGS:** 4; heavily spotted or marked, more so toward larger end; pyriform to oval; 36 x 27 mm; IP 24–26 days (F, M).

Casey McFarland, WA   Matt Monjello, ME

OTHER TEMPERATE-NESTING PLOVERS

All species overlap geographically with Killdeer, and with Spotted Sandpiper and other sandpipers, but little with one another. All but the Wilson's plover are of conservation concern.

SNOWY PLOVER *Charadrius nivosus*

Open sand or pebbly substrate, near the sea or inland water body (often saline). Often beside a prominent object such as driftwood, small plant, etc. Eggs are often partially buried in sandy scrape lined with bits of shell, bone, etc. **EGGS:** 3; spotted and marked more toward larger end; oval to pyriform; 33 x 23 mm; IP 26–33 (F, M). **BEHAVIOR:** Often 2 broods.

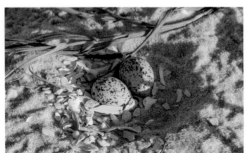

Eric C. Soehren, AL

WILSON'S PLOVER *Charadrius wilsonia*

Open sand or pebbly substrate, or reef, often by a large object such as driftwood or clump of grass. Very similar to nest of Snowy Plover. **EGGS:** 3; *more heavily marked than those of Snowy Plover*; oval; 36 x 26 mm; IP 23–27 days (F, M).

Matt Monjello, NC

PIPING PLOVER *Charadrius melodus*

Open sand along coast or sand and gravel shore of inland rivers, wetlands, and lakes. In area of sparse vegetation but often next to grass tufts, sometimes underneath one. Often in or near tern colonies. Scrape is lined with pebbles and shell fragments. **EGGS:** 4; finely marked, especially toward larger end; oval to pyriform; 32 x 24 mm; IP 26–28 days (F, M).

L. Sanders, ND

MOUNTAIN PLOVER *Charadrius montanus*

Associated with short-grass prairie and, in areas of taller vegetation, heavily grazed areas or prairie dog towns; also areas with short vegetation and open patches of ground, burned grasslands. Nest scrape is often near disturbed soil, lined with lichen,

Mountain Plover.
L. Sanders, WY

plant material, dung. **EGGS:** 3; lightly marked; oval; 37 x 28 mm; IP 28–31 days (M, F). **BEHAVIOR:** May be double-brooded, with 2 clutches laid concurrently, one incubated by male and one by female.

ARCTIC-NESTING PLOVERS

Multiple species with overlapping breeding ranges in Arctic, nesting in open and exposed sites in tundra and along seashore. Four eggs in a complete clutch is typical for all species.

BLACK-BELLIED PLOVER *Pluvialis squatarola*
Tundra, drier upland locations, among heath, willow, or on gravel ridges, often exposed on raised area with view. On moss or lichen, unlined or with minimal lining of pebbles, vegetation. Outside diameter 10–17 cm; depth 3–7 cm. **EGGS:** 4; markings concentrated toward larger end; oval to pyriform; 52 x 36 mm; IP 23–27 days (mostly M).

Chris Smith, AK

AMERICAN GOLDEN-PLOVER *Pluvialis dominica*

Often in areas dense with lichen, other vegetation sparse. Generally in higher and more sparsely vegetated sites than Pacific Golden-Plover. Nest is lined with lichens, sometimes grass, willow, pebbles, etc. *Nest tends to have more lining than that of Pacific Golden-Plover.* Outside diameter 11–13 cm; depth 3 cm. **EGGS:** 3;

markings heaviest on larger end; pyriform; 48 x 33 mm; IP 23–27 days (mostly F). **BEHAVIOR:** Males often return to same breeding territory season after season.

Dan Fontaine, AK

PACIFIC GOLDEN-PLOVER *Pluvialis fulva*

Tundra with abundance of lichen, limited vegetation. Tends to select *lower, more vegetated locations* than American Golden-Plover where they overlap. Scrape is lined primarily with lichens, also other vegetation. *See American Golden-Plover.* Outside diameter 10–13 cm; depth 3–4 cm. **EGGS:** 3; *similar to those of American Golden-Plover;* IP 22–25 days (mostly F).

SEMIPALMATED PLOVER *Charadrius semipalmatus*

Well-drained gravel, shale, or sandy substrates of rivers, tundra, and coastal shorelines. Often near water. Depression in gravel, sand, moss, or lichen is lined with whatever material is most

Mark Nyhof, BC

readily available. Nest cup typically has *straight sides and even depth throughout*, which may help distinguish it from Killdeer and Piping Plover nests. Outside diameter 9 cm; depth 3 cm. **EGGS:** 4; slightly more heavily marked toward larger end; short oval to pyriform; 33 x 23 mm; IP 23–25 days (mostly M).

SANDPIPERS and PHALAROPES

These wading birds generally nest in proximity to water, with some notable grassland-nesting exceptions. All species nest on the ground except Solitary Sandpiper, which uses abandoned nests of other birds in trees. Scrapes are lined sparsely with vegetation, and nests are typically concealed in vegetation or other structures (surrounding grass may be pulled over the nest), features that help distinguish them from plover nests, which are more commonly in the open and lined with minimal plant material.

Most scolopacids are single-brooded, but more broods are possible in polygamous species. Eggs are pyriform to oval and in most species are brown or cream-colored with significant dark markings or spotting. All species typically produce 4 eggs, with rare variations due to destruction of an egg or addition to the clutch from nest parasitism. In many species both sexes incubate, in others only the female. In phalaropes and Spotted Sandpiper, polyandry is common, and the male is the sole incubator. Young are precocial and leave the nest within hours of hatching, led by a parent. For most species, young begin foraging on their own quickly after hatching, following guidance from one or both parents.

Nearly 40 species breed in N. America, most in the Arctic. Some nests, such as those of Spotted Sandpiper, are commonly encountered across a large portion of the continent. Others, such as those of the many Arctic nesters, are rarely observed. Where multiple species overlap, differentiation of nests may be challenging without adult birds present.

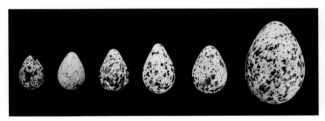

From left to right: Red Phalarope, Sanderling, Pectoral Sandpiper, Lesser Yellowlegs, Wandering Tattler, Long-billed Curlew.

TEMPERATE-BREEDING SANDPIPERS AND ALLIES

(Several of these species also nest in northern boreal and Arctic regions)

SPOTTED SANDPIPER *Actitis macularius*

HABITAT: Wide-ranging sandpiper using a variety of habitats, from sea level to alpine lakes. **LOCATION AND STRUCTURE:** Typically within 100 m of a freshwater stream, lake, or pond, with a mix of open and densely vegetated areas. Scrape is typically constructed among grass, shrubs, or vegetation, partially conceal-ing nest. Lined with grass or other vegetation. Outside diameter 7–12 cm; inside diameter 6–8 cm. **EGGS:** 4; *irregularly spotted*; oval to pyriform; 32 x 24 mm; IP 20–22 days (M, sometimes F). **BEHAVIOR:** Polyandrous where population densities allow for females to attract multiple mates. Females leave one clutch and male to establish others with different males. Often nests near or in colonies of Common Terns.

Note that the nest is built beneath vegetative cover. David Moskowitz, WA

Casey McFarland, WA

UPLAND SANDPIPER *Bartramia longicauda*

HABITAT: Native grasslands and prairies. Occasionally agricul-tural fields and pastures. In other places also mountain mead-ows, tundra, and human-modified landscapes such as airports. **LOCATION AND STRUCTURE:** In grasslands, often in locations with vegetation about 25 cm high, avoiding locations with significantly taller cover. A loose canopy of surrounding grasses may be

Chris Smith, MT

pulled over nest to hide it. Minimal lining. Outside diameter 10–13 cm; depth 3–6 cm. **EGGS:** 4; buff, lightly and *evenly spotted*; oval to subelliptical, *lacking pointed end of many shorebird eggs*; 45 x 33 mm; IP 23–24 days (M, F).

LONG-BILLED CURLEW *Numenius americanus*

HABITAT: Short-grass and mixed prairie, without trees, dense shrubs, or tall grasses. Both wet meadows and drier uplands. **LOCATION AND STRUCTURE:** Ground nest is typically in low grass or other vegetation, often located by distinctive objects in the landscape, such as rocks, dirt piles, or dung. Scrape is lined with local plant materials and other objects, such as droppings, bark, and pebbles. Rim of lining may be raised above ground. Material is more substantial in wet locations than dry. Outside diameter 20 cm; depth 5–7 cm; height of lip raised ~4 cm above ground. **EGGS:** 4; pale beige or olive green, heavily marked; oval to pyriform; 65 x 46 mm; IP 27–28 days (M, F). **BEHAVIOR:** Pairs may make more scrapes than used for nesting. (See page XX for another photo.)

Sarah Hegg,
National Park
Service, WY

MARBLED GODWIT *Limosa fedoa*

HABITAT: In temperate-region prairie population, open short grassland locations. Rarely in cultivated fields. In northern populations, wet tundra or open meadow and wetlands. **LOCATION AND STRUCTURE:** In areas of low, sparse vegetation, not well concealed. Often far from water in prairie habitats, though may be in or adjacent to standing water in northern breeding range. Shallow scrape on ground, often sparsely lined with lichen and grasses. Inside diameter 17 cm; depth 5 cm. **EGGS:** 4; pale buff or olive, lightly spotted and scrawled with dark brown and purplish gray; ovate pyriform; 57 x 40 mm; IP 21–23 days (M, F). **BEHAVIOR:** Monogamous. Often several pairs nest in close proximity.

Hannah Specht,
University of
Minnesota, ND

AMERICAN WOODCOCK *Scolopax minor*

HABITAT: Woodlands and woodland edges, specifically young forests and abandoned agricultural land mixed with forest. Typically by moist ground for foraging. **LOCATION AND STRUCTURE:** A shallow depression on ground lined with plant material and occasionally a rim of twigs. Typically with some sort of vegetation cover around nest. Occasionally eggs are laid directly on ground without creation of depression. **EGGS:** 4; grayish orange, sparsely marked with brown or grayish spots or splotches toward larger end; *short oval to subelliptical, (lacking some of the pointedness of many shorebird eggs)*; 38 x 29 mm; IP 20–21 days (F). **BEHAVIOR:** Polygynous. Males are uninvolved in incubation or rearing. Males engage in distinctive singing and display behavior in areas favorable for nesting. Females may be intolerant of nest disturbance, particularly very early and very late incubation.

David Pierce, MO

WILSON'S SNIPE *Gallinago delicata*

HABITAT: Freshwater wetland habitats, including wet meadows, riparian zones, swamps, and peat bogs. **LOCATION AND STRUCTURE:** Typically concealed in grass or other vegetation, often in

David Moskowitz, MT

wet ground. Depression with woven lining of local vegetation, often more well developed than other shorebird nests. Surrounding vegetation may be woven over nest. Inside diameter 10–17 cm; depth 4 cm. **EGGS:** 4; dark or light olive brown, with markings most abundant toward larger end; ovate pyriform; 39 x 28 mm; IP 19–20 days (M, F). **BEHAVIOR:** Arial "winnowing" performance and associated sounds of male are distinctive part of courtship.

SOLITARY SANDPIPER *Tringa solitaria*

HABITAT: Northern forested bogs near streams, lakes, and ponds. **LOCATION AND STRUCTURE:** The only N. American shorebird that nests in trees, typically conifers. Uses abandoned nests of American Robin, Rusty Blackbird, and others of similar size. Nests are rarely observed. Typically occupies nests 1–9 m aboveground. **EGGS:** 4; pale green to pale brown with chocolate brownish blotches, marked most heavily on larger end; 36 x 26 mm; IP 23–24 days (M, F).

Jason Fidorra, BC

Solitary Sandpiper eggs in the nest of a Rusty Blackbird. Laura McDuffie, AK

WILLET *Tringa semipalmata*

HABITAT: Distinctively different habitat in eastern and western populations. In West: Adjacent to wetlands and ephemeral water in open habitats. Sometimes in grasslands, agricultural fields, or along forested mountain lakes. In East: Strongly associated with

Bri Benvenuti, USFWS, ME

Matt Monjello, NC

tidal wetlands and other coastal habitats. **LOCATION AND STRUC-TURE:** In West: Edges of wetlands and raised ground near water. Distance to water varies from 50 m to almost 3 km. In East: Edges of salt marshes and in coastal sand dunes. Typically associated with a conspicuous rock or other small landscape feature, open bare location, or by tuft of grass, or concealed in low vegetation. Shallow scrape with a variable amount of lining added. Inside diameter 16 cm; depth 6 cm. **EGGS:** 4; color variable, from yellowish to brownish to greenish, speckled, spotted; pyriform to oval; 54 x 38 mm; IP 24–26 days (M, F). **BEHAVIOR:** Monogamous. Male makes several scrapes, female selects one and adds small amount of lining.

WILSON'S PHALAROPE *Phalaropus tricolor*

HABITAT: Wetlands with dense vegetation. **LOCATION AND STRUC-TURE:** In dense vegetation within 100 m of water. Scrape is lined with grass, more lining in wetter areas. Often uses overhead vegetation to conceal nest. Male adds lining after eggs are laid. Outside diameter 9–13 cm; inside diameter 7–9 cm. **EGGS:** 4; buff, *heavily marked*; 37 x 24 mm; IP 20–21 days (M). **BEHAVIOR:** Role reversal: More brightly colored female courts male, lays eggs, then abandons male, leaving him to incubate eggs and tend young.

Scott Somershoe, MT

More than 30 species of sandpipers nest in the Arctic and in open marsh areas in the massive boreal forest zone of N. America. *Many of the species here nest only in the far northern reaches of the continent, and their nests are infrequently encountered. As such, habitat descriptions are generalized.*

All species make simple scrape nests with a variable amount of lining of locally available materials. Depending on the species, the nest may be in the open or tucked within low-growing vegetation. Young are precocial and leave the nest shortly after hatching, tended by adults but foraging for themselves. Most species are monogamous, some are polygynous, and Spotted Sandpiper and phalaropes are typically polyandrous. Many species carry out spectacular annual migrations, some between the Antarctic and Arctic, and thus are commonly seen across temperate N. America in the spring and fall.

DIFFERENTIATING SPECIES: Many of these species target distinctive locations within their breeding range and preferred habitat but there are many locations with multiple overlapping species that are hard or impossible to distinguish without the presence of adults or eggs. Egg color and markings overlap a great deal among and even within species. For the lay observer, besides accompanying adults and geographic and microhabitat location, overall size of the egg is a helpful clue for species identification.

WHIMBREL *Numenius phaeopus*
Nest is in open on a hummock, close to a dwarfed shrub and usually lined with leaves. Arctic coast of AK, w. Canada, w. Hudson Bay. **EGGS:** 58 x 42 mm.

Gerrit Vyn, AK

BRISTLE-THIGHED CURLEW *Numenius tahitiensis*
Small geographic breeding range in w. AK. Dry rocky upland tundra. Well concealed by low vegetation, often underneath willow. **EGGS:** 60 x 42 mm.

BAR-TAILED GODWIT *Limosa lapponica*

Tundra in Arctic AK, often in a raised location and concealed by surrounding vegetation. **EGGS:** *Markings frequently sparse*; 55 x 38 mm.

Gerrit Vyn, AK

HUDSONIAN GODWIT *Limosa haemastica*

In wet tundra or bogs *in forest*. Underneath either a small shrub or grass tuft or on a raised hummock. **EGGS:** *Often sparsely marked, lightly spotted, blotched*; 55 x 38 mm.

RUDDY TURNSTONE *Arenaria interpres*

Variable locations across N. American high Arctic from small islands to river deltas. Typically in open among rocks, occasionally sheltered within vegetation. *Multiple pairs often nest in close proximity.* **EGGS:** 40 x 29 mm.

BLACK TURNSTONE *Arenaria melanocephala*

Coastal areas of AK. Wet tundra, in an open location, typically near a small body of water, river deltas; sometimes beneath willows. **EGGS:** 41 x 29 mm.

Gerrit Vyn, AK

RED KNOT *Calidris canutus*
High Arctic. *Prefers barren ground, rocks.* Nest is lined with lichens. **EGGS:** 43 x 30 mm.

SURFBIRD *Calidris virgate*
Ironically, given its name, nests inland in *open, dry, rocky sites above tree line in mountains.* Nest often has a commanding view. **EGGS:** 43 x 31 mm.

STILT SANDPIPER *Calidris himantopus*
Arctic coast of AK, w. Canada, Hudson Bay. Open sedge tundra near water. Nest is sparsely lined. **EGGS:** 36 x 25 mm.

Kate Fremlin, Amaroq Wildlife Services, NU

SANDERLING *Calidris alba*
High Arctic; coastal tundra. Nest is often tucked against a clump of vegetation, well lined with leaves, etc., close to water. **EGGS:** *More sparsely marked than those of other sandpipers*; 36 x 25 mm.

Garrit Vyn, AK

DUNLIN *Calidris alpina*

Arctic and Bering Sea coasts. Tundra with pools of water or salt marsh. *Multiple pairs often nest in close proximity.* **EGGS:** 36 x 28 mm.

Dan Fontaine, AK

ROCK SANDPIPER *Calidris ptilocnemis*

Coastal AK. Nests in open from coast to mountain ridges. **EGGS:** 39 x 26 mm.

Tim Bowman, AK

PURPLE SANDPIPER *Calidris maritima*

Eastern high Arctic islands, from coastline to ridgeline in open locations. Often lined with leaves and down. **EGGS:** 37 x 26 mm.

BAIRD'S SANDPIPER *Calidris bairdii*

Arctic AK and w. Canada. *Dry tundra locations.* **EGGS:** 33 x 24 mm.

LEAST SANDPIPER *Calidris minutilla*

Across far northern N. America. Wetlands and boggy locations in boreal forest and tundra. In tufts of marsh grass or mossy hummocks. In sand dunes at southern edge of range. **EGGS:** 29 x 21 mm.

Garrit Vyn, AK

WHITE-RUMPED SANDPIPER *Calidris fuscicollis*

Canadian high Arctic. Tundra; *prefers wet locations.* **EGGS:** 34 x 24 mm.

BUFF-BREASTED SANDPIPER *Calidris subruficollis*

Arctic coast. Moist meadows and dry scrubby habitats. Participates in a lek breeding system, in which males defend territories for the purposes of display and mating only, and females take on all other aspects of breeding elsewhere and on their own. **EGGS:** 38 x 27 mm.

PECTORAL SANDPIPER *Calidris melanotos*

Arctic AK and w. Canada. Nest is on slightly raised or well-drained area relative to surrounding wet tundra, typically concealed with vegetation. High degree of sexual dimorphism relative to other *Calidris*, and polygynous behavior. **EGGS:** *Boldly, heavily marked, often appearing as 1 large dark blotch, especially at larger end*; 37 x 25 mm.

SEMIPALMATED SANDPIPER *Calidris pusilla*

Arctic coast. Tundra. Nest is typically in vegetation, often partially concealed. Generally near water. **EGGS:** 32 x 23 mm.

Kate Fremlin, Amaroq Wildlife Services, NU

Susan N. Felage, MB

WESTERN SANDPIPER *Calidris mauri*

Bering Sea and Arctic coast of AK. Tundra, nest often partially concealed. **EGGS:** *Markings "spiral" left to right, are sometimes elongated*; 31 x 22 mm.

Tim Bowman, AK

SHORT-BILLED DOWITCHER *Limnodromus griseus*

Boreal Canada and Pacific Coast of AK. Conifer muskegs, coastal tundra in wet locations, to timberline. Often at base of a tree or shrub, or on top of sedge. Lined with grass and other vegetation, ptarmigan feathers. **EGGS:** 42 x 29 mm.

LONG-BILLED DOWITCHER *Limnodromus scolopaceus*

Coastal AK, YT, NT. Tundra, close to water. A deep depression lined with leaves. Often damp in bottom. **EGGS:** 44 x 29 mm.

WANDERING TATTLER *Tringa incana*

In mountains above tree line in YT and AK. On gravel bar or other open location near water. A natural or created depression, sparsely to elaborately lined. **EGGS:** 43 x 32 mm.

LESSER YELLOWLEGS *Tringa flavipes*

Northern boreal forest bogs near slow-moving streams, lakes, and ponds. Very rare to find. Variable habitats, including open deciduous forests, black spruce muskegs, and open meadows and artificial clearings, such as power lines. Often observed on a hummock next to a shrub. Distance of nest to water ranges from 100 m to 3 km. **EGGS:** 42 x 29 mm.

Laura McDuffie, AK

GREATER YELLOWLEGS *Tringa melanoleuca*

Very rare to find. Muskeg and wet locations along forest edge and high tundra. Often at base of a small conifer, in or on a hummock. **EGGS:** 50 x 33 mm.

RED-NECKED PHALAROPE *Phalaropus lobatus*
Boreal and Arctic N. America. Marsh, wet meadow, tundra and forest edge. Nest is obscured by vegetation pulled over it. Male incubates and rears young. **EGGS:** 30 x 21 mm.

Mark Peck, NU

RED PHALAROPE *Phalaropus fulicarius*
High Arctic. Coastal tundra and offshore islands. Often among sedges. Nest is obscured by vegetation pulled over it. Male incubates and rears young. **EGGS:** 31 x 22 mm.

JAEGERS (Stercorariidae)

These highly pelagic birds spend most of the year on open ocean, returning to land in late spring or early summer to breed in the high Arctic tundra of N. America.

Opportunistic foragers, jaegers feed on a variety of resources, such as mammals, eggs, insects, carrion, and berries. They are also kleptoparasitic, aggressively chasing and forcing other seabirds to give up food. During the breeding season, however, Pomarine and Long-tailed Jaegers rely primarily on a diet of rodents, such as lemmings and voles. This dependency leads to sharp fluctuations in local breeding numbers when rodent population levels either drop or spike. Conversely, Parasitic Jaegers do not depend on rodents for reproductive success, relying instead on birds, chicks, and eggs, and on food stolen from other birds.

Nesting sites are often *exposed*, and found on hummocks, small mounds, low ridges, bare or stony ground, and among small shrubs or moss. Nests are *slight depressions* in the ground, formed by the breast and feet. Both Pomarine and Parasitic Jaegers sometimes include plant material, gathered from the immediate vicinity, lining the depression or placing it around the rim. Long-tailed Jaegers use no nesting material in their construction.

Highly monogamous, pairs may join together in successive years and breed in either isolated pairs or small colonies. Jaegers generally are single-brooded and lay 1–2 sub-elliptical eggs that are smooth and slightly glossy, olive brown, and speckled or blotched with dark brown or gray spots. Both sexes incubate, beginning with the first egg. The semiprecocial young hatch asynchronously in 23–28 days. They may leave the nest within days of hatching but remain in the territory, fed by both parents. They fly in 21–33 days and become independent in 5–8 weeks.

A Long-tailed Jaeger nest, showing typical habitat. Garrit Vyn, AK

POMARINE JAEGER *Stercorarius pomarinus*

HABITAT: Low-lying, wet coastal tundra. Common in marshes surrounded by small bodies of water. *More restricted to wet tundra than Long-tailed Jaeger.* **LOCATION AND STRUCTURE:** Usually slightly elevated from ground. **EGGS:** 2; markings sparse; 64 x 44 mm. **BEHAVIOR:** Both sexes build.

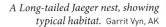

PARASITIC JAEGER *Stercorarius parasiticus*

HABITAT: Arctic tundra. Occupies a broader range of breeding habitats than other jaegers. Heath tundra, coastal and inland marshes. Shows stronger preference for sites near water than Long-tailed Jaeger. **LOCATION AND STRUCTURE:** In marshy areas,

Tim Bowman, AK

nest is generally on a slightly elevated mound. Island nests are often in open away from cover. Along river deltas near colonies of Snow Geese and Brant. **EGGS:** 2; markings may cover egg or be concentrated on larger end; 57 x 40 mm. **BEHAVIOR:** Female builds majority of nest.

LONG-TAILED JAEGER *Stercorarius longicaudus*

HABITAT: Often in tundra habitats far from sea. Will nest in both marshy areas and more exposed and drier sites; in low shrub–dominated habitats and sparsely vegetated barren ground. **LOCATION AND STRUCTURE:** Usually on drier and higher sites than other jaegers. Nest is on a slight slope or elevated spot, surrounded by shrubs, on moss, or in open on bare ground. **EGGS:** 1–2; typical jaeger ground color, rarely light blue, with markings generally on larger end; 55 x 38 mm.

Kate Fremlin,
Amaroq Wildlife
Services, NU

AUKS, MURRES, and PUFFINS (Alcidae)

These seabirds dive for fish beneath the water's surface, and 19 species breed in N. America. With some notable exceptions, they nest along the coast on rocky bluffs and offshore islands (some nest inland in mature trees in old-growth forests). In suitable habitat they are often found with other seabirds seeking similar nesting sites.

Nests are often in a burrow, natural crevice, or occasionally a human structure that mimics this, and cliff ledges. These inaccessible locations protect eggs and young from terrestrial and aerial predators. Some species lay eggs in any available depression in rocks close to the sea, and others establish nests in inaccessible inland locations and commute back to the sea to forage. Some species nest on the surface of islands but are protected by nesting shoulder to shoulder in dense colonies.

Individuals of many species don't breed until they are several years old, and clutch sizes are small, often only 1 or 2 eggs. They are mostly single-brooded, so ensuring the survival of these few eggs is critical. Many species are of conservation concern.

Alcids are monogamous or socially monogamous (varying by species). Young are semiprecocial or precocial and can move around shortly after hatching, but some species may not leave the nesting area for 8 weeks (others leave after just a few days). Both parents bring fish, squid, etc., to the nest to feed young. Many species show strong site fidelity, returning to the exact same location or burrow in successive years.

Several species overlap in breeding range and nest site characteristics (and also overlap with several other groups of seabirds). For burrows, the size and shape of the entrance as well as the specific location can help narrow down the possible maker of the nest. However, burrows are reused often, and not necessarily by the species that constructed it originally.

DOVEKIE *Alle alle*

HABITAT: Most abundant in high Arctic, primarily in rocky and cliffy terrain of coastal mountains, but also on the sides of inland mountains. **LOCATION AND STRUCTURE:** Natural crevices. *A collection of pebbles is often placed conspicuously at entrance*, as well as taken into the crevice for lining. **EGG:** 1; very pale blue or blue-green; *usually unmarked*; 48 x 34 mm; IP 29 days; NP semiprecocial, ~3–4 weeks. **BEHAVIOR:** Nests in large colonies.

COMMON MURRE *Uria aalge*

THICK-BILLED MURRE *Uria lomvia*

HABITAT: Cliffs and ledges. Coastline, rocky offshore islands, tops of seastacks. **LOCATION AND STRUCTURE:** No nest. A single egg is laid on bare ground on an open cliff ledge, occasionally in a covered crevice or cave. Some pebbles may be arranged to keep egg from rolling. Where the 2 species overlap, Common

A Common Murre egg. David Pierce, OR

Common Murres. Peter Hodum, WA

Nesting Thick-billed Murres. Gerrit Vyn, AK

Murres tend to choose wider and lower ledges than Thick-billed Murres. **EGG:** 1; highly variable, from blue-green to turquoise, off-white, beige, or pinkish, marked with speckles, spots, blotches, streaks, and scribbles; ovate-pyriform; 81 x 53 mm (Common), 79 x 50 mm (Thick-billed); IP ~32 days; NP semiprecocial, leave sea in 16–30 days. **BEHAVIOR:** Typically nests in large colonies. Pairs often return to precise nest site year after year. Parents switch off incubation every 12–24 hours. Eggs are placed on parents' feet and covered with belly feathers.

RAZORBILL *Alca torda*

HABITAT: Island and coastal cliffs. **LOCATION AND STRUCTURE:** In a natural crevice or sometimes on an exposed ledge in a cliff above sea. Rarely under shoreline rocks or driftwood, in burrow excavated by another species, or on earthen ledge of steep grassy slope. Prefers locations that are partially or fully enclosed. Makes scrape in soft ground to lay egg in. May import small stones and other objects, in some cases adding no lining at all. **EGG:** 1; variable, from whitish to yellowish, occasionally greenish, lightly or heavily marked in reddish browns, especially on larger end; elliptical to long ovate; 73 x 47 mm; IP 25–36 days; NP ~18 days before leaving nest for sea. **BEHAVIOR:** Colonial, often nesting in hundreds of pairs, less frequently as few as 10 pairs or as many as 10,000. Pairs typically return to same nest location year after year.

Philip Blair

BLACK GUILLEMOT *Cepphus grille*

PIGEON GUILLEMOT *Cepphus columba*

HABITAT: Rocky offshore islands and cliffy coastline. **LOCATION AND STRUCTURE:** Natural cavities or depressions in rocks or cliffs, or under driftwood or tree roots, overhanging grass; essentially any site providing nesting cavity. Also frequently uses human-made structures that provide a suitable cavity, such as under docks (in Arctic AK where there's no rocky shoreline present, all nests are in human-made structures or debris). Occasionally in burrows created by other species (Pigeon Guillemots

A clutch of Black Guillemot eggs from an artificial nest structure. George Divoky, AK

Pigeon Guillemot. Amelia J. DuVall, CA

A Black Guillemot egg in a cavity created by abandoned human debris.
George Divoky, AK

dig their own burrows in sandbanks). Shallow scrape is excavated in loose soil when possible. On rocky surfaces, stones and shells may be gathered around a central depression for eggs. Cavities typically 0.5–1 m deep, and diameter of entrances is typically 13–29 cm; large range reflects wide variety of structures used. **EGGS:** 2 (1–3; all 3-egg clutches are the result of 2 females); whitish or pale whitish green, with dark spots, blotches; short elliptical. Black: 58 x 40 mm. Pigeon: 61 x 41 mm. IP ~28–32 days; NP semiprecocial, ~30–40 days before leaving nest for sea. **BEHAVIOR:** *Cepphus* are the only alcids that regularly nests as single pairs as well as in small to medium-sized colonies, and are typically considered to be semicolonial.

MURRELETS

Murrelets nest either in a burrow or natural crevice close to the ocean on offshore islands, or well inland in a natural crevice or in old-growth conifers, and have also been documented in red alder and big-leaf maple. They are monogamous or socially monogamous (varying by species), and often exhibit high fidelity over years to a specific nest area and sometimes to one specific nest. Males and females both incubate, switching daily or every several days, often at night. Eggs are long elliptical and vary among species from greenish to pale blue, olive, or brown, or brownish yellowish, lightly spotted. Some species are semiprecocial and remain at the nest for long periods of time. Others are precocial and just a few days after hatching leave the nest for open ocean under the cover of darkness; they are reared at sea by both parents for more than 1 month, possibly 2.

MARBLED MURRELET *Brachyramphus marmoratus*

HABITAT: Mature forest stands with trees often older than 200 years, up to 12.4 km or more from ocean. **LOCATION AND STRUCTURE:** Usually 10–82 m high in a mature tree, typically in the top half to third of canopy. Nest is commonly located on a horizontal branch close to tree trunk, but broken tree top, burl, dwarf mistletoe, or other structural features are possible. Egg is laid on moss or other substrate, with no other material added. After 30–40 days, fecal ring forms around edge of nest cup. In CA, OR, and WA, nests in trees (1 cliff nest in WA). From BC to AK, also nests on ground or in crevices in tundra on scree slopes, boulder-strewn terrain, and at base of dwarfed patches of spruce, alder, or in forests at tree bases near ledges. Tree nests: average outside diameter about ~10 cm, depth about ~4 cm. **EGGS:** 1; light greenish yellow or pale olive, variably marked with fine specks or spots; long elliptical; 59 x 37 mm; IP 27–30 days; NP semiprecocial (remaining at nest site), 30–40 days. **BEHAVIOR:** Occurs in forests or areas with other Marbled Murrelets but not in dense colonies.

A Marbled Murrelet ground nest.
Katherine Hocker, AK

A Marbled Murrelet nest atop a limb, high in an old-growth tree. Nick Hatch, USDA Forest Service, Pacific Northwest Research Station, WA

A Marbled Murrelet on its nest high in a Douglas-fir. J. Brett Lovelace, OR

A classic tree nest of a Marbled Murrelet.
J. Brett Lovelace, OR

KITTLITZ'S MURRELET *Brachyramphus brevirostris*

HABITAT: Glaciated areas along AK coast. Wide variety of steep slopes up to 2,500 m elevation and 0.25–75 km from ocean. **LOCATION AND STRUCTURE:** Rocky ridgeline or mountain summit, scree slopes above tree line. Open nest is on bare ground, often

Look closely for the Kittlitz's Murrelet sitting hidden atop her egg on mountainside tundra.
Robb Kaler, USFWS, AK

A Kittlitz's Murrelet egg in situ.
Robb Kaler, USFWS, AK

A Kittlitz's Murrelet egg.
Robb Kaler, USFWS, AK

located on downhill side of a large rock. Little or no lining is added. **EGG:** 1; *similar to egg of Marbled Murrelet*; long elliptical; 60 x 36 mm; IP 30 days; NP semiprecocial (remaining at nest site), varies, average 26 days. **BEHAVIOR:** Occurs in areas with other murrelets (Marbled or Kittlitz's) but not in dense colonies.

SPECIES THAT NEST ON OFFSHORE ISLANDS

SCRIPPS'S MURRELET *Synthliboramphus scrippsi*
(formerly Xantus's Murrelet)

HABITAT: Islands in Pacific off coast of Baja and s. CA. **LOCATION AND STRUCTURE:** Natural crevice or cave on steep slope or cliff, occasionally under thick vegetation on similar terrain. Occasionally in a burrow excavated by another species. **EGGS:** 1–2; 53 x 36 mm; IP 34 days; NP precocial, leave nest after 2 days and are reared at sea, presumably for 40–60 days. **BEHAVIOR:** Colonial.

Amelia J. DuVall, CA   Dan Fontaine, CA

ANCIENT MURRELET *Synthliboramphus antiquus*

HABITAT: Offshore islands, coastal cen. BC to Aleutian Is. **LOCATION AND STRUCTURE:** In forest in southern part of range and in densest vegetation possible in north. Typically within 300 m of sea, on well-drained slopes rather than flat or wet locations. Nests in burrows are most common; birds may dig their own burrow or occupy one dug by another species. Also natural crevices and cavities in rocks or under tree roots, etc. Burrow entrance diameter 8–14 cm; length 0.6–1.5 m. **EGGS:** 1–2; 61 x 39 mm; IP 35 days; NP precocial, leave nest after 1–3 days and are reared at sea for ~40–60 days. **BEHAVIOR:** Colonies of 1,000–10,000 pairs, sometimes in association with Cassin's and Rhinoceros Auklets and storm-petrels. New pairs prefer to create their own burrow rather than use an unoccupied burrow, leading to many unused burrows in active colonies.

AUKLETS

Auklets nest on offshore islands, either in a burrow that each pair digs, in a natural cavity or one dug by other species, or in crevices under rocks. The nest contains little or no lining. Auklets are colonial, with some species nesting in massive, dense colonies (more than 1 million pairs), and others in smaller numbers and more widely spaced. Auklets are often found in association with other species of auklets, as well as other seabirds. They are monogamous or socially monogamous (varying by species) and often exhibit high fidelity to a specific nest site over years. They lay a single subelliptical white or off-white egg (sometimes bluish), usually with little or no markings. Young are semiprecocial and leave the nest in 27–52 days, varying by species. Unlike some other auks, young are independent upon leaving.

BURROW-NESTING SPECIES (EXCAVATORS)

CASSIN'S AUKLET *Ptychoramphus aleuticus*

Uses either a natural crevice or an excavated burrow, usually within a few hundred feet of shoreline. Little to no lining in nest chamber. Burrow is typically 0.7–1 m long with entrance 13 cm high x 15 cm wide. **EGG:** 1; 47 x 34 mm; IP 37–42 days. **BEHAVIOR:** Nest density is high in big colonies, with every available space used. Adults visit nest primarily at night. Only alcid to produce 2 broods a year, which it may do at southern end of range. Opportunistic in choosing nest sites; will nest under boards and in nest boxes.

RHINOCEROS AUKLET *Cerorhinca monocerata*

Digs burrow into a grassy slope (level ground may be used but is less preferred); may be among tree roots or logs or under bushes, up to 200 m from shoreline. Nest chamber with some lining of vegetation and other materials. Burrow density up to 1

Rhinoceros Auklet burrows on an island hillside. Peter Hodum, WA

Rhinoceros Auklet burrow entrance and tracks. Peter Hodum, WA

Rhinocerous Auklet burrow on an offshore island.
David Moskowitz, BC

per square meter. Burrow diameter 13 cm; length 0.3–5 m. **EGG:** 1; 69 x 46 mm; IP 45 (39–52) days. **BEHAVIOR:** Incubation shifts typically occur every 24 hours but may be up to 4 days. Food is often delivered to young at night.

NATURAL CAVITY-NESTING SPECIES

PARAKEET AUKLET *Aethia psittacula*
Natural crevices or burrows (typically modified burrows originally constructed by other seabirds) on islands. No nest material is added. **EGG:** 1; 54 x 37 mm; IP 35 days.

Brendan Higgins,
USFWS, AK

LEAST AUKLET *Aethia pusilla*
Natural crevices or cavities under rocks on islands. Large colonies are typical (from several hundred to more than 1 million pairs). **EGG:** 1; 39 x 29 mm; IP ~30 days.

WHISKERED AUKLET *Aethia pygmaea*

Limited range in Aleutian Is. and coast of Asia. **EGG:** 1; 44 x 31 mm; IP ~35 days.

CRESTED AUKLET *Aethia cristatella*

Natural crevices or cavities under rocks on islands. Large colonies are typical (from several hundred to more than 1 million pairs). No nest material is added where egg is laid. **EGG:** 1; 54 x 38 mm; IP ~34 days.

PUFFINS

Three colonial species nest on rocky headlands and offshore islands, either in an earthen burrow dug below the sod by the puffin itself, in a burrow excavated by another species, or in a natural crevice in boulders or cliffs along the ocean. Tunnels range in length from .6 to 3 m. Suitable nesting habitat is limited for all 3 species.

Where Tufted and Horned Puffins overlap, their burrows are indistinguishable, though Horned Puffins use burrows less frequently. All puffins can be found nesting in proximity to other seabirds that use burrows or crevices, and they have larger-diameter burrows than those constructed by smaller murrelets and shearwaters. (See also Murrelets; Northern Storm-Petrels; and Shearwaters and Petrels.) Puffins lay a single subelliptical egg and incubate for ~40–44 days. Young are semiprecocial, leave the nest in about 38–51 days, and are independent thereafter.

ATLANTIC PUFFIN *Fratercula arctica*

LOCATION AND STRUCTURE: Burrows excavated by birds themselves or made by shearwaters or rabbits, often at the top of rocky cliffs on small islands. Also crevices on cliffs or under large rocks, especially in northern part of range. Slopes are preferred to flat ground. Pairs tend to reuse burrows for multiple years. Chamber at the end may include some lining of grass and twigs. Burrow diameter 13–18 cm; length 70–110 cm. **EGG:** 1; dull whitish, occasionally heavily spotted; subelliptical; 63 x 44 mm.

Earl Deickman, NL

An Atlantic Puffin egg in a nest cavity.
Sigurður Stefnisson

HORNED PUFFIN *Fratercula corniculata*

LOCATION AND STRUCTURE: Tends to use natural crevices in cliffs or under rocks in talus field or along beach more often than burrows. Natural crevice often has dry grass added as well as feathers. **EGG:** 1; rough surface, whitish or cream, unmarked or with gray or lavender spots or scrawls; 67 x 46 mm.

Horned Puffins nest in natural cavities.
Bill Bumgarner, AK

TUFTED PUFFIN *Fratercula cirrhata*

LOCATION AND STRUCTURE: Burrows excavated into soil, often on a densely vegetated steep slope or edges of cliffs. Sometimes natural crevices. Burrow entrances can be densely packed together in some colonies. Nest chamber in back of burrow may have substantial lining of grasses, vegetation, and items collected at sea, or be minimally lined. Burrows are typically wider than high: 16–19 cm wide x 13–18 cm high. Typically a little less than 1 m deep, though some are much longer. Burrow entrances of 14.5 cm are used as a cutoff to separate burrows of smaller murrelets and storm-petrels from those of larger Tufted Puffin. **EGG:** 1; similar to Horned Puffin; 72 x 49 mm.

Tufted Puffin nest burrows.
Peter Hodum, WA

Peter Hodum, WA

GULLS, TERNS, and SKIMMERS (Laridae)

The 34 species of these familiar medium-sized to large seabirds breed in N. America, along coasts and on offshore islands, interior lakes, rivers, open areas, and marshes. Many species use simple scrape nests, but some nest over water, on cliffs, and even in trees. They are generally monogamous, with many species building strong pair bonds that last for successive years or the life of a partner.

Most species are colonial. They may breed in single or mixed-species aggregations, in small groups or in colonies numbering in the tens of thousands. Some species, such as the tree-nesting Bonaparte's Gull, may also breed in isolated pairs. Within colonies, larids are highly territorial and defend areas that range from the immediate space surrounding a nest (kittiwakes, terns, skimmers) to over 100 meters (Glaucous Gull). In colonies where females outnumber males, females may develop pair bonds with one another and lay their eggs in the same nest, resulting in larger than average clutch sizes.

GULLS

The 18 species of gulls that breed in N. America mostly nest on the ground close to water in a variety of locales, including beaches, cliff ledges, freshwater and coastal islands, rock stacks, and dunes. They also use human-made structures. Two species build *nests on water* in wetlands, and two others will *nest in trees* (see categories below). Both sexes generally build nests, which vary from sparsely lined shallow scrapes to large bulky mounds or mats with well-defined, bowled cups. A given species can produce either variation. Gulls are single-brooded, though some replace failed clutches. Many species usually lay 3 eggs, though the number can vary per clutch and across species (noted below). Eggs are subelliptical, smooth to slightly glossy, and well camouflaged. They range in color from pale to dark olive, olive-buff, whitish blue, cream, or greenish and are speckled, scrawled, or blotched with olive brown, brown, gray, or blackish brown markings. Eggs of many species can be similar, and wide variation within a single species is possible. Descriptions for species are not included here; use geographic location and accompanying adults.

Both parents incubate for 20–33 days, varying by species. Young are semiprecocial and may leave the nest within a few days but remain nearby, where they're fed by both the male and female. They fledge and are able to fly within 3–7 weeks, but parents may continue to tend them for longer periods.

BLACK-LEGGED KITTIWAKE *Rissa tridactyla*

HABITAT: Cliffs on offshore islands, sea stacks, and along mainland coast. **LOCATION AND STRUCTURE:** Commonly on ledges, but also will nest on steep ground, boulders, human structures. Will use caves on nesting grounds that lack cliff sites. Nest is a mound with a shallow cup, though *cup is often deeper than in ground nests of other gulls*. Made from mud, vegetation,

occasionally seaweed and feathers. Outside diameter 20–50 cm.
EGGS: 2; 56 x 41 mm. **BEHAVIOR:** Colonial, ranging from small groups to hundreds of thousands.

Black-legged Kittiwake nest colony.
Douglas Mason, AK

Black-legged Kittiwake nests.
Douglas Mason, AK

RED-LEGGED KITTIWAKE *Risa brevirostris*

HABITAT: Limited range (Aleutian Is., AK). **LOCATION AND STRUCTURE:** Often with Black-legged Kittiwakes, but *on narrower ledges. Nest is similar to Black-legged Kittiwake's, but smaller and generally with more mud.* **EGGS:** 1; 56 x 41 mm.

GULLS THAT NEST ONLY IN THE ARCTIC OR FAR NORTH

IVORY GULL *Pagophila eburnea*

HABITAT: High Arctic islands. **LOCATION AND STRUCTURE:** On coastal and inland cliff ledges, scree or rock slopes, boulder-strewn flat ground. Nests vary from shallow hollows with little to no lining to large bulky mounds of moss, dry grass, bits of driftwood, feathers, seaweed, mud. **EGGS:** 1–2; 61 x 43 mm. **BEHAVIOR:** Colonial, sometimes with other seabirds.

ROSS'S GULL *Rhodostethia rosea*

HABITAT: Limited range, n. MB. High Arctic on wet, boggy tundra ponds and marshes. **LOCATION AND STRUCTURE:** On an island or mound, next to water. Nest is a shallow scrape lined with dried grass, willow leaves, sedges, moss, other plant material. **EGGS:** 3; 42 x 33 mm. **BEHAVIOR:** Colonial, often with Arctic Terns.

SABINE'S GULL *Xema sabini*

HABITAT: In boggy tundra near coast and fresh water: ponds, marshes, lakes, tidal flats, lagoons, salt marshes, coastal shorelines, and islands. **LOCATION AND STRUCTURE:** On ground, often along edges of fresh or brackish water, on small islands in fresh water. Infrequently on saltwater islands. Nest is a shallow scrape, usually unlined. Sometimes lined with grass stems, sedges, seaweed, feathers. **EGGS:** 3; 44 x 32 mm. **BEHAVIOR:** Colonial, often with Arctic Terns.

ICELAND GULL *Larus glaucoides*

HABITAT: Coastal Arctic. On cliff ledges facing sea, occasionally on rocky sea islands; rarely interior freshwater lakes. **LOCATION AND STRUCTURE:** Nest is a bulky cup of grasses, moss, feathers, other surrounding material. **EGGS:** 2–3; 69 x 48 mm. **BEHAVIOR:** Colonial. May nest with Glaucous Gulls and Black-legged Kittiwakes, but often separated into distinct colonies. Little is known about this species.

GLAUCOUS GULL *Larus hyperboreus*

HABITAT: Mostly coastal. Lagoons, estuaries, sea cliffs, offshore islands, ice edges. Also inland lakes, ponds, rivers. **LOCATION AND STRUCTURE:** Variable. On ledges, rocky bases and vegetated tops of cliffs, coastal and freshwater islands, in tundra along pond shorelines. Sometimes on ice edges. Nest is a *mound of vegetation with a shallow bowl in center and no lining*. Mound consists of grass, moss, twigs, sedges, occasionally feathers. Sometimes nests are simple scrapes lined with seaweed. Outside diameter 30–55 cm; height up to 76 cm; inside diameter 25–40 cm; depth 10 cm. **EGGS:** 2–3; 77 x 54 mm. **BEHAVIOR:** Colonial on coast, often mixed with other species. In isolated pairs inland.

Chris Smith, AK

GULLS WITH MORE SOUTHERN-REACHING BREEDING RANGES

GREAT BLACK-BACKED GULL *Larus marinus*

HABITAT: Along coast on natural and dredge-spoil islands, sea stacks, barrier beaches and dunes, salt marshes. Also inland on freshwater lake and river islands. **LOCATION AND STRUCTURE:** On ground, sheltered by an object (log, rock, etc.). Nests may be on rocky outcroppings, in low vegetation or sandy soil. Also under bushes in salt marsh. Nest is a scrape lined with grass, sticks, rubbish, seaweed, moss, some feathers, or a bulkier, bowled nest of similar materials. *May have little to no lining when located*

Great Black-backed Gull.
Matt Monjello, NC

in grassy area. Outside diameter 20–56 cm; height 5–12 cm; inside diameter 25–33 cm; depth 6–10 cm. **EGGS:** 2–3; 77 x 54 mm. **BEHAVIOR:** Colonial, often with Herring Gulls, but in more open areas at site.

LAUGHING GULL *Larus atricilla*

HABITAT: Broad variety of coastal habitats, including salt marshes, dunes, beaches, natural and dredge-spoil islands, also inland sites in FL and Salton Sea (s. CA). **LOCATION AND STRUCTURE:** Variable. Nest is on ground on vegetative mats, among grasses, rushes, and shrubs. Structure varies with site. May be a simple scrape, sparsely lined with dry grass stems and plant material, *or a bulkier nest* of similar material and lined with finer grasses. Outside diameter of NJ nests ~28 cm; height 28 cm; inside diameter 14 cm; depth 6 cm. **EGGS:** 3; 54 x 38 mm. **BEHAVIOR:** Colonial, ranging from small group to tens of thousands. Will nest with other species, including gulls, terns, rails, Black Skimmer, etc. Favors elevated sites but occupies lower and more open portions of habitat when nesting with larger gulls.

Bill Summerour, AL

GLAUCOUS-WINGED GULL *Larus glaucescens*

HABITAT: Pacific Northwest coast. On low rocky or vegetated off-shore islands. Also coastal cliffs, human-made structures. **LOCATION AND STRUCTURE:** On ground among low herbage, on rock or bare soil. Also on cliff ledges, roofs, and other artificial structures near water. Nest is a scrape lined with grasses, weeds, feathers, seaweed, moss, roots, bones, debris. Outside diameter 38.6 cm; inside diameter 21.6 cm; depth 10.6 cm. **EGGS:** 2–3; 73 x 51 mm. **BEHAVIOR:** In small to large colonies. *Hybridizes with Western Gull in southern range and with Herring and Glaucous Gulls in northern range.*

Brendan Higgins, AK

Peter Hodum, WA

Hybrid of Western Gull x Glaucous-winged Gull. Timothy Laws, OR

MEW GULL *Larus canus*

HABITAT: Coastal and freshwater areas: sea cliffs, along lakes, ponds, marshes, rivers, tundra, grass-covered meadows, wooded and rocky islands. **LOCATION AND STRUCTURE:** Variety of substrates, including on large rock, grass clumps, mounds, tree stumps. Also *nests in coniferous trees* and on human-made structures. Nest is a *substantial bowl* of dry grasses and other vegetation: twigs, mosses, seaweed, etc. Dimensions vary by placement and materials. Outside diameter ~25–56 cm; height 16–24 cm; depth 9–16 cm. **EGGS:** 3; 57 x 41 mm. **BEHAVIOR:** Breeds in small groups or isolated pairs. Female mainly builds nest. Often reuses previous year's nest where rearing was successful.

Rick Bowers,
Bowersphoto.com,
AK

RING-BILLED GULL *Larus delawarensis*

HABITAT: Both inland and along coast. On small, sparsely vegetated islands. Also beaches, bays, estuaries. Sometimes lakeshore peninsulas, infrequently on river islands. Common near human habitation and agricultural areas (especially western population). **LOCATION AND STRUCTURE:** On ground, often in open, shielded by scant low vegetation, sometimes among rocks. Nest is usually a shallow scrape lined with small sticks, twigs, grasses, dead plants, lichens, moss, feathers possible. Amount of material used varies, with some nests are sparsely lined. Outside diameter 25–63 cm; height 7.6–10 cm; inside diameter 15.2–23 cm; depth 5 cm. **EGGS:** 2–4; 59 x 42 mm. **BEHAVIOR:** Colonial. Will share habitat with Herring Gulls (Northeast), terns (Great Lakes), California Gulls (West). In colonies with more females than males, females may lay their eggs in the same nest, resulting in clutch sizes greater than 4 eggs in a single nest.

Thomas Erdman, WI

WESTERN GULL *Larus occidentalis*

HABITAT: Pacific Coast. On offshore rocky islets and islands, coastal cliffs, old piers, salt flats. **LOCATION AND STRUCTURE:** Nests on ground, sheltered from wind by an object (shrub, rock, etc.). Nest may be in rock crevice or on exposed cliff ledge, in sand or soil or on piers or dikes; sometimes grassy hillsides. Nest is a shallow scrape lined or substantially filled with grasses, seaweed, and other vegetation, feathers; rarely debris. Outside diameter 23–35 cm; inside diameter 9–20 cm; depth 5–10 cm. **EGGS:** 2–3; 72 x 50 mm. **BEHAVIOR:** Colonial. *Hybridizes with Glaucous-winged Gull in northern part of range.* In colonies with more females than males, females may lay eggs in the same nest, resulting in clutch sizes greater than 3 eggs. Often returns to nest site and may reuse nest from previous year.

CALIFORNIA GULL *Larus californicus*

HABITAT: Commonly on islands of interior freshwater lakes, marshes, rivers; on levees at alkaline lakes. **LOCATION AND STRUCTURE:** Usually near and surrounded by water. Nests are on ground, in open areas or protected by short vegetation. A shallow depression with variable amount of accumulated material. A simple depression that is sparsely lined, or a large cup with material rising slightly up and off ground. Lining of bones, feathers, grasses, rubbish, other surrounding material. Outside diameter 25.4–30.5 cm; inside diameter 17.8 cm; depth 5 cm. **EGGS:** 3; 52 x 37 mm. **BEHAVIOR:** Colonial, sometimes in the thousands and with other gull species.

Bird Research
Northwest, BC

HERRING GULL *Larus argentatus*

HABITAT: Wide range, often surrounded by or near water: on boulders in tundra lakes; on interior lake or ocean islands; mounds in marshes, sand or pebbly substrate, promontories; on sea cliffs, gravel rooftops. **LOCATION AND STRUCTURE:** Nests on ground. Prefers areas with scattered vegetation that provide good view of surroundings. Nest is commonly placed near cover such as a rock or log that provides concealment from nearby nesters and protection from wind. Sometimes will nest under small trees or shrubs, under buildings. At artificial sites, nests may be fully exposed. Nest is usually a shallow bowl lined with grass and other plant material, seaweed, feathers, debris. May have little to no lining. *Can also build more substantial nest.* Outside diameter 30–70 cm; height 12.7–25.4 cm; inside diameter 25.4 cm; depth 7.62 cm. **EGGS:** 2–3; 70 x 48 mm. **BEHAVIOR:** Colonial along coast; in solitary pairs inland. May reuse old nest.

Herring Gull. Thomas Erdman, WI

Nest of a "Chandeleur" Gull, a hybrid of Kelp Gull x Herring Gull. Bill Summerour, AL coastal islands

TREE-NESTING GULLS

(See also Mew Gull)

BONAPARTE'S GULL *Larus philadelphia*

HABITAT: Boreal forests. Common along wooded margins of inland lakes, bogs, ponds, marshes, also tree-covered lake islands. **LOCATION AND STRUCTURE:** Typically within 100 m of water, nest well hidden and *built on horizontal branch of a coniferous tree (usually spruce)*; infrequently on a mound of wetland vegetation; 1.2–17 m off the ground. Nest is a platform of small sticks and twigs with a sturdy open cup lined with mosses, grasses, lichen. Outside

Christian Artuso, MB

George Peck, MB

diameter 22–33 cm; height 8–13 cm. **EGGS:** 3; 49 x 35 mm; IP 23–24 days (F, M). **BEHAVIOR:** Will breed in loose groups or isolated pairs and may return to same area each year. Both sexes build.

GULLS THAT OFTEN BUILD NESTS ON WATER

LITTLE GULL *Hydrocoloeus minutus*

HABITAT: Breeding range limited to pockets around Great Lakes region, s. MN, along Hudson Bay. Freshwater marshes, low marshy shorelines along lakes, rivers, and on offshore islands. **LOCATION AND STRUCTURE:** Nests on bulrush or grass tussocks, muskrat mounds, mudflats, floating islands. Nest is built of dead vegetation (cattail, reeds, rushes, grasses) with a shallow depression. Little data; outside diameter 16–20 cm. **EGGS:** 3; 42 x 33 mm. **BEHAVIOR:** Both sexes build. Usually in small colonies or isolated pairs; occasionally with other gulls, terns.

Thomas Erdman, WI

FRANKLIN'S GULL *Leucophaeus pipixcan*

HABITAT: Interior freshwater marshes. **LOCATION AND STRUCTURE:** In thinly dispersed stands of emergent vegetation near open water, sometimes in denser stands of cattail. Nest is commonly a floating mat of bulrush, cattail, or similar (often cut green by gulls) attached to rooted vegetation, *with a ramplike "porch" leading to nest* and a slightly hollowed cup. New materials are added regularly owing to material decay and nest sinking. Outside diameter 30–76 cm; height 10–20 cm; inside diameter 13 cm. **EGGS:** 3; 52–36 mm. **BEHAVIOR:** Usually in large colonies, and material theft from other nests is common. Young will also add material to nest.

TERNS and SKIMMERS

These seabirds—16 species of which breed in N. America—nest mostly on the ground, usually creating a shallow scrape with little or no lining, but 2 frequently build *nests on water*, and some others in bushes, rock crevices, and cliffs.

Species are single-brooded and typically lay 1, 2, or 3 subelliptical, smooth, nonglossy, well-camouflaged eggs. Color ranges from very pale to pale blue, creamy or pinkish buff, buff-olive, or brownish olive. Eggs are variably spotted, speckled, and blotched with brown, black, or gray markings. Within a few days of hatching, young of some species gather in creches. Parents and offspring recognize each other by voice, enabling adults to find their chicks among the raucous crowd. Both parents incubate for 19–36 days, varying by species. Young hatch asynchronously, are semi-precocial, and may leave the nest within days, but remain nearby and continue to be tended by parents. Young are generally able to fly in 19–30 days, but some species, such as Bridled and Sooty Terns, may take as long as 2 months. Parents may continue to tend young for longer periods after fledging.

Royal Terns, pictured here, and many other terns nest colonially in close proximity.
Bill Summerour, AL

BROWN NODDY *Anous stolidus*

HABITAT: Dry Tortugas, FL. **LOCATION AND STRUCTURE:** *On ground or in shrubs, trees, or cacti.* From ground to 10 m. On ground, no nest to simple collections of shells, etc. In trees and bushes, a *shallow platform of twigs, debris, seaweed,* sometimes lined with shells and coral, bones. Old nests may be repaired and reused, resulting in larger structures. Tree nests to 30 cm in outside diameter. **EGGS:** 1; whitish buff, sparingly marked with spots, speckles, blotches, frequently at larger end; 52 x 35 mm. **BEHAVIOR:** Colonial, from small to large aggregations.

SOOTY TERN *Onychoprion fuscatus*

HABITAT: Small offshore islands. **LOCATION AND STRUCTURE:** On ground, usually in open and sparsely vegetated areas. Occasionally will nest under shrubs or bushes. Nest may be a shallow scrape or just a clutch laid on ground. May be unlined or lined with shell and vegetative fragments, pebbles. **EGGS:** 1; whitish or pale pink, heavily speckled and spotted; 50 x 35 mm. **BEHAVIOR:** Colonial. Often in the tens of thousands (Dry Tortugas, FL).

Bill Summerour, AL

BRIDLED TERN *Onychoprion anaethetus*

HABITAT: FL Keys. **LOCATION AND STRUCTURE:** Usually well concealed among rubble, under rock overhangs, in rock crevices and limestone cavities, under bushes or trees. Occasionally on artificial structures. Rarely in open sandy areas with no vegetation. Nest is a small depression, typically unlined but may contain rock or limestone fragments, dry vegetation, leaves. May reuse scrape from previous year. **EGGS:** 1; creamy or pinkish white, heavily marked with spots, speckles, and blotches; 46 x 33 mm.

ALEUTIAN TERN *Onychoprion aleuticus*

HABITAT: Coastal: on vegetated islands, sandbars, beaches, in bogs and marshes, along pond edges, lush meadows. **LOCATION AND STRUCTURE:** In moss, matted grass, or low vegetation. Occasionally in taller grass. Nest is a slight depression with or without a lining of dry grass stems. **EGGS:** 2; clay yellow to olive-buff, heavily marked with blotches and small spots, which may be wreathed or scattered over entire egg; 42 x 29 mm. **BEHAVIOR:** Colonial, commonly with Arctic Terns.

LEAST TERN *Sterna antillarum*

HABITAT: Coastal: Beaches, lagoons, estuaries, natural and dredge-spoil islands. Inland: Along freshwater lakes and sandbars of large rivers. Occasionally on gravel rooftops and parking lots. **LOCATION AND STRUCTURE:** On open, flat, sandy or shelly ground with sparse vegetation, usually close to shallow water. Nest is a shallow scrape, usually unlined but sometimes with shell and wood fragments, plant stems, pebbles. **EGGS:** 2–3; creamy to pale olive brown, variably marked with spots, speckles, and blotches; 32 x 23 mm. **BEHAVIOR:** Colonial, occasionally in solitary pairs. Will nest with other terns, Black Skimmer, small plovers.

Least Tern.
Matt Monjello, NC

GULL-BILLED TERN   *Gelochelidon nilotica*

HABITAT: Usually on sparsely vegetated coastal beaches, lagoons, and offshore islands. Also on gravel rooftops, inland on small islands in Salton Sea. Traditionally nested on saltmarsh islands but has abandoned those sites, most likely because of hunting. **LOCATION AND STRUCTURE:** Prefers to nest in most elevated spot within site. Nest is a shallow scrape lightly lined, materials placed particularly around nest rim. Coastal nests are often composed of shells and other debris; in Salton Sea, nests may be composed of pebbles, fish bones, vegetation, exoskeletons, plastic. Outside diameter 19–29 cm. **EGGS:** 3; pale buff, marked; 49 x 35 mm. **BEHAVIOR:** Colonial, often in small groups.

Bill Summerour, AL

CASPIAN TERN *Hydroprogne caspia*

HABITAT: Broad variety of open habitats, including natural and dredge-spoil islands, coastal beaches, salt marsh. Also inland on freshwater lake and river islands, saline lake islands. **LOCATION AND STRUCTURE:** Usually in open areas with little vegetative growth. May nest among driftwood or logs. Nest is a shallow scrape or natural hollow and may or may not be lined with shells, pebbles, plant material, seaweed. Nest rim may be built up. Sometimes builds more substantial nest of sticks, etc. Outside diameter 19.5 cm; inside diameter 16 cm; depth 4.5 cm. **EGGS:** 2–3; pinkish buff, lightly marked; 64 x 45 mm. **BEHAVIOR:** Colonial and isolated pairs. Usually near other seabirds.

Bill Summerour, AL

Peter Hodum, WA

ROSEATE TERN *Sterna dougallii*

HABITAT: Low, rocky, sandy or coral islands, barrier beaches. Infrequently on islands in salt marsh. Often with Common Terns (Northeast), but also with Least Terns (FL Keys). **LOCATION AND STRUCTURE:** *Nests are often better concealed than those of Common Tern, and are usually under cover (grass, driftwood, rock).* Sometimes in open areas. Also gravel rooftops (FL). A shallow scrape in sand, shell, occasionally on matted grass. Sometimes clutch is laid on rock, driftwood, other hard substrates. Usually lined with surrounding material. **EGGS:** 1–2; *similar to Common Tern's but usually slightly darker and more pointed*; 42 x 30 mm.

Roseate Tern nest tucked beneath a rock overhang. Matt Monjello, ME

Roseate Tern nest concealed in vegetation. Matt Monjello, ME

COMMON TERN *Sterna hirundo*

HABITAT: Coastal and inland lake islands, on beaches, in salt marsh. Sometimes in freshwater marsh. **LOCATION AND STRUCTURE:** On ground, usually in open sites with low vegetation close to water. Sometimes on heaps of sea wrack, on artificial structures (old piers, rooftops, etc.), in freshwater marshes on muskrat lodges or floating vegetation. Nest is a slight scrape, with or without lining of dead vegetation, seaweed, or can be substantial. Occasionally trash, shells, and other nearby material. Very variable, depending on build and substrate. Outside diameter of built-up nests ~16–18 cm; height ~10 cm. **EGGS:** 2–3; pale buff, cinnamon brown, greenish, or olive, lightly marked with spots, blotches, fine lines, sometimes concentrated at larger end. **BEHAVIOR:** Colonial, often with other terns.

Matt Monjello, ME

ARCTIC TERN *Sterna paradisaea*

HABITAT: Open, often treeless areas with little to no vegetative growth: small coastal and offshore islands, barrier beaches, sand or gravel promontories, meadows, inland lakes, rivers, wetlands. **LOCATION AND STRUCTURE:** Often close to water. In sand, shell, moss, among or on bare rock, in short grass. Sometimes uses old nests of other birds. A shallow scrape, unlined or lined to form a thin, bowled mat. May or may not have material placed around rim. Material includes dead plants, twigs, occasionally shells, pebbles, bones, trash. Size variable; built-up nests to 15 cm or more in outside diameter. **EGGS:** 2; *similar to Common Tern's but usually darker.* **BEHAVIOR:** Colonial (coastal) or in small groups (inland), sometimes with gulls, ducks, other terns.

ROYAL TERN *Thalasseus maximus*

HABITAT: Atlantic and Gulf Coasts, on low barrier and dredge-spoil islands. **LOCATION AND STRUCTURE:** In open, usually on sand, sometimes in shelly or silty areas or among sea wrack. A shallow scrape, usually unlined, but may contain debris. Feces are deposited around scrape. **EGGS:** 1; whitish to dark brown, heavily spotted at larger end; 63 x 45 mm. **BEHAVIOR:** Often in large, dense colonies. Young form a creche. Adults may continue to feed young well after hatching (5–8 months).

A Sandwich Tern egg, next to the larger egg of a Royal Tern, from a dredge-spoil island where these species nest side by side.
Matt Monjello, NC

SANDWICH TERN *Thalasseus sandvicensis*
HABITAT: Atlantic and Gulf Coasts. On low, sandy barrier islands or dredge-spoil islands. **LOCATION AND STRUCTURE:** Often among Royal Terns. Nest is a shallow scrape, with or without lining of nearby material. **EGGS:** 1; pinkish, pale buff, or creamy white, with wide variety of size and dispersal of markings; 51 x 36 mm. **BEHAVIOR:** Young form a creche.

ELEGANT TERN *Thalasseus elegans*
HABITAT: Limited to isolated areas in southern coastal CA: small, sandy islands and dikes separating salt ponds. **LOCATION AND STRUCTURE:** Nest is a shallow depression (in smooth depressions in hard surfaces where occurring) or scrape. May or may not have material placed in scrape or around rim. Material includes shells, pebbles, bones, feathers, twigs. *Occasionally feces are deposited around scrape, forming a rim.* **EGGS:** 1; pinkish buff to white, spotted or blotched; oval to subelliptical; 53 x 38 mm. **BEHAVIOR:** *Often with Caspian Terns.* Young may gather in a creche.

BLACK SKIMMER *Rynchops niger*
HABITAT: Sandy or dredge-spoil islands, remote beaches, shell banks, saltmarsh islands, highway embankments, occasionally rooftops (FL); inland at Salton Sea (CA). **LOCATION AND STRUCTURE:** Favors open, sparsely vegetated, sandy substrates. Also nests on piled-up seawrack mats. Avoids nesting on pressed-down saltmarsh grass. Nest is a shallow, unlined scrape. **EGGS:** 4–5; white to pale creamy white, infrequently blue-green or pinkish, marked with blotches and spots. **BEHAVIOR:** Often in mixed colonies with Common Terns.

Black Skimmer.
Bill Summerour, AL

TERNS THAT BUILD NESTS ON WATER

BLACK TERN *Chlidonias niger*

HABITAT: On shallow freshwater ponds and marshes in both open and wooded areas. Also marshy shorelines of lakes, rivers, and freshwater islands; occasionally in wet meadows, rice fields. **LOCATION AND STRUCTURE:** Over still water in marsh; on low, floating mats of dead emergent vegetation. Also on muskrat platforms, wood, root tangles, abandoned grebe nests, or on ground close to open water. A platform of emergent vegetation with or without a depression at center. Depression may be damp. Sometimes a simple scrape with little lining. Platform nest: Outside diameter 10–25 cm; height 2–6 cm; depth 9 cm. **EGGS:** 3; pale buff to deep olive, marked with spots, blotches, scrawls, typically concentrated at larger end, occasionally wreathed; 34 x 25 mm. **BEHAVIOR:** Often in loose colonies.

Thomas Schultz, WI

Damon Calderwood, OR

FORSTER'S TERN *Sterna forsteri*

HABITAT: Freshwater, brackish and saltwater marshes; on river islands (Colombia R., WA). **LOCATION AND STRUCTURE:** In marshy areas along coastal beaches, islands and estuaries, interior freshwater lakes and marshes. Nest may be placed on mats of floating vegetation, muskrat houses, or on ground. Sometimes uses old grebe nests and artificial floating platforms. Structure varies from a simple scrape with little to no lining to a platform of marsh vegetation. Size variable. Outside diameter ~8–40 cm; inside diameter 11–14 cm; depth 3 cm. **EGGS:** 2–3; olive to buff or pinkish buff with heavy spots, blotches, scrawls, often wreathed at larger end; 43 x 31 mm. **BEHAVIOR:** Colonial, often in loose colonies. Occasionally nests near Black Tern or Yellow-headed Blackbird colonies. Prefers higher and drier sites than Black Tern.

Floating nest built atop a mat of emergent vegetation. Paule Hjertaas, SK

Nest built at water's edge.
Thomas Erdman, WI

LOONS (Gaviidae)

These 5 species of diving birds breed in protected, undisturbed fresh waters throughout northern N. America. They can be found along a variety of waterways, including large to small lakes, reservoirs, ponds, the Arctic coast, and shorelines of rivers. Having evolved to a life on water, loons are exceptional swimmers. Their feet, located posteriorly on the body, help propel them through water with ease. Conversely, this makes their ability to move on land more difficult. Nests, therefore, are often placed adjacent to the water's edge, where they are more readily accessible. With one exception (Red-throated Loon), loons cannot take flight from land. They require a relatively large body of water with enough of a "runway" to gain the speed necessary for takeoff.

Loons select the nest location, often on small islands, shores, and promontories that provide easy access to and from the nest, protection from wind and waves, and good visibility of the surrounding area. When water levels in a breeding area drop, nests may end up high above and farther from the water's edge, making access difficult and causing adults to abandon the nest. A significant rise in water levels may flood nests, leading to clutch failure. Nests are variable: an onshore mound or depression, a heap of vegetation with a damp inner cup built in shallow water, or a clutch of eggs on bare rock. Both sexes share in nest building, incubation, and rearing young.

Socially monogamous, loons form strong pair bonds that can last the lifespan of the pair. These species often return annually to the same breeding habitat, where they maintain and defend their territory from other loons and other species that use similar habitat. Single-brooded, loons typically lay 2, sometimes 1 or 3, olive brown eggs, which are subelliptical to oval, glossy to slightly glossy, and speckled and blotched with blackish brown markings.

Pacific Loon nest built at water's edge.
Kate Fremlin, Amaroq Wildlife Services, NU

Yellow-billed Loon eggs. Chris Smith, AK

Common Loon nest built in shallow water. Matt Monjello, ME

RED-THROATED LOON *Gavia stellata*

HABITAT: Favors low-lying freshwater wetlands along coastal tundra regions, but also in northern forests and up to 1,070 m elevation in montane areas. Prime habitats are shallow ponds with a silty and decayed vegetated bottom, usually 0.3–1 m deep. May also occupy slightly deeper ponds, larger lakes, and shorelines of marshes. Generally occupies small ponds where range overlaps with Pacific Loon. **LOCATION AND STRUCTURE:** More common along shorelines of islands than on mainland. Nest is a depression on ground lined or rimmed with grasses and moss, vegetation; also a built-up mass of wet, decayed vegetation and moss in shallow water. Outside diameter 40–50 cm; height 0–15 cm; inside diameter 20–28 cm; depth 2.5–5.5 cm. **EGGS:** Color variable, usually olive buff speckled with irregular blackish brown spots; 74 x 45 mm; IP 24–31 days (F, M); NP leave nest at 1 day, fly in 38–50 days (F, M). **BEHAVIOR:** May build new nests on top of older ones from previous seasons.

ARCTIC LOON *Gavia arctica*

PACIFIC LOON *Gavia pacifica*

HABITAT: Once considered a single species, Pacific Loon, now designated as two closely related species. Arctic Loon: In nw. AK, along coast on brackish lakes. Pacific Loon: On lakes in tundra and in forested landscape. *Pacific Loon inhabits larger, deeper waters where range overlaps with Red-throated Loon.* **LOCATION AND STRUCTURE:** Nests both on shorelines of islands and mainland. A shallow depression in ground with small amounts of lining, or a large mound built from bottom up with wet, decayed vegetation and mud. Rarely builds a floating nest within reeds. Outside diameter 46–71 cm; height 8.2–9.8 cm; inside diameter 23–36 cm; depth 4–10 cm. **EGGS:** Color variable, usually olive brown, speckled with blackish brown spots. Arctic: 76 x 47 mm; IP 27–30 days (F, M); NP 60–65 days (F, M). Pacific: 76 x 47 mm; IP 23–25d (F, M); NP 60–65 days (F, M).

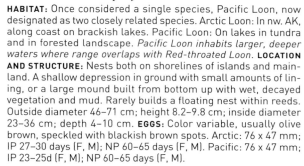

COMMON LOON *Gavia immer*

HABITAT: Clear freshwater lakes with small islands and coves creating a sheltered shoreline. Small fish must be plentiful. From wooded country to past tree line in subarctic and Arctic regions; occasionally breeds on slow-moving water around oxbows on rivers. **LOCATION AND STRUCTURE:** Site varies. Nest is always built in or very close to water, including on protected sides of islands, sheltered shorelines, in marshy or reedy areas, on floating bogs, muskrat houses, logs, similar. Readily uses artificial nest platforms and rafts at management sites. Nest structure varies. More commonly a large circular vegetative mound of various surrounding plant materials; also a simple scrape or just a clutch on bare ground or rock. Outside diameter 55.9–56.9 cm; inside diameter 24.4–33 cm. **EGGS:** Olive brown with blackish brown spots or blotches; 90 x 57 mm; IP 27–31 days (F, M); NP 2 days, fly at 10–11 weeks (F, M). **BEHAVIOR:** May take more than a month from arrival on territory before building nest and laying clutch. Will reuse and refurbish nest from previous year. Within 1–2 days of hatching, young accompany parents to nursery areas: often shallow protective coves with an abundant food supply of plants, invertebrates, crayfish, small fish.

YELLOW-BILLED LOON *Gavia adamsii*

HABITAT: Commonly near coast on low-lying lakes in Arctic tundra and inland along shorelines of larger lakes and torpid rivers. **LOCATION AND STRUCTURE:** Most commonly occupies elevated promontories along mainland shoreline and also small islands when available. Nest is typically a shallow depression, occasionally a mound of mud or moss. Lining is usually absent but may contain sedges, grasses, willow, moss. Outside diameter 38–82 cm; height 8–33 cm; inside diameter 24–37 cm; depth 1–7.5 cm. **EGGS:** *Similar to Common Loon's*; 89 x 56 mm; IP 27–28 days (F, M). **BEHAVIOR:** Frequently reuses nests. Highly territorial; pairs will defend against other loons as well as diving ducks.

NORTHERN STORM-PETRELS (Hydrobatidae)

Northern storm-petrels are colonial nesters on offshore islands, or occasionally in inaccessible cliffs on mainland coasts, and are found with other seabirds with similar nesting habits. They are the smallest seabirds, and all nest in burrows or natural crevices; some species may use the abandoned burrows of other seabirds, and pairs often return year after year to the same nest burrow. Lining material is occasionally added to the nest chamber.

Northern storm-petrels are monogamous (some species only presumed so). Adults are typically absent from the nest during the day and come and go from the colony at night to feed the chick. Young are often fatter and larger than adults while being fed; when the NP nears its end, adults usually abandon the nest entirely, and the young lose weight for several days before finally fledging (typically at night). IP 37–53 days, NP ~5–70 days. Female lays a single round-ended elliptical to subelliptical white egg that occasionally has very light markings. Least Storm-Petrel *(Oceanodromo microsoma)*, breeding in Baja CA, is not included here.

FORK-TAILED STORM-PETREL *Oceanodroma furcate*

In open rather than treed areas. Uses both burrows and natural crevices; depending on the colony, one or the other may be more common. When burrows are used, they are often preexisting, made by other seabirds. There is little to no lining material in the nest chamber. Colonies range from a few hundred pairs to several hundred thousand. Burrow diameter 12 x 14 cm; length 1 m. **EGG:** 34 x 26 mm.

LEACH'S STORM-PETREL *Oceanodroma leucorhoa*

Excavates a burrow, about 1 m long, in forest, shrubs, or grass cover. Multiple pairs sometimes share one burrow with separate nest chambers. A shallow scrape in burrow is lined with vegeta-

Leach's Storm-Petrel burrows, excavated on a grassy island slope.
Alix d'Entremont, NS

Leach's Storm-Petrel burrow on a forested island.
Alix d'Entremont, NS

tion and other items. Sometimes nests in natural crevices. Colonies may be as small as 7 pairs to more than 3 million. **EGG:** 33 x 24 mm; IP 37–50 days.

ASHY STORM-PETREL   *Oceanodroma homochroa*

Uses natural cavity or a burrow dug by another bird. May share nesting chamber with other species. Asynchronous breeding at colonies. Fewer than 10,000 birds remain in the entire population. Introduction of mammalian predators to nesting islands by humans has significantly affected this species. **EGG:** 30 x 23 mm; NP can be as long as 119 days.

An Ashy Storm-Petrel broods a chick in its rock-crevice nest.
Amelia J. DuVall, CA

BLACK STORM-PETREL   *Oceanodroma melania*

Around Santa Barbara I., CA. Uses natural crevice or abandoned burrow of another species (often Cassin's Auklet). No lining is added to nest chamber. Occasionally shares nesting burrow or crevice with conspecifics or other seabird species. Entrance to natural crevice may be as small as 5 cm, which would exclude larger seabirds. **EGG:** 37 x 27 mm.

SHEARWATERS and PETRELS (Procellariidae)

Most shearwaters and petrels nest in the Southern Hemisphere and wander to coastal N. America after the breeding season. Two species nest in N. America, however, mostly on coastal islands in inaccessible locations. As in the alcids, with which procellarids share nesting habitats and strategies, nests are either on bare rock ledges, in natural crevices, or in excavated burrows. Both species below are colonial and monogamous. They produce a single subelliptical, dull white egg, which both the male and female incubate. Both sexes also feed. Nestling period 46–62 days (Manx Shearwater at the longer end). Black-vented shearwater *(Puffinis opisthemelas)*, which breeds on a few islands offshore of Baja CA, is not included here.

NORTHERN FULMAR *Fulmarus glacialis*

HABITAT: Cliffs on islands and rocky coastline. **LOCATION AND STRUCTURE:** On sea stacks, islands, occasionally onshore. Typically a cliff ledge with a wall on 1 or 2 sides, occasionally in an overhang. Sometimes on moderate slopes with grass and other vegetation. On rock, no nest is built. A slight depression is

formed in softer substrate with sparse lining if material is available within reach. **EGG:** 71 x 50 mm; IP 55–57 days. **BEHAVIOR:** Adults switch off incubating every 5–7 days. Monogamous; long-term pair bond is associated with extreme site fidelity. When an individual survives its mate, it will retain "ownership" of nest site until a new, previously unattached, mate is found, often with a delay of multiple years. When a pair bond is formed without a nest site, it may take years for the new pair to claim a nest location.

Brendan Higgins, AK

Garrit Vyn, AK

MANX SHEARWATER *Puffinus puffinus*

HABITAT: Although worldwide range is extensive, in N. America known from only 1 small colony on an island off the coast of NL. **LOCATION AND STRUCTURE:** Constructs burrow or uses natural rock crevice. Nest chamber is large enough to accommodate both adults. Has lining of plant and other materials or bare dirt. Burrow diameter 20 cm; length 0.4–3 m. **EGG:** 61 x 42 mm; IP 52–54 days. **BEHAVIOR:** Adults switch off incubating every 4–5 days.

Manx Shearwater nests.
Jens Kirkeby

STORKS (Ciconiidae)

WOOD STORK *Mycteria americana*

HABITAT: Forested wetlands, cypress swamps, flooded impoundments, and any area with trees above standing water. **LOCATION AND STRUCTURE:** Nests in trees close to or in fresh or brackish water. Historically, storks preferred large cypress trees but now often use a variety of nonnative species in locations with artificially impounded water. Nest is a large stick platform over standing water with a lining of finer material (twigs, leaves, etc.). Typically in treetops, occasionally just a meter or so above water. Outside diameter 1–1.5 m; depth 18–25 cm. **EGGS:** 3–4; 68 x 46 mm; IP 28–32 days (F, M); NP ~75 days (F, M). **BEHAVIOR:** Apparently monogamous. Colonial nesting at various heights in stands of trees. Nests may be within 1 m of one another. In FL breeds in late winter, matching the dry season there when fish become concentrated in shrinking pools of water.

Matt Monjello, GA

FRIGATEBIRDS (Fregatidae)

MAGNIFICENT FRIGATEBIRD *Fregata magnificens*

HABITAT: Mangrove cays, low trees and shrubs on islands. Islands off coast of s. FL and occasionally farther north on East Coast; also on West Coast in Baja CA and on coast of s. CA. **LOCATION AND STRUCTURE:** Loosely constructed stick platform is built on top of shrubs or in a tree, typically within several meters of ground. Sometimes on ground. Has slight hollow in center, occasionally with dry grass lining. Outside diameter 25–35 cm. **EGGS:** 1; 68 x 47 mm; IP 56 days (F, M); NP 150 days (F, M). **BEHAVIOR:** Variation on classic lekking behavior; multiple males display in close proximity to attract females, and newly established pair builds nest at male's display site (male collects materials, female builds). Male may remain with female (and chick) for 3 months before he leaves, possibly to breed with another female. It takes ~6 months for young to be ready to fly, and female will continue to feed young for several months after this. Colonial in s. FL, from as few as 10 pairs to several thousand.

Judd Patterson,
Birds In Focus, FL

BOOBIES and GANNETS (Sulidae)

NORTHERN GANNET *Morus bassanus*

HABITAT: Breeds in large colonies in cold temperate waters of N. Atlantic, on small offshore islands and sea stacks; occasionally on protected cliffs of mainland coast or large islands. **LOCATION AND STRUCTURE:** Prefers sites facing prevailing winds on cliff tops, ledges, steep slopes; occasionally on more even ground. Nests are often no more than 60–90 cm apart. May be located from just above splash zone to over 200 m high. Nest is a shallow cupped mound built with mud, seaweed, grass, debris, human-made materials, feathers, flotsam, adhered with excrement. Typically 30 cm in diameter x 20 cm high. **EGGS:** 1; initially smooth in texture and pale blue or greenish, becoming rough, flaky, and chalky white with deep brown stains; long subelliptical to oval; 82 x 49 mm; IP 43–46 days (F, M); NP 90 days (F, M). **BEHAVIOR:** Monogamous. Male selects nest site and is primary builder. Construction continues through incubation. New nest starts as scrape with rim of material that gradually becomes a taller mound as more material gets added. Pairs often reuse same nest for years.

Jean Knowles,
Nova Scotia

CORMORANTS and ANHINGA
(Phalacrocoracidae and Anhingidae)

The 7 species in this group (6 cormorants in Phalacrocoracidae and Anhinga in Anhingidae) all nest colonially. Although most are associated with marine environments, Double-crested and Neotropical Cormorants and Anhinga will also nest near and use freshwater habitats; these 3 species typically nest near bodies of water large enough to support their fishing activities but will occasionally nest on small wetlands, flying many miles to forage.

Nests may be constructed on the ground, in trees, or on raised platforms such as cliffs and ledges. Nests on ledges and rocks are often composed primarily of grass, seaweed, or other found materials from the environment, and the size of the nest may be dictated by the size of the ledge it is built on. Tree nests are constructed primarily with sticks. Trees supporting cormorant colonies are often dead or dying because of the guano from the birds. Nest structure is similar in all species; use geographic location, placement, size, and microhabitat to aid identification.

These species are primarily socially monogamous, but some species may re-pair in consecutive years. In most species the male picks the nest location and collects most of the nest material while the female weaves the nest. Male cormorants will steal material from other nests for their own. Young cormorants and Anhingas are altricial. Older young may leave their nest to congregate with other young but will return to their nest to be fed by their parents; they typically leave the nest permanently after 5–8 weeks, varying by species. Eggs of all species are unmarked and white to pale blue.

BRANDT'S CORMORANT *Phalacrocorax penicillatus*

HABITAT: Cliffs and ledges in marine environments, primarily islands. Slopes, often windward. **LOCATION AND STRUCTURE:** On cliff ledges or flattish areas of ground or rocks. Nest is a large and untidy mass of matted plant material, marine debris. Often shares locations with other seabirds. *Where Brandt's and Double-crested Cormorants overlap, Double-crested is found on steeper terrain.* Outside diameter 35–80 cm; height 15–36 cm; inside diameter 19–25 cm; depth 8–10 cm. **EGGS:** 1–6; 61 x 39 mm; IP 29 days (F, M).

Bird Research
Northwest, OR

NEOTROPIC CORMORANT *Phalacrocorax brasilianus*

HABITAT: Limited breeding range in s. U.S. Inhabits coastal marshes and inland lakes, **LOCATION AND STRUCTURE:** Nests in small living or dead trees. Nest is usually less than 4 m high. Occasionally on human structures, rocks, or bare ground when trees are absent. Nest is a platform of sticks, sometimes with a lining of grass, seaweed, and debris, sometimes cemented together with guano. Outside diameter 29–40 cm; height 9–19 cm; inside diameter 16–26 cm; depth 4–9 cm. **EGGS:** 3–4 (2–6); 55 x 34 mm; IP 25 days (assumed M, F).

DOUBLE-CRESTED CORMORANT *Phalacrocorax auritus*

LOCATION AND STRUCTURE: On ground on islands and in inaccessible locations. Elsewhere in trees or shrubs, emergent vegetation, or other raised structures. Guano often kills nesting trees. Nest is a bulky structure of finger-sized sticks, often with seaweed, human debris, and parts of dead birds. Outside diameter 45–90 cm; depth 10–43 cm. Ground nests are wider and flatter than tree nests. **EGGS:** 3–4; 61 x 38 mm; IP 25–28 days (F, M). **BEHAVIOR:** Ground nests have occasionally been found to include rocks, apparently cared for like eggs.

Double-crested Cormorant ground nest.
Thomas Erdman, WI

Colonially nesting Double-crested Cormorants, ground site.
Thomas Erdman, WI

The colonial tree nests of Double-crested Cormorants.
Scott Somershe, TN

GREAT CORMORANT *Phalacrocorax carbo*

LOCATION AND STRUCTURE: Coastal cliffs and rocky islands on ne. Atlantic Coast. Nest is a solid mass of seaweed and sticks in a tree or on rocks. Often in colonies with Double-crested Cormorants and other seabirds. Outside diameter 49 cm; height 22 cm. **EGGS:** 3–5; bluish green with chalky coating; 63 x 41 mm; IP 28–31 days (F, M).

RED-FACED CORMORANT *Phalacrocorax urile*

HABITAT: Nests in colonies on ledges of cliffs on islands and rocky headlands. **LOCATION AND STRUCTURE:** Often nests with other seabirds, selecting for steeper and higher parts of cliffs. *Not found on flat ground, or human structures where similar Pelagic Cormorants sometimes nest.* Prefers ledges 30–60 cm wide. Constructs compact oval nest with grass, seaweed, and other soft material as well as guano. Outside diameter 50 cm; height 15 cm. **EGGS:** 2–4; 61 x 37 mm; IP 31–38 days (F, M).

PELAGIC CORMORANT *Phalacrocorax pelagicus*

HABITAT: Primarily on narrow ledges or caves on cliffs along exposed shorelines. **LOCATION AND STRUCTURE:** *Not in trees or shrubs, as is possible for Double-crested Cormorant.* Nest generally appears *more compact than that of Brandt's Cormorant*, constructed with seaweed and marine debris often cemented to cliff with guano. Size of nests varies to match size of cliff area it is built into. Outside diameter 44–48 cm; height 26–41 cm. **EGGS:** 3–4; greenish white to blue; 59 x 37 mm; IP 30 days (F, M). **BEHAVIOR:** Least colonial of all cormorants. May be in colonies of thousands but more often in small numbers or isolated pairs. Where it overlaps with Double-crested and Red-faced Cormorants, Pelagic is found on narrower, more seaward ledges.

Brendan Higgins, AK

Mark Nyhof, BC

ANHINGA *Anhinga anhinga*

HABITAT: Shallow and slow-moving fresh water with an abundance of areas for perching. Rarely near salt water. Feeds almost exclusively in fresh water but may nest in colonies along salt water and fly inland to forage, often nesting with other waterbirds, including cormorants. **LOCATION AND STRUCTURE:** In fork of a tree or shrub close to or overhanging water. Up to 30 m high but usually less than 5 m. Nest is relatively compact, built with sticks and lined with green leaves and other soft material. *Presence of*

Mary Alice Tartler, SC

green vegetation in lining is helpful for distinguishing from nests of herons; nest is also less bulky than those of herons. Occasionally uses nests originally constructed by egrets or storks. Outside diameter 29 cm; depth 16 cm. **EGGS:** 3–5; pale blue-green with chalky coating, distinctively pointed at one end; 53 x 41 mm; IP 26–30 days (F, M). **BEHAVIOR:** May nest alone or in colony of up to several hundred pairs. Male starts building nest, female finishes.

PELICANS (Pelecandiae)

Two species of pelicans breed in fresh- and saltwater habitats in N. America's interior as well as on its coastlines. Highly gregarious, they nest in colonies that in some areas can number in the thousands. Pelicans prefer to nest in protected areas that have low predator populations. Nests are placed on the ground or elevated in bushes or mangrove trees. Size and structure vary, ranging from simple scrapes with little lining to a bulky ground nest of sticks and vegetation, or a sturdy elevated platform of sticks secured into surrounding branches.

Monogamous and single-brooded, pelicans lay 2–3 elliptical to long sub-elliptical, chalky white eggs. Rough in texture when laid, eggs become smooth and stained over time by the parents' feet and guano. Both parents incubate for 29–30 days. Young fledge in anywhere from 64–70 days (American White Pelican) to 77–84 days (Brown Pelican).

American White Pelicans in a large, mixed nesting colony with Double-crested Cormorants. Thomas Erdman, WI

AMERICAN WHITE PELICAN *Pelecanus erythrorhynchos*

HABITAT: Remote islands on interior freshwater and saline lakes. Also on islands susceptible to flooding in wetlands (n. Great Plains) and in coastal waters (TX). **LOCATION AND STRUCTURE:** Often next to or among low herbage or shrub, sometimes under trees in forested areas. Ground nest is a shallow, unlined depression with a rim of soil, gravel, and vegetation, scraped up by the sitting adult. Feathers possible. Outside diameter 61–89 cm; height 20–30 cm; depth up to 5 cm. **EGGS:** 2; 87 x 56 mm.

An American White Pelican's nest composed of only a small collection of material. Timothy Lawes, OR

BEHAVIOR: Both sexes build in 2–9 days. Distinct subcolonies become evident at breeding sites due to pairs establishing nesting groups with others at similar stages of the reproductive cycle.

BROWN PELICAN *Pelecanus occidentalis*

HABITAT: Atlantic and Gulf Coasts: Artificial and natural islands in shallow coastal waters, estuaries, and tidal rivers; mangrove islands (FL). Pacific Coast: Bare, rocky coastal islands. **LOCATION AND STRUCTURE:** Nest site varies. Often on ground among herbage or low shrubs, occasionally in open sand or shelly ground or aboveground near top of thickets or low trees. Ground nest may be a simple scrape lined with grasses or a large bulky mound of sticks, leaves, reeds, seaweed. Elevated nests typically are sturdy platforms of sticks secured into supporting branches, lined with grasses, leaves. **EGGS:** 3; 87 x 56 mm. **BEHAVIOR:** Male selects nest site and gathers material. Female builds. Construction takes 4–10 days. May steal material from unattended nests. Chicks dismantle nests in mangrove colonies, leaving little of the original structure at time of fledging.

A Brown Pelican nest elevated in shrubbery. Bill Summerour, AL

A Brown Pelican nest consisting of heaped vegetation. Matt Monjello, NC

BITTERNS, HERONS, and EGRETS (Ardeidae)

These 12 small to large wading birds can be found occupying nearly every aquatic habitat in fresh, brackish, and salt water. Nests are commonly found near water, on or above ground in a variety of sites: tall grasses or emergent vegetation, shrubs or bushes, tall trees, or rock ledges. They may be well hidden in foliage, set close to a tree trunk, or far out on a fork of a horizontal branch. Some nests are constructed from reeds and grasses, but most are made of sticks, twigs, and stems, woven into a platform that contains little to no lining. Platform nests may be small, flimsy, and see-through, or sturdy and thickly built. Discrepancies in size and structure are often attributed to whether the nest is newly built or has been reused and added to over multiple breeding seasons. Many ardeids also add material to nests throughout the NP.

For many species, the choice of nest site and its initial construction or repair are made by males. The male may begin the platform, and then the female typically completes the nest, with the male providing materials. Exceptions occur among both bittern species, Great Egret, Yellow-crowned Night-Heron, and possibly others. Nests are usually built within a week.

Typically seasonally monogamous, most ardeids breed in small or large colonies, either interspecific or conspecific. Initial territories at large colonial sites will decrease in size as more birds arrive at the breeding grounds, resulting in a territory that surrounds only the nest. Single-brooded, ardeids lay 2–7 elliptical to long subelliptical, pale bluish green unmarked eggs, except for American Bittern, whose eggs are buff brown to olive buff. Both parents incubate, generally beginning with the first or second egg and lasting 16–28 days. The altricial young hatch asynchronously, are tended by both parents, and depending on the species, may fledge in 1 week or in up to 3 months. The NP listed for herons and egrets refers to the time in the nest itself, after which the young begin to leave for the surrounding branches until they are ready to fly.

AMERICAN BITTERN *Botaurus lentiginosus*
HABITAT: Predominately in large, shallow freshwater marshes, bogs, swamps with tall emergent vegetation. Less abundant in smaller wetland areas. Rarely in salt marshes and unspoiled,

R. Wayne Campbell, BC

elevated grassy fields around wetlands. **LOCATION AND STRUC-TURE:** Commonly 7.6–20 cm above shallow water in dense emergent vegetation; occasionally on ground in grassy areas with tall, thick herbage. Nest is a platform of nearby material, including cattails, grasses, and sedges, lined with fine grasses. Heavily used runways often provide approach to nest. Outside diameter 25–40 cm; height 15–25 cm. **EGGS:** 2–5; smooth and slightly glossy, buff brown to olive buff, unmarked; elliptical to short sub-elliptical; 49 x 37 mm; IP 24–29 days (F); NP 7–14 days (F). **BEHAVIOR:** Monogamous and likely also polygynous; may nest singly, or multiple females may nest in 1 male's territory. Female apparently chooses nest site, builds, and tends young alone. Will renest if first clutch is lost.

LEAST BITTERN *Ixobrychus exilis*

HABITAT: Primarily in freshwater and brackish marshes with a mixture of tall, thick emergent vegetation, shrubby growth, and open water; sometimes in salt marshes or mangroves. **LOCATION AND STRUCTURE:** Well concealed, in marsh or shrubby growth, usually within 10 m of open water, including waterways and openings associated with muskrat activity. Occasionally on top of an old nest. Nest is a platform, with a canopy built by bending down emergent vegetation, adding sticks and stems on top. Sticks are often arranged in a radiating pattern, 15–76 cm above water. **EGGS:** 4–5; 31 x 24 mm; IP 16–19 days (F, M); NP ~2 weeks (F, M). **BEHAVIOR:** Male selects nest site and is primary builder. Normally solitary but will nest in loose colonies, sometimes in close proximity to Boat-tailed Grackles (along Atlantic and Gulf coastlines).

Eric C. Soehren, AL

GREAT BLUE HERON *Ardea herodias*

HABITAT: Highly adaptable. Breeds in fresh- and saltwater habitats, including swamps, on islands, upland deciduous and coniferous forests, along wooded lakes and ponds, and in riparian woodlands. **LOCATION AND STRUCTURE:** Generally in trees, up to

40 m above ground or water, but also on bushes, nesting platforms, cattails, in mangroves ("Great White" Heron), cliff ledges, on ground (especially predator-free islands). Nest is a large platform of sticks and twigs; shallow cup is lined with moss, pine needles, reeds, leaves, grasses, and finer twigs. Ground nests may be constructed with grassy vegetation. Outside diameter 0.5–1.2 m. Older nests become quite substantial in size. **EGGS:** 3–5; smooth or slightly rough and nonglossy; oval to long oval, long elliptical or subelliptical; 64 x 45 mm; IP 25–30 days (F, M); NP 7 weeks (F, M), depart nest in 64–91 days **BEHAVIOR:** Seasonally monogamous. Nests colonially both with conspecifics and other wading birds; sometimes with eagles, hawks, and owls. Construction takes 3–14 days.

Great Blue Heron nesting colony in a group of dead trees. David Moskowitz, WA

Great Blue Heron ground nest. Bird Research Northwest, OR

GREAT EGRET *Ardea alba*

HABITAT: Near fresh, brackish, and saltwater wetlands, including hardwood and cypress swamps; marshes, lakes, ponds, thickets along streams, rivers, and glades; islands, coastal bays and estuaries. **LOCATION AND STRUCTURE:** Nest height varies. Generally high up in a tree or shrub, on or close to canopy. Often exposed from above and sides. Typically 4.5–12 m above ground or water, sometimes lower or up to 30 m high; occasionally on ground or artificial platform. Nest is *similar to Great Blue Heron's but can be less bulky.* Platform nest: Outside diameter 50–100 cm; height 12.5–29.9 cm. Ground nest: Outside diameter 43.3–55 cm; height 8.3–12.7 cm. **EGGS:** 3; smooth and nonglossy; elliptical to subelliptical; 56 x 41 mm; IP 23–26 days (F, M); NP 21 days

(F, M), fly in 42–49 days. **BEHAVIOR:** Nests both singly and in colonies. Both male and female (or only male) may complete nest. Often constructs new nests each year. Aggression is common among nestlings, and younger or smaller siblings often do not survive.

Great Egret nest built close to the ground.
Matt Monjello, NC

Great Egret tree nest with Wood Stork in background.
Matt Monjello, GA

SNOWY EGRET *Egretta thula*

HABITAT: Wide variety of sites near fresh and salt water. Along the East Coast: In cedar swamps, artificial and saltmarsh islands, mangroves, freshwater wetlands. In Texas: On dry scrubby, coastal islands. In Midwest: Marshes and swamps, lake shorelines, low-lying rivers, reservoir groves, flooded lowland fields. In West: Along large willow-strewn rivers, lakes, irrigation canals, wet meadows, grass-covered marshes, bay islands. **LOCATION AND STRUCTURE:** Nests are usually placed on outer margins of trees, on top of shrubs and low woody tangles; prickly pear (TX); on ground in reeds. Typically 1.5–3 m aboveground, occasionally lower and up to 9 m high. Nest is a flat and shallow round or elliptical platform of loosely woven sticks, lined with finer sticks and twigs and other plant material. Outside diameter 30.5–61 cm; height 20.5–33.5 cm. **EGGS:** 3–5; 43 x 32 mm; IP 20–24 days (F, M); NP 20–25 days (F, M). **BEHAVIOR:** Colonial, sometimes nesting in the thousands. Often in mixed-species colonies. Construction takes 5–7 days.

Matt Monjello, NC

LITTLE BLUE HERON *Egretta caerulea*

HABITAT: Breeds near fresh, brackish, and salt water: swamps, lakes, ponds, streams, water impoundments, inundated fields, mangrove bays, upland islands. **LOCATION AND STRUCTURE:** Typically nests below canopy in small trees, low shrubs and bushes, 0.9–7.6 m above ground or water. Nest is a flat elliptical platform of loosely woven twigs and sticks; a slight depression is lined with finer twigs and green vegetation. Outside diameter 30–50.8 cm; height 15.2–20.3 cm. **EGGS:** 3–5; 45 x 33 mm; IP 20–23 days (F, M); NP 14–21 days (F, M), fly in 28 days, independent in 42–49 days. **BEHAVIOR:** Nesting colonies may be conspecific or interspecific. Construction takes 3–5 days, sometimes 6–8.

Matt Monjello, GA

TRICOLORED HERON *Egretta tricolor*

HABITAT: Primarily along coast in tidal marshes, estuaries, mangrove swamps. Also inland at freshwater sites: marshes, rivers, lakes. Prefers islands, but also nests on areas of high ground with suitable nest substrate surrounded by open area. **LOCATION AND STRUCTURE:** Usually well concealed and shaded in dense vegetation: small trees or shrubs, mangroves, in marsh thickets

or dense cattail stands, rushes and grasses on saltmarsh islands; dry scrubby islands (TX), 0.14–4 m aboveground. A shallow platform of twigs is lined with finer twigs, grass stems. **EGGS:** 3–4; 45 x 33 mm; IP 21–25 days (F, M); NP ~21 days (F, M), fly in ~35 days. **BEHAVIOR:** Colonial, often with other species, sometimes in conspecific colonies or isolated pairs. In mixed-species assemblages, typically nests along edges in densest vegetation. Construction takes 4–5 days.

Tricolored Heron, showing the worn path to and from its nest. Matt Monjello, NC

Tricolored Heron nest close up. Matt Monjello, NC

REDDISH EGRET *Egretta rufescens*

HABITAT: The rarest N. American ardeid. Coastal areas near salt water, including mangrove bays and islands (FL); dry coastal islands (TX). **LOCATION AND STRUCTURE:** In trees, bushes, on ground, up to 10 m high. TX islands: In low scrub, cacti, or herbage; on ground in sand. FL: Commonly over water in a tree fork (especially mangrove). Outside diameter 30–66 cm; height 20–25 cm; inside diameter 25–30 cm; depth 7.6–10 cm. **EGGS:** 3–4; elliptical to short elliptical; 50 x 36 mm; IP usually 26 days (F, M); NP ~21 days (F, M), fly in ~35 days. **BEHAVIOR:** Breeds singly or in mixed colonies, sometimes ranging in the thousands.

CATTLE EGRET *Bubulcus ibis*

HABITAT: Near freshwater, saltwater, and terrestrial habitats: swamps; riparian or dry upland forests or groves containing or absent the undergrowth layer; treed and brushy interior and coastal islands; edges of reservoirs, lakes, and wetlands; pastures, farms, agricultural and irrigated fields. **LOCATION AND STRUCTURE:** In large aggregations, nests are placed on any stable area, including on ground when elevated sites are taken. Usually in trees or bushes aboveground or over water; in both live and dead vegetation. Nest is a shallow, bowl-shaped platform of heavy sticks, lined with thin twigs and vines, sometimes other herbage. May use green leafy twigs for nest construction. Chick excrement seeps into gaps in nest, solidifying the structure. Outside diameter 17.8–61 cm; height 5–27.9 cm; depth 1.3–7.6 cm. **EGGS:** 2–4; *very pale blue or whitish blue*; 46 x 34 mm; IP 21–26

Cattle Egret ground nest.
Scott Somershoe, TN

Elevated Cattle Egret nest.
Scott Somershoe, TN

days (F, M); NP 15–20 days, fly in 29–35 days (F, M), independent in ~45 days. **BEHAVIOR:** Colonial, often found with other herons. Construction occurs 5–6 days before first egg is laid, continues through incubation and after hatching, with nest growing considerably in size.

GREEN HERON *Butorides virescens*

HABITAT: Common in a variety of habitats, often near water, including marshes, swamps, ponds, lakes, impoundments, thickets, and mangroves, but also terrestrial areas away from water, such as dry groves and orchards. **LOCATION AND STRUCTURE:** Typically found over water, in trees and bushes, from ground level to 10 m up, occasionally higher. Often in a large fork, with overhanging branches that provide concealment. Nest is a

Matt Monjello, NC

shallow, round or elongated platform of thin sticks, occasionally reeds, unlined. Nest may be thin and see-through to substantial. Outside diameter 20–30.5 cm; inside diameter 10–13 cm. **EGGS:** 3–5; *green to pale greenish blue*; 38 x 29 mm; IP 19–21 days (F, M); NP 16–17 days (F, M), fly in 21–23 days. **BEHAVIOR:** Single- or double-brooded. Solitary or in loose colonies; often apart from other species. Construction takes 3–4 days. May reuse old nests.

BLACK-CROWNED NIGHT-HERON *Nycticorax nycticorax*

HABITAT: Wide variety of fresh, brackish, and saltwater habitats, including marshes, swamps, rivers, lakes, ponds, canals, wet fields. Colonies are often on islands, swampy areas, or above water. **LOCATION AND STRUCTURE:** Wide variety of substrates, including trees, shrubs, thickets, emergent vegetation, tall grass, cliff ledges, artificial structures, up to 30 m high. On islands, nests may be among rocks on ground. Structure varies with available material; usually a shallow platform of sticks, twigs, reeds, or similar material. May be very thin and see-through or substantial. Outside diameter 30–61 cm; height 20–45.7 cm. **EGGS:** 3–4; 53 x 37 mm; IP 21–26 days (F, M); NP 4 weeks (F, M), fly in 6–7 weeks . **BEHAVIOR:** Colonial, often with other herons or other waterbirds, occasionally with Franklin's Gulls in w. Great Plains. Nesting sites free from predation and disturbance may support colonies for decades. Will reuse nests. Construction takes 2–7 days.

Eric C. Soehren, AL

YELLOW-CROWNED NIGHT-HERON *Nyctanassa violacea*

HABITAT: Limited to areas near water. Breeds in both fresh- and saltwater habitats, including coastal islands, cypress swamps, bayous, densely covered lowlands, wooded riparian uplands, mangroves, edges of bays. Common in open-wooded neighborhoods. **LOCATION AND STRUCTURE:** Usually in a tree or shrub 2–18 m above ground or water, sometimes lower. Nests are generally in a fork along outer margins of large trees or on lowest available tree branch and usually over water in habitats with dense canopy cover; in shrubs in bays and on islands. Nest is usually a thick, slightly concave platform of heavy, dead sticks, lined with fine twigs, rootlets, leaves; occasionally unlined. Outside diameter 50–102 cm. **EGGS:** 2–4; 51 x 37 mm; IP 21–25 days (M, F); NP 30–43 days (F, M). **BEHAVIOR:** Nests singly or in small to medium-sized colonies (more than 50 pairs), sometimes with other heron species. Pairs may build multiple nests until one is completed. May reuse nests. Construction takes 10 days.

IBISES and SPOONBILLS (Threskiornithidae)

Four species of these long-legged wading birds nest in N. America, and 3 of them overlap geographically during the breeding season (White Ibis, Glossy Ibis, and Roseate Spoonbill). Although the birds themselves are clumsy in appearance, their nests, usually near freshwater or saltwater wetlands, are solidly constructed. Either on the ground or up to 4 m above-ground on the top of shrubs or low trees, nests are a well-woven platform of sticks, twigs, and leaves. The cup, which tends to be deeper than in other wading- or waterbird species, is typically well lined with green leaves or other soft material. Monogamous or socially monogamous (varying by species), ibises and spoonbills are typically colonial nesters, often in mixed colonies with herons and other waterbirds. White and White-faced Ibises also nest in "neighborhoods" within the greater colony, and egg laying within subcolonies is highly synchronized. IP varies by species, 21–24 days (F, M). Young leave the nest for surrounding branches in ~2–3 weeks but are unable to fly for ~5–8 weeks. Both sexes incubate and care for young.

WHITE IBIS *Eudocimus albus*

HABITAT: Fresh- and saltwater marshes, mangrove swamps. **LOCATION AND STRUCTURE:** Often in fork of a tree or shrub, but nests are very close together, so many are not supported by a

White Ibis nestlings.
Jason Fidora, FL

*White Ibis
tree nests.*
Jason Fidora, FL

forked branch. Ground nests are in cattail, sawgrass, or other
herbaceous vegetation. Nest is a sturdy but messy construction
of sticks, leaves, grass, etc. Ground nests are often made of only
the herbaceous vegetation in which they are built. Outside diameter 25–29 cm; depth 6–9 cm. **EGGS:** 3–4; pale tan to blue-green,
speckled irregularly; subelliptical; 58 x 39 mm. **BEHAVIOR:** Colony
location often changes annually. Intolerant of nest disturbance
by humans.

GLOSSY IBIS *Plegadis falcinellus*

HABITAT: Usually in freshwater wetlands, but also found in saltwater marshes, mangroves, and other areas of still water. **LOCA-
TION AND STRUCTURE:** Often on islands where available. In small
trees, shrubs, or on ground or above water in reeds or grass in
areas of dense vegetation. Often close to water and up to 4 m
high. Nest is a well-constructed platform of sticks, reeds, and
other vegetation. Outside diameter 30–40 cm; height 4–27 cm.
EGGS: 3–4; solid deep greenish blue; subelliptical to long subelliptical; 52 x 37 mm. **BEHAVIOR:** In mixed-species colonies, often
nesting lower than other species.

The nest of a Glossy Ibis built in emergent vegetation.
Joel Jorgensen/ Nebraska Game and Parks Commission, NE

WHITE-FACED IBIS *Plegadis chihi*

HABITAT: Inland freshwater marshes in West, coastal freshwater marshes and salt marshes in Southeast. Not in cypress swamps. **LOCATION AND STRUCTURE:** Usually less than 1 m over water in emergent vegetation, shrubs, or small trees. Occasionally on ground, on islands. Supportive platform is woven from bent-over stalks of surrounding vegetation. Nest is constructed from nearby available material on top of platform and is well lined with fine grass and parts of rushes. Outside diameter 27–39 cm; height 16–26 cm. **EGGS:** 3–4; green-blue; subelliptical; 58 x 39 mm. **BEHAVIOR:** In small colonies.

ROSEATE SPOONBILL *Platalea ajaja*

HABITAT: Coastal and inland waterways and marshes. **LOCATION AND STRUCTURE:** On islands or over water with nearby shallows. In shrubs and trees, up to 5 m high, sometimes on ground. Almost always in dense cover. In mixed-species colonies, *often found grouped together toward middle of mixed colony in shaded areas.* Bulky, sturdy stick nest is lined with twigs and leaves. Outside diameter 40–60 cm; depth 11–18 cm. **EGGS:** 2–3; white with fine speckles; subelliptical; 65 x 44 mm.

Bettina Arrigoni, TX

NEW WORLD VULTURES (Cathartidae)

There are 3 N. American cathartids: 2 species of vultures and California Condor. No nest is built, but adults may make a subtle scrape and arrange bits of material from a sitting position. Nest sites may be reused; multiple-use sites can accrue uric splatter and other waste and may have a pungent odor. Sites are located in a variety of recesses: caves, covered cliff ledges, thickets, hollowed treetops, and similar. New World vultures are monogamous, forming long-term pair bonds that generally endure until one mate dies. Courtship consists of pairs flying in close formation, and variations of perched wing-lifting and head bobbing. Condors lay a single egg, vultures 2, and all are single-brooded. Raising young is a long affair, requiring much energy from both parents; both incubate eggs and feed the young by regurgitation. Vultures fledge in about 8–13 weeks, condors in 6 months. New World vultures are quite social, and feed, roost, and sometimes nest in close association.

BLACK VULTURE *Coragyps atratus*

HABITAT: Dense woodlands or in open habitat with available secluded sites. **LOCATION AND STRUCTURE:** Nests are typically well hidden, dark, and sheltered: caves, brush piles, dense thickets, beneath or in deep crevices of boulders, in stumps or hollow logs, abandoned buildings. **EGGS:** Usually 2, *larger than Turkey Vulture eggs*; smooth, light gloss, pale gray-green or bluish to dull white, with dark brown or lavender spots or blotches, usually concentrated on larger end, or wreathed; subellipitcal to blunt oval; 76 x 51 mm. **BEHAVIOR:** Highly social. Nest sites are very sensitive to human disturbance.

David L. Sherer, FL

TURKEY VULTURE *Cathartes aura*

HABITAT: Wide variety, from mixed farmland and woodland to swampy hills to arid rangelands; typically chooses forested or somewhat wooded zones with available nesting sites. May be found in urban and suburban areas, unlike Black Vulture. **LOCATION AND STRUCTURE:** Like that of Black Vulture; may also nest in

Turkey Vulture.
Mark Nyhof, BC

stands of cactus, chaparral in West. **EGGS:** Usually 2, *smaller than Black Vulture eggs*; smooth, slight gloss, creamy or dull white, marked with spots, sometimes blotches of bright reddish browns, mostly on larger end; subelliptical to long subelliptical; 71 x 49 mm.

CALIFORNIA CONDOR *Gymnogyps californianus*

HABITAT AND LOCATION: Landscapes with cliffs or steep mountainsides providing overhangs, ledges, caves, or crevices among boulders. Sometimes in a hollow or dead top in massive trees (sequoia). Often south-facing at high elevations, north-facing at lower. Often at very back of recess, generally requiring some layer of soft substrate. **EGGS:** 1; smooth, glossy, finely pitted surface, white with very subtle wash of green or blue; long subelliptical; 110 x 67 mm. **BEHAVIOR:** Wild populations were completely extirpated by the mid-1980s due to shooting and ingestion of lead bullets. Adults do not breed every year, and condors require 6 years to sexually mature, making their population recovery slow and difficult (not yet self-sustaining).

A California Condor nesting site was placed high in these desert cliffs.
Matt Monjello, AZ

HAWKS, KITES, EAGLES, and ALLIES
(Accipitridae, Pandionidae)

Twenty-four species in this group of medium-sized to very large birds of prey nest in N. America; Osprey (Pandionidae), and twenty-three hawks, kites, and eagles (Accipitridae). Hook-billed Kite *(Chonrohierazx uncinatus)*, however, rare in s. TX, is not included below.

Females are typically larger than males, and the bird-hunting species are the most sexually dimorphic. Members of this group are generally monogamous, but some species are occasionally polyandrous or polygynous. Courtship may consist of impressive flight displays, with dramatic climbs and dives, and pairs falling toward the ground in tandem. Nests are large, bulky, somewhat messy masses of sticks that are loosely piled and interlocked into platforms and generally lined with finer materials. Green material is commonly added during the breeding season and may serve as an antifungal or insect repellent, as insulation, and may also indicate to others that the nest and territory are in use. Some nests are moderately cupped, though the nest bowl is often flattened with use. Most hawks and allies are tree nesters, but variation occurs: Golden Eagles and Rough-legged Hawks often use cliffs (as will other species), and multiple species will nest on the ground in some regions. Nests take considerable time and energy to build, usually constructed by both sexes over weeks or months. Identifying inactive hawk nests can be difficult; it is important to know what species occur in the area and to note habitat preference, site location, and nest differences where possible. The eggs of hawks and allies usually hatch asynchronously, and brood reduction is common—the smaller, latest hatchlings may starve if food sources are scarce. In some species (e.g., Golden Eagle), the larger chick simply kills and consumes the smaller. Both males and females incubate, though in some species males are much smaller than females and can only crouch awkwardly over the eggs, likely serving better as guards than effective incubators. Eggs vary in color and size across species and typically have rounded ends. Males are typically the primary hunters during the fledgling stage and bring food to the female, who feeds the chicks and shields them from the elements.

From left to right: Sharp-shinned Hawk, Ferruginous Hawk, Golden Eagle.

OSPREY *Pandion haliaetus*

HABITAT: Near fresh or salt water: lakes, ponds, rivers, estuaries, coasts, and marshes, from mangroves to boreal forest. **LOCATION AND STRUCTURE:** Wide variety of sites (requiring only a solid foundation) generally overlooking or close to water. In live or dead trees, cliffs, pinnacles, atop massive rocks, sometimes on ground (islands without terrestrial predators), and many artificial sites (platforms, towers, channel buoys, windmills, poles, etc.). Up to 30 m high, or quite low. Old nests are massive, made of small to large sticks, debris (sod, seaweed, etc.), and often human-made materials such as rope, twine, clothing, garbage. Lined with finer grasses, algae, mosses, similar. Outside diameter 1–2 m; height can grow to 3–4 m. New nests are a thin flat platform ~0.7 m in diameter. **EGGS:** 3; creamy white to pinkish, yellowish, variably and heavily marked with reddish brown, especially on larger end; short subelliptical to long oval. 61–45 mm. **BEHAVIOR:** Single-brooded. Both sexes build, and construction happens throughout nesting cycle, with flurry of activity right after hatching. Active builders even as nestlings, young will rearrange nesting materials and even maintain a nest rim. Acclimates easily to human activity.

David Moskowitz, WA

David Moskowitz, WA

KITES

Kites build nests with smaller materials than many other hawks (except small accipiters), primarily composed of *twigs and plant vegetation*, and often including items such as epiphytes (lichens and Spanish moss). Nests are fairly small, similar in size across species, and typically built by both sexes (but see Snail Kite). Kites collect materials on the wing or sometimes from the ground. Kites are quite social and often nest in loose colonies. They are primarily monogamous, though Snail Kite differs uniquely in that both sexes may abandon a mate (and offspring) to pair with another. Courtship displays include food exchanges in flight, dive-chases, paired soaring, and hanging from one leg above the nest site. Kites are known to aggressively defend nests. Eggs are dissimilar in appearance but are generally short subelliptical, smooth and nonglossy or with a slight gloss. Generally single-brooded.

WHITE-TAILED KITE *Elanus leucurus*

HABITAT: Low-elevation, open country: grasslands and savanna, cultivated lands, marshes and wetlands, oak woodland. **LOCATION AND STRUCTURE:** In a tree (oak, eucalyptus, willow, cottonwood, other deciduous) 3–50 m aboveground, in forest or in isolated stand, on habitat edge. Often in upper two-thirds of tree. Nest is well made of small twigs and lined with dry grasses, leaves, forbs, rootlets, stringy lichen. *Deeply hollowed.* Outside diameter 53 cm; height 21 cm; inside diameter 18 cm; depth 9 cm. **EGGS:** 4–5; whitish base, but mostly covered or washed with heavy reddish brown marks; 42 x 32 mm; IP 28–32 days (F); NP 35–40 days (F, M).

SWALLOW-TAILED KITE *Elanoides forficatus*

HABITAT: Humid woodlands; wooded tracts or edge habitat of pines, cypress, mangrove, hardwoods at swamp or marsh forest; clearings; wet prairie. **LOCATION AND STRUCTURE:** Often near the

Mac Stone, FL

top of prominent trees in area, commonly near edge with open area, but may be lower in other substrates such as mangrove. Usually high, 23–29 m or more. Nest is made of small sticks and Spanish moss or other stringy lichens; moss provides a padded cup and helps bind the stick structure. *Dimensions are similar to those for Snail Kite,* but can be larger; often shallow. **EGGS:** 2; white but boldly marked with dark and light browns; 47 x 37 mm; IP 28 days (mostly F); NP 36–42 days (F, M).

MISSISSIPPI KITE *Ictinia mississippiensis*

HABITAT: Wide variety but favors tall trees. Bottomland and riparian woodlands, rural woods, shelterbelts, mixed short-grass prairie, large forest tracts adjacent to open habitat. Also in mesquite, oak scrubland; in towns in parks, golf courses, similar. **LOCATION AND STRUCTURE:** Typically high up, to 41 m, but as low as 3 m. In trees often supported by 3 or 4 branches in limb forks or crotch of tree. Nest is circular or *oval*, made of dead twigs, and is a shallow saucer, mostly flat on top, thickly lined with a mat of green vegetation. Size varies by location. Outside diameter 27–35 cm long x 24–29 cm wide; height 13–15 cm. **EGGS:** 2; whitish to blue white, *usually unmarked*; 41 x 34 mm; IP 3–32 days (F, M); NP ~34 days (F, M).

Mathew Jung, TX

SNAIL KITE *Rostrhamus sociabilis*

HABITAT: Permanent open marshes in FL Everglades. **LOCATION AND STRUCTURE:** Large, bulky, but compact platform is made of sticks and dry plant material, lined with other dry vegetation, including leaves, vines, grasses. Outside diameter 25–58 cm; height 8–44 cm; inside diameter 7–20 cm; depth 2–11 cm. **EGGS:** 3–4; whitish, but covered in dark reddish brown markings, and

large blotches; 44 x 36 mm; IP 27 days (F, M); NP 23–24 days (F, M). **BEHAVIOR:** Male is primary builder. Often nests in association with Anhinga, herons, ibises.

EAGLES

Bald and Golden Eagles build large, bulky stick nests composed of sizeable branches and lined with finer materials. Bald Eagles nest in trees, occasionally on cliffs, ridgelines, and on ground (treeless regions). Golden Eagles mostly use cliffs but will also use trees, ground sites, and towers, varying geographically. Eagles often reuse nests (sometimes for decades), and both species may have alternate nests within their territories that they concurrently work on while building their chosen nest of the year. Construction may occur throughout the year, by both sexes, and new nests may take 1–3 months to build. Eggs are short subelliptical to short oval, whitish and nonglossy, with a granulated texture. Golden Eagle eggs are often marked with brownish spots, blotches, and specks, while those of Bald Eagle are usually unmarked. They measure 77 x 59 mm and 71 x 54 mm, respectively. There are usually 2 eggs per clutch, and both species are single-brooded. Young may not establish breeding territories until at least 4 years of age.

GOLDEN EAGLE *Aquila chrysaetos*

HABITAT: From sea level to high-elevation tundra, rangelands, shrublands, riparian zones, and mountain conifer habitats, but primarily in western mountainous canyon country in deserts, grasslands, and riparian corridors in Great Plains. **LOCATION AND STRUCTURE:** Most commonly on cliffs to 100 m and higher, but also in trees. *Nest is like that of Bald Eagle*, often smaller but dimensions can overlap. Nests on cliffs are thinner than those in trees. **EGGS:** IP 43–45 days (mostly F); NP ~30–70 days (F, M).

Two Golden Eagle nestlings in a cliff nest. Neal Wight, CO

Golden Eagle tree nest. Jonah Evans, WY

BALD EAGLE *Haliaeetus leucocephalus*

HABITAT: Mostly in wooded or forested areas (though nest trees do not require contiguous stands) near lakes, shorelines, rivers. **LOCATION AND STRUCTURE:** Usually in tree (coniferous or deciduous) larger than others, with features that will support the nest foundation, and generally built near top, at highest point where branches are strong and stable. Nest is large (one of the largest of all birds, and *generally larger than Golden Eagle nest, especially on cliffs*), made of interwoven sticks and lined with grasses, pine needles, forbs, moss, sod, debris. May include down feathers, and greenery is added during nesting season. Nests grow massive over the years, to 2.7 x 3.6 m. Outside diameter 2 m; inside diameter 0.4 m; depth 0.1 m. **EGGS:** IP 35–46 days (F, M); NP 10–11 weeks (F, M).

David Moskowitz, WA

NORTHERN HARRIER *Circus cyaneus*

HABITAT: Fresh- and saltwater marshes, wet meadows and pastures, sloughs; also in drier grassland and prairie, croplands, riparian woodlands. **LOCATION AND STRUCTURE:** *On ground,* in open area, usually in tall, dense vegetation. Nest may be thin, or thick if on wetter ground (sometimes in standing water). A platform of reeds, small sticks (willow, etc.), and water plants, with an inner cup of finer rushes, grasses, similar. Outside diameter 39–72 cm; height 4–20 cm; inside diameter 20–25 cm; depth ~5–10 cm, sometimes deeper. **EGGS:** 4–6; bluish white, unmarked but occasionally blotched or spotted; short subelliptical; 47 x 36 mm; IP 29–39 days (F); NP ~37 days (F, M). **BEHAVIOR:** Single-brooded. Males are mostly monogamous but may be bigamous, or occasionally have more than 2 mates. Male often initiates nest building; female is primary builder. Construction takes 7–14 days.

David Moskowitz,
WA

ACCIPITERS

The 3 accipters in N. America each build stick nests. The largest, Northern Goshawk, makes a large structure similar to that of many buteos, but the other 2 species build smaller nests composed of finer materials. They are all unique, however, in their use of *outer bark fragments to line the nest*, unlike the more common use of "softer" or green items used by other hawks. Accipiters may reoccupy nest sites, and alternate nests are common, but unlike in many other species, reuse of nests in consecutive years is rare. They are aggressive and known to dive at or attack nearby persons who venture too close to eggs or chicks. Eggs are short subelliptical to elliptical, nonglossy, and pale bluish white; those of Cooper's Hawk are marked. Single-brooded.

SHARP-SHINNED HAWK *Accipiter striatus*

HABITAT: Wide range, from sea level to high country. In a variety of forests, preferring those with conifers (aspen-conifer, hardwood-pine). Prefers dense groves and cover, heavy canopy. **LOCATION AND STRUCTURE:** Often on a horizontal limb against trunk but also in crotch or out in branches, in robust canopy but well below treetops. Often in conifers, but also in deciduous trees, varying regionally. 2.4–19 m aboveground, often 9–10 m. Broad platform of small sticks and twigs is lined with bark chips. Large for bird's size: outside diameter at widest point 35–60 cm; height 10–17 cm; inside diameter 15 cm; depth 5–7.6 cm. **EGGS:** 4–5; pale bluish white, variably marked with splotches of deep browns, violets, grays, at either end or middle; 38 x 30 mm; IP 30–35 days (F, M); NP ~23 days (F; M hunts). **BEHAVIOR:** Both sexes collect materials; female is primary builder. Secretive, often nesting in areas where it is difficult to find nest or bird. Nest reuse is rare.

Sharp-shinned Hawk high in a conifer tree.
Matt Monjello, AZ

COOPER'S HAWK *Accipiter cooperii*

HABITAT: Wide variety: deciduous and evergreen forests, scrub in treeless regions. Common now in urban and suburban habitats, and often close to edges of roads, development, active recreation areas, etc., *unlike Sharp-shinned Hawk and Goshawk. Sharp-shinned usually nests in conifers.* **LOCATION AND STRUCTURE:** In trees, typically 8–15 m aboveground, often in 2–6 horizontal limbs against trunk, or in crotch. Sometimes atop old nest of a bird, woodrat, or squirrel. *Like Sharp-shinned nest but larger,* always lined with bark chips, sometime evergreen springs. Often broader and flatter in conifers. Outside diameter 61–76 cm;

height 15–43 cm. **EGGS:** 4; pale bluish or dirty white, rarely marked with dark browns; 49 x 38 mm; IP 36 days (F); NP 30–34 days (F, M). **BEHAVIOR:** Male is primary builder. Young males leave nest first, followed 4 days later by females. Nests sometimes are reused or built on.

View from below of a Cooper's Hawk nest.
Matt Monjello, AZ

Coopers hawk nest viewed from above.
Jack Bartholmai, WI

Cooper's Hawk eggs in a nest.
Mark Nyhof, BC

NORTHERN GOSHAWK *Accipiter gentilis*

HABITAT: Mostly in fairly remote tracts of various forests types (coniferous, deciduous, mixed), generally with robust canopy, often in small stands. **LOCATION AND STRUCTURE:** In a coniferous or deciduous tree, typically the largest in stand, often in lower third or just below canopy. Height varies by trees in region, to 23 m but as low as 6 m. *Nest is much larger than those of other accipiters*, on limbs against trunk, in fork, or sometimes out on a limb. Made with thin sticks, lined with bark chips, greenery. Outside diameter 88–121 cm; height 46–66 cm. **EGGS:** 2–3; rough textured, pale bluish white; 59 x 45 mm; IP 36–41 days (F); NP 40 days (F, M). **BEHAVIOR:** Female is primary builder. May have up to 8 alternate nests in territory. Nest reuse is rare.

Northern Goshawk nest high in a conifer tree.
Jonah Evans, WY

Note the lining of bark chips in this Northern Goshawk nest, common to accipiters.
Thomas Erdman, WI

HAWKS

These 12 raptors mostly nest in trees, but Ferruginous, Red-tailed, and Zone-tailed Hawks may use cliffs. Rough-legged Hawk can nest on the ground in its northern reaches. Nests are commonly reused, growing in size over time, and are typically large, bulky structures of sticks and twigs. Eggs vary by species in size, color, marking, and shape (they are generally short subelliptical, subelliptical, or sometimes elliptical). Most species are single-brooded, both sexes incubate, and both care for young; exceptions are noted below.

COMMON BLACK HAWK *Buteogallus anthracinus*

HABITAT: *Restricted to riparian habitat,* primarily in mature forest near permanent streams. **LOCATION AND STRUCTURE:** In groves, in a tree, usually in crotch of trunk, sometimes on branches. Reuses nest; starts smaller and grows large over time; sticks used are usually similar to those of nest tree. Flattish top, lined with green leaves. Outside diameter ~0.4–1 m; height 0.2–1 m. **EGGS:** 1, occasionally 2; granular, grayish white to dull white, marked with brownish to purplish specks or smudges, sometimes unmarked; 57 x 45 mm; IP 38 days; NP 43–50 days.

HARRIS'S HAWK *Parabuteo unicinctus*

HABITAT: Dry, open habitats. Savanna and grassland, scrub (mesquite, similar), riparian and wetland woodlands, palo verde and saguaro desert. Access to open water may be important. Increasingly in urban and suburban areas. **LOCATION AND STRUCTURE:** Requires a somewhat tallish, strong structure: cactus, palo verde, yucca, oak, etc.; sometimes on mistletoe. Usually low, 2.7–5.8 m, occasionally higher. Nest is built of live or dead sticks, lined with various materials, varying geographically: fresh sprigs, sage, grass, cholla bones, etc. In NM, feathers and grasses are common. Nest is round to elliptical, shallow to fairly bowled. Outside diameter 47–59 cm; height 23 cm. **EGGS:** 2–4; whitish blue or dull white, usually unmarked; 53 x 42 mm; IP

Karen McCrorey, AZ

Mary Jo Bogatto, TX

This nest has been repaired and reused in consecutive years. Mary Jo Bogatto, TX

31–36 days; NP 38 days. **BEHAVIOR:** Double-brooded, sometimes treble. Monogamous and polyandrous. Female is primary builder. May nest and breed year-round. Renowned for its unique behavior: cooperatively hunts and nests in small social units of up to 7 members. Young may stay for 3 years, helping rear subsequent offspring, and unrelated birds may also contribute.

WHITE-TAILED HAWK *Geranoaetus albicaudatus*

HABITAT: Open or semiopen country; low scrub, grassland and prairie, savanna. **LOCATION AND STRUCTURE:** Usually at crown of bushes, low tree, yucca. Low, 1–3.5 m. Nest is usually elliptical in shape, made of small sticks and twigs (commonly thorny), grass tufts, lined with leaves, forbs. Outside diameter of larger nests ~54 cm; height 46 cm; depth 8 cm. **EGGS:** 2; white, unmarked, may have a few reddish or brown specks; elliptical; 59 x 47 mm; IP 31 days; NP 49–53 days.

GRAY HAWK *Buteo plagiatus*

HABITAT: Uncommon, estimated at 100 nesting pairs or fewer. Riparian areas, generally in mesquite country. **LOCATION AND STRUCTURE:** Usually in a lone tree (commonly cottonwood), typically 12–19 m high, usually away from main trunk. **EGGS:** 2; white or bluish white, usually unmarked; 51 x 41 mm. Female incubates.

RED-SHOULDERED HAWK *Buteo lineatus*

HABITAT: Generally in mature, moist, extensive woodlands. In West often in riparian, oak, eucalyptus woods. May nest farther from forest edges, openings, agriculture, and development than Red-tailed Hawk, though also nests in suburban settings. **LOCATION AND STRUCTURE:** In a deciduous tree, sometimes a conifer, often in larger-sized tree than surrounding, commonly with water nearby. Usually in crotch in main trunk, or against trunk in conifer, usually 10–24 m up. *Nest is like that of Red-tailed,* but dimensions are like those of Harris's Hawk. Nest is made of sticks, dead leaves, bark strips, rubbish; lined with inner bark strips, moss, stringy lichens, fresh sprigs. **EGGS:** 2–3; dull white or bluish white, blotched with browns, light purples; smooth, with slight gloss; 55 x 43 mm; IP 23–25 days (F, M); NP 35–42 days. **BEHAVIOR:** Other conspecific pairs may reuse nests as well.

Matt Monjello, CA

John Jacobs, WI

BROAD-WINGED HAWK *Buteo platypterus*

HABITAT: Large tracts of deciduous or mixed forest, usually near openings, water. *Red-shouldered Hawk usually nests in more closed, older forest.* **LOCATION AND STRUCTURE:** Placement is similar to that of Red-shouldered, but *usually in smaller trees, in lower third of canopy and closer to edges*, openings. Usually in primary crotch of a deciduous tree or placed against trunk on limbs in conifers. Variety of fresh twigs, dead sticks, *commonly lined with bark chips* (accipiter-like), but other materials may be included: vines, red cedar, moss. Fresh sprigs are added before laying. May be placed on an old hawk, crow, or squirrel nest. Dimensions are similar to Red-shouldered's but can be slightly smaller. Nest is flattish (in conifers) or well cupped, to 8 cm deep. **EGGS:** 2–3; dull white, creamy, or bluish white, varyingly marked, with blotches of lavenders overlaid with speckles of browns, or can be mostly unmarked; 49 x 39 mm; IP 28–31 days (mostly F); NP 29–30 days. **BEHAVIOR:** Forms large flocks during migration.

Matt Monjello, ME

SHORT-TAILED HAWK *Buteo brachyurus*

HABITAT: Rare. Restricted in U.S. mostly to FL peninsula. In woodlands, near or in swamps, sometimes in mangroves. **LOCATION AND STRUCTURE:** Near crown of tall, straight trees, in trunk fork or on horizontal branch. Usually 8–29 m aboveground but can be lower. Typical buteo nest, lined with leaves and often cypress sprigs. Outside diameter 60–90 cm; height 30 cm. **EGGS:** 2; bluish white or whitish, unmarked or with sparse speckling of reddish browns, or splotches, usually mostly on larger end; sometimes oval; 53 x 43 mm; IP 34–39 days (F). **BEHAVIOR:** Builds 1–3 nests each year, usually within a few hundred meters of each other.

SWAINSON'S HAWK *Buteo swainsoni*

HABITAT: Open country: prairies, plains, open woodlands, shrublands, agricultural areas. *Usually in smaller groves and smaller trees than Red-tailed Hawk.* **LOCATION AND STRUCTURE:** Typically in scattered trees or lone tree (shelterbelts, narrow riparian belts, similar), but also in saguaro, Joshua trees. Will use utility poles and towers. Usually near treetop in crown, often built on small limb. Nest is often bulky but in some instances appears *thinner, more flimsy, and messier* than those of other hawks. May use wool in lining, in addition to greenery. Outside diameter ~60 cm; height ~32 cm; inside diameter up to 20 cm; depth 6–7 cm. **EGGS:** 2; white, mostly unmarked but may have sparse brownish speckles; 56 x 44 mm; IP 28 days (mostly F); NP 30–35 days. **BEHAVIOR:** Makes one of the longest migratory journeys among raptors, with some birds traveling 10,000 km each way. Gregarious, traveling in flocks, sometimes by the thousands. Populations are declining in parts of range.

Swainson's Hawk. David Pierce, OK

Nest on a utility pole, showing the relatively shallow nest profile common to Swainson's Hawks. David Moskowitz, WA

A large Swainson's Hawk nest high in a tree. David Moskowitz, WA

ZONE-TAILED HAWK *Buteo albonotatus*

HABITAT: Sw. U.S., in riparian woodland, canyons, scrubby cliffs, montane mixed forests (primarily pine-oak). **LOCATION AND STRUCTURE:** In deciduous or coniferous tree (oak, sycamore, cottonwood, ponderosa pine) or on cliff ledges of shrubby canyons. Usually in crotch of main trunk. Typical buteo nest, lined with green mosses, juniper, leaves (oak, cottonwood often with twig

Note the fresh greenery recently added to this Zone-tailed Hawk nest.
Chris McCreedy, NM

attached), pine needles. May be well cupped. Dimensions vary, are similar to those for Swainson's Hawk. **EGGS:** 2; dull white or bluish white, smooth or finely granular, unmarked or rarely with fine brown spots; 55 x 43 mm; IP 35 days; NP 35–42 days. **BEHAVIOR:** Performs spectacular aerial courtship displays, and readily attacks nest intruders.

RED-TAILED HAWK *Buteo jamaicensis*

HABITAT: Common in a broad variety of habitats. Open to semiopen country, from desert to high mountains. **LOCATION AND STRUCTURE:** Forest and tree types vary; nest is often in mature woodlands, in trees taller than or above others with good views, and next to expansive openings, usually near crown or in top quarter of tree. Also in smaller trees (pinyon-juniper, etc.),

saguaro, and on cliffs where trees are few. Adapted to human structures: towers, power lines, sometimes skyscrapers. *Nest site is often in more open habitat and in taller trees than those of Broad-winged, Harris's, and Red-shouldered Hawks. Swainson's and Ferruginous Hawks often choose lower nest substrates and may nest farther from water sources.* Typical buteo nest. Outside diameter ~71–76 cm; inside diameter 35–37 cm; depth 10–13 cm. Can grow larger with reuse. **EGGS:** 2–3; dull white to bluish white, smooth or finely granular and marked with varying brown spots, specks, scrawls; 59 x 47 mm; IP 28–32 days (mostly F); NP 30–35 days. **BEHAVIOR:** Many pairs remain together throughout year.

David Moskowitz, WA

ROUGH-LEGGED HAWK *Buteo lagopus*

HABITAT: An Arctic species, breeding only in far north. Open habitat: tundra, boreal forest edges, open areas with scattered woods. **LOCATION AND STRUCTURE:** Primarily on protected ledges on cliffs, escarpments, banks, occasionally in tree, usually near top. Typical buteo nest, large for bird's size. Lining of fur, feathers from prey accumulates over NP. Dimensions similar to those of Red-tailed Hawk, or slightly larger or smaller. **EGGS:** 2–3, or 5–7 when food is abundant; white, marked with red, brown, or pale purple streaks or blotches; 55 x 44 mm; IP 28–31days or more (F,M, or just F); NP ~40 days. **BEHAVIOR:** Nests on cliffs are often near those of other species, such as Gyrfalcon, Peregrine Falcon, Common Raven.

Kate Fremlin,
Amaroq Wildlife
Services, NU

FERRUGINOUS HAWK *Buteo regalis*

HABITAT: Open country; shrub-steppe, grassland, sparsely wooded riparian belts, open canyon with cliff or rocky features. **LOCATION AND STRUCTURE:** Wide variety: often in a lone tree or shrub, but also on cliff ledges, on elevated land features (boulders, pinnacles, hoodoos, etc.), human structures, artificial platforms, on ground (usually in rocky, brushy hillsides, rises)

Casey McFarland, WA

A dead nestling from a failed nesting attempt. David Moskowitz, WA

0–20 m or higher. Considered a low-nesting buteo among western species, but this may be because of nest substrate options; apparently favors elevation in all regions. Typical buteo nest, but tends to be *very large, and deeply cupped to protect young from strong, open-country winds.* Made of sticks (sagebrush, etc.), twigs, ground debris (sod, dung), lined with shredded bark, dung. Outside diameter and height usually at least 1 m, up to 1.85 x 1.5 m; depth 17–32 cm. **EGGS:** 3–4; white to bluish white, marked irregularly, boldly with speckles, spots, blotches of rich browns, pale purples; smooth or finely granulated; 62 x 49 mm; IP 32–36 days; NP 44–48 days. **BEHAVIOR:** This large, eaglelike buteo is secretive, wary, and often roosts on ground. Despite its size, it is often subordinate to Red-tailed and Swainson's Hawks. Passerines have been known to nest within active nests. Population is declining in parts of range.

OWLS (Strigidae and Tytonidae)

Nineteen species of owls—18 in the family Strigidae (typical owls) and 1 in the family Tytonidae (barn owls)—breed in N. America. Most of these small to large predatory birds are nocturnal. They are mostly monogamous, and some species maintain long-term pair bonds, though polygyny has been reported in a few species and many others are understudied. Courtship is largely vocal, though aerial displays do occur in some species, as well as interactions such as allopreening, feeding, and bill rubbing. With the exception of Short-eared Owl, owls *do not build nests*, and limitations on nest-site availability pose considerable conservation challenges for several species. Finding where owls raise their young can be difficult, but in some cases whitewash, pellet accumulations, and other clues help indicate the presence of a nest. Owls can be loosely divided into three nesting categories: (1) *secondary cavity or small natural tree-cavity users* (holes in rotted limbs, old tree wounds, etc.); (2) *those that use nests of other species, or natural cavities or ledges*; and (3) *ground nesters*. Larger-bodied owls commonly use nests of raptors, corvids, or squirrels, among other sites. Many of the small owls are secondary cavity nesters, using holes created by woodpeckers; in the desert Southwest some may share the landscape with as many as 15 other cavity-nesting species, of which a few are also owls. Eggs hatch asynchronously (with the exception of Northern Pygmy Owl), and smaller young are often outcompeted by their larger siblings. Eggs range from short subelliptical to elliptical, though the ends are usually quite rounded, as in many

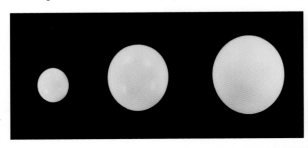

Left to right:
Elf Owl, Spotted
Owl, Snowy Owl.

other cavity-nesting species. Barn Owl eggs are more tapered. Eggs are white or off-white, typically smooth and slightly glossy, but may have a granular texture. Most species are single-brooded. Some owls have small clutches (2–3), while some are quite large (4–10), and many species can lay significantly more eggs when food is abundant (10–15 at the high end). In most species the female incubates and is fed by the male for ~21–24 days to upward of 30–36 days. Cavity-nesting females often reside inside a chosen site for several days or longer before laying, fed by the male. Young of most species leave the nest in about 1 month but are usually weak flyers, vulnerable, and entirely dependent on their parents for long periods of time (1.5–5 months). Large species are usually slowest to develop into independent hunters.

BARN OWL *Tyto alba*

HABITAT: Highly flexible in habitat use; well distributed across numerous landscape types, both natural and human-made, which contributes to the species' impressive global range. **LOCATION AND STRUCTURE:** Tree cavities, crevices or cracks in rocks or cliffs, wide variety of human structures, buildings. May dig nest burrow into arroyo wall or cut bank if substrate is soft enough, using its feet. *A layer of shredded pellets* may be prepared for eggs, or eggs are laid on bare surface. **EGGS:** 4–7; more pointed than those of typical owls; 42 x 33 mm. **BEHAVIOR:** Often double-brooded, sometimes treble; in warmer climates can breed year-round. Usually monogamous, but polygamy occurs. Chosen nest sites are often similar to natal site of the adult bird.

A Barn Owl nest in a basalt cliff.
David Moskowitz, WA

Bill Summerour, AL

OWLS THAT TYPICALLY USE WOODPECKER HOLES (SECONDARY CAVITY) OR SMALL NATURAL TREE CAVITY

FLAMMULATED OWL *Psiloscops flammeolus*

HABITAT: Mature conifer forests, usually high elevation. Ponderosa pine, fir, aspen. **LOCATION AND STRUCTURE:** Primarily in woodpecker hole, sometimes in natural cavity. Prefers larger woodpecker holes (Pileated Woodpecker, Northern Flicker; occasionally Hairy Woodpecker, Yellow-bellied Sapsucker). Often in live trees, but uses dead snags. Occasionally uses nest boxes. **EGGS:** 3–4; 29 x 26 mm. **BEHAVIOR:** Highly migratory. Male likely shows female multiple nest site options, from which she chooses. This tiny owl has a surprisingly deep voice, and males often sing from large trees on ridges or slopes to best broadcast their presence.

WESTERN SCREECH-OWL *Megascops kennicottii*

HABITAT: A common owl, with a wide variety of habitats from desert to an array of low-elevation woodlands; often in deciduous woods and riparian areas. Quite tolerant of humans, and will nest in suburban neighborhoods, parks, farms, etc. **LOCATION AND STRUCTURE:** Across range, generally in woodpecker holes of large species (including flicker cavities in saguaro, cardon cactus), but also commonly in natural cavities. Readily uses nest boxes. **EGGS:** 2–4; 38 x 42 mm. **BEHAVIOR:** Nonmigratory. Pairs usually remain in territory year-round. In Pacific Northwest, Barred Owls are expanding into Western Screech-Owl habitat, causing population declines. Screech-owls defend alternate nesting sites.

Western Screech-owl nestlings in a natural tree cavity.
Mark Nyhof, BC

EASTERN SCREECH-OWL *Megascops asio*

HABITAT: Common in diverse habitats: coniferous, deciduous, or mixed forest of lowlands or low mountainsides (to only ~1,500 m in elevation across range). Common in rural, urban, and suburban environments. **LOCATION AND STRUCTURE:** Often in a deep natural tree cavity with wide floor. Also in holes of large woodpecker species. Some natural cavities used are enlarged first by squirrels. Readily uses nest boxes and other artificial sites. **EGGS:** 4–5; 36 x 30 mm. **BEHAVIOR:** Permanent residents that usually form lifelong pairs. Nest site is often at great distance from larger nesting owls, but sites reported in close proximity to other species (American Crow, Northern Flicker, and others).

Eastern Screech Owl nestlings in a natural cavity nest.
Christian Artuso, MB

WHISKERED SCREECH-OWL *Megascops trichopsis*

HABITAT: Common within its small U.S. range. Montane landscape in coniferous, deciduous (oak), or mixed (pine-oak) forests, riparian canyons, 1,000–2,900 m in elevation. Nests at higher elevation than Western Screech-Owl typically does, though overlap occurs at lower elevations. **LOCATION AND STRUCTURE:** Similar to other screech-owls. In natural cavity (commonly rotted limb holes) or woodpecker hole, usually Northern Flicker. **EGGS:** 3, occasionally 4; 33 x 28 mm.

Lois Manowitz, AZ

NORTHERN HAWK OWL *Surnia ulula*

HABITAT: A circumpolar species, breeding only as far south as n. MN. Coniferous or mixed forests, often at clearing edges of marshes, logging cuts, burns (fire is apparently important for creating nesting sites). **LOCATION AND STRUCTURE:** In natural, often shallow cavities in trees (storm-damaged trees; broken, rotted treetops; fire-hollowed stumps; etc.), woodpecker holes.

Northern Hawk owl nest in a natural tree cavity.
Rick Bowers, Bowersphoto.com, MB

Sometimes uses old stick nests. **EGGS:** 3–10; 40 x 32 mm. **BEHAVIOR:** Monogamous, but polygynous with abundance of prey or nesting sites.

NORTHERN PYGMY-OWL *Glaucidium gnoma*

HABITAT: Lower-elevation deciduous to high-elevation coniferous and mixed forests, 1,525–3,050 m. **LOCATION AND STRUCTURE:** Woodpecker cavity or natural cavity in tree. Possibly adds material on occasion. **EGGS:** 4–6; 29 x 24 mm; IP 28-29 days; NP 28-29 days. **BEHAVIOR:** The only N. American owl demonstrating synchronous incubation and fledging. A very small owl, but may aggressively defend territory, chasing other species, and sometimes locking talons midair with intruding conspecifics in fierce grappling matches.

Garrit Vyn, OR

FERRUGINOUS PYGMY-OWL *Glaucidium brasilianum*

HABITAT: Small year-round range. *Rare and endangered in U.S.* Saguaro, riparian areas in AZ; mesquite-oak, mesquite-ebony, riparian areas in TX. **LOCATION AND STRUCTURE:** Woodpecker cavities, natural cavities in trees, stumps, etc., nest boxes. **EGGS:** 3–4; 29 x 24 mm. **BEHAVIOR:** Like Elf Owl, may remove some of nesting material from prior bird inhabitants.

Thomas P. Brown, AZ

ELF OWL *Micrathene whitneyi*

HABITAT: Sw. desert to higher montane desert: in wooded desert washes, riparian canyon woodlands, coniferous woodland. Also in areas of human habitation. **LOCATION AND STRUCTURE:** Almost always in woodpecker holes: in saguaro cactus (often in Gila Woodpecker or Gilded Flicker cavities) in lower deserts, and wide variety of deciduous trees, conifers (variety of woodpecker species) in other habitats. Also fenceposts, utility poles, nest boxes.

Bettina Arrigoni, AZ

EGGS: 3; 27 x 23 mm. **BEHAVIOR:** Migratory. The world's smallest owl, likely able to use holes created by a wider variety of woodpecker species than other secondary-cavity-nesting owls. Will nest in same tree simultaneously with other cavity nesters, such as flycatchers, trogons, woodpeckers, and other owl species, though Elf Owls are sometimes displaced or displace others. May remove preexisting nest material from cavity.

BOREAL OWL *Aegolius funereus*

HABITAT: Circumpolar. Coniferous, mixed forests, often mature. **LOCATION AND STRUCTURE:** Often in larger woodpecker holes (like Northern Saw-whet Owl, frequently in Pileated Woodpecker and Northern Flicker holes). Also in natural tree cavities, and readily uses nest boxes. **EGGS:** 3–6; 32 x 27 mm. **BEHAVIOR:** Uncommon, secretive, and little studied in N. America.

NORTHERN SAW-WHET OWL *Aegolius acadicus*

HABITAT: Common and widespread. Wide variety of habitat types, where nesting and roosting sites are available: deciduous forests (including swamp, riparian, hardwoods), coniferous forests (often near riparian areas). In general, appears drawn to thick coniferous forest for roosting and mix of abundant deciduous forest for nesting. **LOCATION AND STRUCTURE:** Primarily in woodpecker holes, often those of larger species (Pileated Woodpecker, Norther Flicker). Also occurs in natural cavities, nest boxes, 2.5–25 m high. **EGGS:** 5–6; 30 x 25 mm. **BEHAVIOR:** Females are thought to sometimes be polyandrous and mate with a second male while prior male is left to care for female's first brood.

OWLS THAT TYPICALLY USE OTHER SPECIES' NESTS OR LARGE NATURAL CAVITIES, LEDGES

GREAT HORNED OWL *Bubo virginianus*

HABITAT: An impressive range geographically and ecologically, from arid desert to boreal landscapes. Typically in open woodland habitat, usually with some matrix of openings: agricultural lands, fields, meadows, sagebrush flats, wetlands, etc. In treeless deserts on cliffs. Also comfortable near human habitations.

David Moskowitz, WA   David Moskowitz, WA

LOCATION AND STRUCTURE: Frequently in nests of other bird species: commonly Red-tailed Hawk nests but also those of eagles, other hawks, corvids, herons, squirrels. Also in tree or snag cavities, rocky ledges in boulders or cliffs, caves, holes in earthen banks (arroyos), and in empty buildings. Rarely on ground. EGGS: 2–3; 56 x 47 mm. BEHAVIOR: Very territorial, and possibly kills and consumes conspecifics. Birds without territories (floaters) live quietly and inconspicuously within an established territory of other Great Horned Owls.

SPOTTED OWL *Strix occidentalis*

HABITAT: Scattered and rare across a wide range, divided into 3 subspecies. Usually in coniferous forests, commonly old growth. Also in mostly oak woodlands at lower elevations, in Southwest in mountainous conifer or pine-oak, also steep canyons, often wooded but not always, with coniferous or riparian woodland. LOCATION AND STRUCTURE: Often in tree cavities, hollows from broken treetops, also uses platforms (old raptor, corvid, or squirrel nests, mistletoe, etc.). May use ledges or other features on cliffs, particularly in southwestern canyons. May add feathers to scrape. EGGS: 2; may be slightly granular; 50 x 41 mm. BEHAVIOR: Some Spotted Owls have been known to reuse same nest over a period of years. This endangered species not only face declines due to severe logging and deforestation of native habitat, but also from expanding populations of more aggressive Barred Owl, with which it can also hybridize.

Spotted Owl nest in a natural cavity of an old growth tree.
Garrit Vyn, OR

BARRED OWL *Strix varia*

HABITAT: Wide range of forest types (coniferous, hardwood, mixed) from swamp and riparian to mountainous regions, usually in old-growth or mature stands (but does well in second-growth, though not young, forest), favoring large tracts of unfragmented landscape. LOCATION AND STRUCTURE: Often nests in a deciduous tree, commonly where understory is robust.

Mostly in cavity, similar to other larger-bodied owls: broken treetops, broken branch holes, disease malformation, etc. Also in old nests of other species. Will use nest boxes. May add sprigs, feathers, lichen. Upper surface of old nests of other species may be modified by removing material. *Nest can look identical to those of Great Horned, Snowy, and Boreal Owls.* **EGGS:** 2–3; 49 x 42 mm. **BEHAVIOR:** Maintains territory throughout year, and pairs may hold a territory for many years. Like Spotted Owl, relies on healthy mature or old-growth forest. Territories tend to be large, with pairs spaced far apart. May nest near Red-shouldered Hawks.

Barred Owl nest in broken top of a dead tree.
Jack Bartholmai, WI

GREAT GRAY OWL *Strix nebulosa*

HABITAT: Circumpolar. Old-growth coniferous boreal forest, also coniferous and mixed montane and occasionally riparian forest, often adjacent to forest edges at muskeg bogs, mountain meadows, wetlands, similar. **LOCATION AND STRUCTURE:** Commonly in old stick nests (accipiter, Red-tailed Hawk, Common Raven), often with dense canopy cover, and natural platforms (commonly in tops of broken-off snags; mistletoe). Readily takes to human-made nest platforms, which has had positive effects on breeding and nesting success. **EGGS:** 3–5; 54 x 43 mm. **BEHAVIOR:** Very secretive (quiet) while nesting. Chicks are killed and consumed by Great Horned Owls and Northern Goshawks, which can nest in close proximity to Great Grays. They keep nests extremely clean. Commonly use good nesting sites for multiple breeding seasons. Males may feed young well into autumn.

Steven Poole, WY

LONG-EARED OWL *Asio otus*

HABITAT: Nests and roosts in dense, concealing vegetation, but with plentiful open habitat for hunting nearby: coniferous or deciduous woods near meadows, mixed woodland adjacent to shrub-steppe, similar. Wooded parks, orchards. **LOCATION AND STRUCTURE:** Primarily on other bird nests, *commonly corvids* (Black-billed Magpie, American Crow, or Common Raven). Also in accipiter and hawk nests, occasionally squirrel nests. Rare in tree cavity, cliffs. **EGGS:** 4–5; 41 x 33 mm. **BEHAVIOR:** Male shows nesting sites to female by performing aerial courtship displays above potential area. Pairs seem tolerant of other individuals, primarily defending only immediate area around nest. Will roost communally in large numbers during nonbreeding season. Females not easily flushed from nest. Adults and young alike are killed by many species, including Great Horned Owls, hawks, Golden Eagles, and mammals.

Marcus Reynerson, WA

Mark Nyhof, WY

GROUND-NESTING OWLS

SNOWY OWL *Bubo scandiacus*

HABITAT: Circumpolar, and the world's northernmost owl. Variety of Arctic tundra habitats from coast to tree-line: open, sparse vegetation/shrub, boggy meadows, salt marsh. **LOCATION AND STRUCTURE:** On ground, usually on a rise with commanding view (mounds, hummocks, hill, occasionally on stone topped with sod, mosses), mostly on grounds cleared of snow by winds. Female scrapes into soil; nest is usually unlined, or with small amount of feathers, perhaps mosses. **EGGS:** 4–10, as many as 15; 57 x 45 mm. **BEHAVIOR:** N. America's largest owl by weight, and one of the largest on the planet. Some remain in far north through the long darkness of winter. Diurnal. (See nest photo on page 35.)

SHORT-EARED OWL *Asio flammeus*

HABITAT: Wide variety. Tundra, prairie, dunes, grasslands (including coastal), shrub-steppe, forest clearings, agricultural lands. **LOCATION AND STRUCTURE:** *On ground*, often within grass or other vegetative concealment, but sometimes more exposed, usually on slightly raised dry area (mounds, natural berms, hummocks). This species is unique in that it builds a nest, creat-

Short-eared Owl.
R. Wayne Campbell, BC

ing a scrape and lining it with grasses, forbs, down feathers. Outside diameter 23–25 cm; height 4–5 cm. **EGGS:** 4–8 (as many as 14 when food is abundant); 40 x 31 mm. **BEHAVIOR:** Diurnal. Nests are often challenging to find, and as with Long-eared Owl, females are not easily flushed. Also roosts in large numbers during nonbreeding season.

BURROWING OWL   *Athene cunicularia*

HABITAT: Prairies, grasslands, deserts, shrub-steppe; also in human habitations, including vacant lots, agricultural areas, golf courses, similar. Usually in area with burrowing rodents (prairie dogs, ground squirrels) or other mammals such as badgers, armadillos, skunks, kangaroo rats. **LOCATION AND STRUCTURE:** Usually in a mammal burrow (but on rare occasions dug by owls if soils permit and other burrows are lacking), often in concentration of multiple burrows. Minimum entrance size 11 cm; to about 25 cm high. Tunnel floors are gently sloped, not steep. Cavity and occasionally entrance tunnel are lined with dry manure chips, and *collected items may decorate the entrance:* pellets, desiccated small rodents, carnivore scats, etc. Use of

An active Burrowing Owl nest in an old badger burrow. Note the "decorations"— a desiccated kangaroo rat, owl pellets, and bits of an old coyote scat.
David Moskowitz, WA

underground nest boxes has been successful. **EGGS:** 5–6; can be quite round; 31 x 26 mm. **BEHAVIOR:** May nest in loose colonies, and young and adults often roost together on mounds. Nonmigrant populations use nest burrow year-round. Populations have seen dramatic declines. Sexual behavior varies across range, from changing mates yearly to long-term bonds.

An adult Burrowing Owl purchases on a post set up next to an artificial nest constructed specifically for this species.
David Moskowitz, WA

TROGONS (Trogonidae)

ELEGANT TROGON *Trogon elegans*

HABITAT: Sycamore canyons, generally through pinyon, pine, juniper, oak woodlands. **LOCATION AND STRUCTURE:** In a woodpecker cavity, or natural cavity (may enlarge those in decayed wood), usually with no nesting material. Often in sycamore, but also in other deciduous or coniferous trees, 2.4–13.7 m high. Holes average 14 cm wide. **EGGS:** 2; dull white to pale bluish

Scott Olmstead, AZ

Elegant Trogon egg.

white, nonglossy; 28 x 23 mm; IP 17–22 days (F, M); NP 20–23 days, possibly shorter (F, M). **BEHAVIOR:** Apparently monogamous. Males announce potential nest sites to females by staying near nest tree through day, continuously calling.

KINGFISHERS (Acedinidae)

Three species of these small to medium-sized birds breed in N. America. Belted Kingfisher often migrates to warmer climates during winter, while Ringed and Green Kingfishers are permanent residents. Associated with perennial freshwater and coastal habitats, kingfishers can be found along shorelines of clear lakes, estuaries, and slow-moving waterways, where they excavate nest burrows along earthen and dirt banks. Both sexes partake in nest construction, chipping away at the tunnel with their bills and ejecting loose material with their feet. Generally taking a week to complete, the tunnel angles upward from the entrance hole, terminating at an unlined nest chamber, which over time accumulates fish bones, exoskeletons, and excrement. Active burrows can show parallel scrapes at the entrance from the bird's feet as they enter and exit the hole.

Belted Kingfisher egg.

Seasonally monogamous, kingfishers are usually single-brooded with an occasional second brood (Belted Kingfisher) if the first nest is destroyed. Eggs are elliptical to short elliptical, white, smooth, and glossy. Both parents feed young, which, depending on the species, take 22–38 days to fledge.

This close-up of an active Belted Kingfisher burrow shows the small parallel trenches made by the birds' feet upon entry and exit.
Nicholas J. Czaplewski, OK

RINGED KINGFISHER *Megaceryle torquata*

HABITAT: Limited to southern tip of TX but expanding northward. *Habitats similar to those of Belted Kingfisher.* Generally more common along inland freshwater sites than coastal areas. Prefers clear waters unobstructed by overgrown vegetation and with banks containing perches suitable for hunting. **LOCATION AND STRUCTURE:** Typically over water in earthen banks devoid of vegetation, 3–4 m from top of bank, and on upper half of bank wall, extending inward 2.3–2.7 m. Tunnel entrance width 15 cm; height 10 cm. **EGGS:** 4–5, sometimes 3–7; 44 x 34 mm; IP not well known but thought to be 22 days (F, M); NP 35–38 days (F, M). **BEHAVIOR:** Unlike Belted, Ringed rarely hovers for prey, especially for extended periods of time.

BELTED KINGFISHER *Megaceryle alcyon*

HABITAT: Wide variety. Prefers clear, open waters unobstructed by overgrown vegetation: rivers, streams, ponds, sheltered lake shorelines; estuaries, coastal waters, up to 2,743 m elevation. **LOCATION AND STRUCTURE:** Prefers earthen banks clear of vegetation near feeding territory. May breed away from water in other

David Moskowitz, WA

suitable sites, including road and railroad cuts, ditches, sand and gravel pits; rarely in tree cavity, dredge spoils, agricultural trenches, dunes. Burrow entrance is generally located 35–64 cm down from top of bank, extending inward 0.3–2.5 m, infrequently as far as 4.5 m. Tunnel entrance width 7.6–12.6 cm; height 7.6–12.6 cm. Egg chamber width 20–30 cm; height 15.2–17.8 cm. **EGGS:** 6–7; 34 x 27 mm; IP 22–24 days (F, M); NP 27–29 days (F, M). **BEHAVIOR:** Pairs may have multiple burrows along nesting site, but choose one in which to nest. Occasionally reuses nests over subsequent years. Generally constructs new nests in 3–7 days, but may take 21 days to complete. Commonly hovers above water before plummeting to capture prey.

GREEN KINGFISHER *Chloroceryle americana*

HABITAT: Low shrubby banks with overhanging branches over clear torpid or still fresh water, including streams, brooks, ponds, lakes. *Unlike Belted and Ringed Kingfishers, may nest in much narrower and densely vegetated streams and waterways.* At elevations up to 2,100 m. **LOCATION AND STRUCTURE:** *Usually concealed by overhanging vegetation* in a vertical dirt bank along a river or stream. Burrow is typically unlined, 1.8–2.4 m up from bank bottom, extending inward 0.7–1 m. Tunnel entrance width 5.5 cm; height 5 cm. **EGGS:** 5, sometimes 3–6; 24 x 19 mm; IP 19–21 days (F, M); NP 26–27 days (F, M). **BEHAVIOR:** Apparently practices true monogamy.

WOODPECKERS (Picidae)

Twenty-two species of woodpeckers nest in N. America, 1 or more of which can be found in every wooded habitat from coast to coast. They are also well adapted to other environs, including savanna and desert habitats, where they excavate their cavities in large standing cactuses. All use tree (or cactus) cavities for nest locations. The vast majority of these are drilled by the woodpeckers themselves into live or dead trees, depending on the environment and species. They may create more than 1 cavity prior to the breeding season and use only 1; most species construct new cavities each year. Both sexes often participate in excavation, but in many species males do most of the work. Cavities frequently take 2–3 weeks to complete, but sometimes require less time, or as long as 6 weeks (or even longer for Red-cockaded Woodpecker). Nest cavities are unlined. Woodpeckers provide an essential service to many other species of birds (from small owls to ducks) and mammals, which rely on these ready-made, well-constructed, and well-protected nest sites for their own reproduction. Differentiating vacant woodpecker holes is challenging. Know the species in the area and note habitat, tree species, condition of wood, and size, shape, and height of the cavity entrance, as all are helpful clues for narrowing down the potential maker.

Woodpeckers are monogamous or socially monogamous (varying by species), and several species (e.g., Acorn Woodpecker) have complex social structures and mating systems. Young are altricial and are fed in the nest cavity. The excited, shrill chatter of nestlings—often somewhat cricketlike in sound—can help locate active nests. Males engage in "drumming" as a courtship behavior, in which they bang on wood (or metal in the modern world) to produce a repetitive pattern of noise that is readily distinguished from the more irregular sound of drilling into wood for foraging or cavity excavation. Drumming may also be used by either sex as an indication of selected nest sites. Many species also use tree cavities as roosts, and Pileated Woodpeckers are known to excavate cavities solely for this purpose. Eggs vary in shape but generally are ovate or elliptical. They are white and described as flat or glossy, depending on species. Egg size varies with size of species. Clutch size commonly 4–6, but varies by species. Incubation period is typically ~12–13 days and NP typically ~26–31 days, varying by species; significant differences are noted below. Males of many species may spend more time attending to eggs than females do, and incubate throughout the night as well as taking part during daylight hours; both sexes care for young.

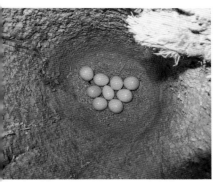

The unlined nesting cavity of a Northern Flicker, typical among woodpeckers.
Mark Peck, ON

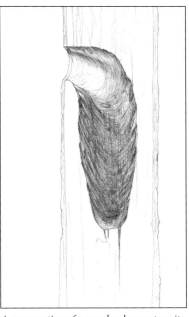

A Pileated Woodpecker excavating a nest cavity in an aspen tree. David Moskowitz, MT

A cross section of a woodpecker nest cavity. Woodpeckers typically use their excavation for a single season, but the sturdy cavity goes on to provide a nesting location for many other species, including owls, passerines, and mammals. As important providers for a wide variety of life in forests and deserts, woodpeckers are considered keystone species.

LEWIS'S WOODPECKER   *Melanerpes lewis*

HABITAT: Open ponderosa pine and riparian woodlands, often associated with previously burned forests. Occasionally other mixed-coniferous woodland, oak woodland, and nut orchards. **LOCATION AND STRUCTURE:** Cavity is often just below a limb in heavily rotted trunk of a large standing dead tree. Often reuses cavities made previously by Lewis's or other woodpeckers and also uses natural cavities. Cavity height 2–30 m; entrance diameter 5–7.5 cm. **EGGS:** 6–7; 26 x 20 mm. **BEHAVIOR:** May return to same nest location in successive years. At times multiple pairs nest in same tree.

David Moskowitz, WA

RED-HEADED WOODPECKER   *Melanerpes erythrocephalus*

HABITAT: Variety of open woodlands with standing dead trees, including landscapes heavily influenced by humans. Especially in forest edges and disturbed areas. **LOCATION AND STRUCTURE:** Typically in a location with open understory; cavity is excavated in a dead tree or limb, often one retaining little of its bark. Also in utility poles. Occasionally uses live trees, buildings, nest boxes. Although sometimes low to ground, nest cavity is typically toward top of a tree or pole. Excavation often begins with a natural crack and is found in trees with well-decayed center. Cavity height 7–12 m (2–25 m); entrance round, diameter 5.6 cm. **EGGS:** 5; 25 x 19 mm. **BEHAVIOR:** Male selects nest site and does most or all of excavation. Reuse of old nest cavities varies among populations. Sometimes takes over in-use cavities or uses old nest cavities of other species.

ACORN WOODPECKER   *Melanerpes formicivorus*

HABITAT: Oak and pine-oak woodlands. **LOCATION AND STRUCTURE:** Cavity in trunk or limb of a large tree, typically oak. Nest tree or limb may be living or dead, and may or may not also be a "granary tree." Multiple cavities are often present in territory of a family group, and nest cavities are often reused for many years. Cavity height 8 m (2–18 m); entrance diameter 3.8 cm. **EGGS:** 4–6;

25 x 19 mm; IP 11–12 days (M, F). **BEHAVIOR:** Breeding strategy is both communal and cooperative, with specifics varying widely among family groups. Family groups, rather than breeding pairs, maintain territories. Acorns are collected and stored in granary trees, managed communally. Two or 3 breeding individuals in a group are common, but there may be up to 7 breeding males and up to 3 breeding females. Breeding females lay their eggs in the same cavity, but an individual will destroy eggs deposited in the cavity before she begins to lay. Once all females are laying, eggs are no longer destroyed. In addition to breeding adults, there may be up to 10 nonbreeding helpers in the family group, typically offspring from previous years. Helpers rarely incubate, but help care for nestlings and fledglings.

An Acorn Woodpecker cavity. Also note the active granary. Matt Monjello, CA

GILA WOODPECKER *Melanerpes uropygialis*

HABITAT: Desert areas with saguaro and other large cactus, riparian woodlands, human-dominated suburban landscapes. **LOCATION AND STRUCTURE:** In saguaro cactus where available, also in mesquite, cottonwood, willow. Rarely in oaks, palo verde. Nest cavity in cactus is often excavated months prior to use,

allowing the plant to form a casing around excavated cavity. *Typically found lower on cactus than Gilded Flicker, whose nests are typically close to top.* Cavity height 6–7 m (1–11 m); entrance round or oval, diameter 5 cm. **EGGS:** 3–4; 25 x 19 mm.

Scott Olmstead, AZ

RED-BELLIED WOODPECKER *Melanerpes carolinus*

HABITAT: Wide variety of woodlands. **LOCATION AND STRUCTURE:** Cavity is often on underside of a leaning dead tree or limb. Readily uses fenceposts, utility poles, and similar structures, as well as cavities of Red-cockaded Woodpeckers. Often returns to same nest tree in subsequent years, but excavates a new cavity directly below old one. Cavity height 1.5–12 m; entrance slightly elliptical or round, diameter 5.7–5.9 cm. **EGGS:** 4–5; 25 x 19 mm. **BEHAVIOR:** Sometimes double- or treble-brooded. Recorded taking over and reusing cavities excavated by Red-cockaded Woodpecker.

GOLDEN-FRONTED WOODPECKER *Melanerpes aurifrons*

HABITAT: Open and semiopen woodlands, riparian areas, oak-juniper savanna, and urban and suburban environments. In arid brushland, closely associated with mesquite. In other parts of range, confined to riparian cottonwood groves. **LOCATION AND STRUCTURE:** Cavity is in trunk or limb of a live or dead tree. Also occasionally in fenceposts, utility poles, and nest boxes. Cavity height 2–9 m; entrance oval, diameter 5.5 x 4.9 cm. **EGGS:** 4–5; 25 x 19 mm.

WILLIAMSON'S SAPSUCKER *Sphyrapicus thyroideus*

HABITAT: Montane coniferous and mixed coniferous-deciduous forests. Particularly associated with ponderosa pine, western larch, Douglas-fir and true firs, and quaking aspen. **LOCATION AND STRUCTURE:** Excavates cavity in a live but well-decayed aspen, in live western larch with some heart rot, or in snag of other conifer species. Selects for trees that are larger than average in any given forest. *Tends to select sites in more coniferous-dominated forests with fewer aspens than does overlapping Red-naped Sapsucker, which tends to favor aspen-dominant areas for nesting.* Cavity height 1.5–18 m; entrance round, diameter 4 cm. **EGGS:** 5–6; 24 x 17 mm; NP 31–32 days (F, M). **BEHAVIOR:** Often uses same tree year after year, typically drilling a new cavity each nesting season.

Matt Monjello, AZ

YELLOW-BELLIED SAPSUCKER *Sphyrapicus varius*

HABITAT: Various forest types but typically associated with early successional forests. Often along aspen and birch riparian zones in mixed-conifer forests. **LOCATION AND STRUCTURE:** Often in heartwood fungus–infected aspens, also in other deciduous trees, rarely conifers. Cavity is drilled into main tree trunk, occasionally a dead stob. Entrance is small for the size of bird. Cavity height 9 m (3–19 m); entrance round, diameter 3.2–4 cm. **EGGS:** 5–6; 22 x 17 mm. **BEHAVIOR:** Often reuses cavity tree, and sometimes same cavity, for multiple years.

Paul Suchanek, MN

RED-NAPED SAPSUCKER *Sphyrapicus nuchalis*

HABITAT: Variety of deciduous and mixed forests. Avoids oak and pine-oak forests. **LOCATION AND STRUCTURE:** In aspen most often, also in pines, birch, larch. In aspens, nest cavity is usually in a live tree with heart rot; pair may use same tree multiple years, with cavity of each year higher than the ones before as heart rot moves upward. In conifers, cavity is more often in a dead tree. Cavity height 3–6 m, occasionally up to 21 m; entrance round or oval, diameter 3.8–4.6 cm. **EGGS:** 4–5; 23 x 17 mm.

David Moskowitz,
WA

RED-BREASTED SAPSUCKER *Sphyrapicus ruber*

HABITAT: Coniferous and mixed forests; sometimes riparian areas, orchards, edges, and burned areas. **LOCATION AND STRUC-TURE:** Nest cavity is in a dead tree or dead part of a live tree. Cavity height up to 15 m; entrance round, diameter 3.8–4.6 cm. **EGGS:** 4–5; 23 x 17 mm; IP 12–14 days (M, F); NP 23–28 days (F, M).

AMERICAN THREE-TOED WOODPECKER *Picoides dorsalis*

HABITAT: Boreal or montane coniferous forests, occasionally aspen or birch. Associated with bark beetle infestations and other forest disturbances resulting in dead and dying trees. **LOCATION AND STRUCTURE:** Usually in trunk of a dead tree and usually in a conifer. Entrance hole may have sloped lower portion, like a ramp, *making it taller than wide at entrance*, tapering to a round hole within. Cavity height 5.5 m (1–16 m), rarely higher. Entrance irregular, diameter 3.8–4.5 cm. **EGGS:** 4 (3–7); 23 x 19 mm. **BEHAVIOR:** In good habitat is loosely social with numerous nests in proximity to each other. Commonly excavates new cavity each year, but will sometimes reuse old ones.

Douglas Mason, AK

BLACK-BACKED WOODPECKER *Picoides arcticus*

HABITAT: Boreal and montane coniferous forests, closely associated with bark and wood-boring beetle infestations that follow fires. **LOCATION AND STRUCTURE:** In either a live or dead tree, with tendency toward one or the other, varying by region. Relative to other woodpeckers, Black-backed nests in stands with higher tree density and smaller-diameter trees. Bark is often, but not always, removed from around nest cavity, causing sap to flow in live trees. Bottom of entrance may be strongly beveled so that hole slants upward into tree, *similar to Hairy Woodpecker*. Cavity height 5–10 m; entrance irregular, 3.3 x 4 cm. **EGGS:** 4–5; 23 x 18 mm; NP ~24 days (F, M). **BEHAVIOR:** Inhabits suitable burned forests within 1 year post-fire, and will remain for up to 10 years, often less.

A Black-backed Woodpecker nest in a live tree with the bark removed from around the nest cavity.
David Moskowitz, ID

DOWNY WOODPECKER *Dryobates pubescens*

HABITAT: Variety of forests and woodlands, though more strongly associated with deciduous than coniferous forest. Readily uses human-created landscapes such as parks and orchards. **LOCATION AND STRUCTURE:** Cavity is excavated in rotten wood in a dead tree or in dead stob of a live tree. Cavity entrance is often on underside of a limb leaning away from vertical. Lined only with wood chips. Occasionally nests in siding of buildings. Cavity height 2–15 m; entrance circular, diameter 2.8–3.8 cm. **EGGS:** 4–5; 19 x 15 mm; NP 18–21 days (F, M).

NUTTALL'S WOODPECKER *Dryobates nuttallii*

HABITAT: Primarily live oak woodlands, but also riparian woodlands and rarely coniferous forests. **LOCATION AND STRUCTURE:**

Allen J. Vernon, CA

Cavity is in a soft, often dead, trunk or limb in a variety of tree species within preferred habitat. Cavity height 1–20 m; diameter 5 cm (4–8 cm). **EGGS:** 4–5; 22 x 16 mm; IP 14 days (M, F).

LADDER-BACKED WOODPECKER *Dryobates scalaris*

HABITAT: Arid scrublands, pinyon-juniper or juniper woodlands, and thorn forests. Also in riparian woodlands, especially where it does not overlap with Nuttall's Woodpecker. **LOCATION AND STRUCTURE:** Cavity is in a tree or occasionally a cactus, yucca, or fencepost. Cavity height 1–10 m; entrance round or oblong, diameter 3.2–4 cm. **EGGS:** 4–5; 21 x 16 mm. **BEHAVIOR:** Although often in conflict with similar Nuttall's Woodpecker where ranges overlap, these two species are also known to hybridize.

Ladder-backed Woodpecker.
Lois Manowitz, CA

RED-COCKADED WOODPECKER *Dryobates borealis*

HABITAT: Mature open pine forests. Rarely in younger forests, forests with notable hardwood presence, or areas of cypress adjacent to pines. **LOCATION AND STRUCTURE:** Cavity is in trunk of a mature live longleaf pine, occasionally other pine species, usu-

Note the bark chipped away around the cavity entrance. Jake Scott, FL

A Red-cockaded Woodpecker nest box, installed into the tree to create more nest locations for this endangered species.
Liz Tymkiw, SC

ally with decaying heartwood. Cavity is typically just below lowest branch on tree, usually on south or west side of tree. Occasionally in vertical branch of tree. *A bare area is made around cavity, causing sap to flow around entrance of nest, and is an indicator of long-term use.* Interior of cavity extends both upward as well as downward, thought to provide a refuge from potential predators of adult birds. Cavity height 1–30 m, typically 10–13 m. Entrance round, possibly irregular, diameter 5–7 cm. **EGGS:** 3–5; 24 x 17 mm; IP 10–11 days (F, M). **BEHAVIOR:** Highly social, traveling and foraging in small family groups. The cluster of cavity trees is occupied by a single breeding pair and 1–4 male helpers, young from previous year, that assist with incubation as well as care of nestlings and fledglings. Occasionally other individuals that are not breeders or helpers are present in the group. Individual cavities are excavated or reused over multiple years, with the best cavity often used as the nest hole and also the breeding male's roost hole. Endangered and declining because of direct habitat destruction and degradation from fire suppression.

HAIRY WOODPECKER *Dryobates villosus*

HABITAT: Variety of woodlands. More associated with mature forests than with small woodlots or human-created environments, although present in all types. **LOCATION AND STRUCTURE:** Cavity is excavated in trunk of a living tree with heart rot, or a dead trunk or stob. Preference for these sites seems to vary regionally. Entrance cavity is often on underside of a stob, large branch, or trunk that leans away from vertical, thought to be a deterrent against flying squirrels and sapsuckers. Cavity height 1–19 m; entrance oblong, 4.8–5 cm x 3.8–4.8 cm. **EGGS:** 4; 25 x 19 mm; IP 11–15 days (M, F).

Matt Monjello, ME

WHITE-HEADED WOODPECKER *Dryobates albolarvatus*

HABITAT: Mature montane mixed-coniferous forests dominated by pines, especially ponderosa. Most abundant in forests with 2 or more pine species, but avoids those dominated by small or close-coned pine species. **LOCATION AND STRUCTURE:** Cavity is in trunk of a large, often dead, tree; often ponderosa pine or other

*White-headed
Woodpecker
feeding young
in the nest.*
David Moskowitz,
WA

conifer, rarely in a deciduous tree. Also occasionally in a fence-post, stump, or fallen log. *Cavity is often lower than that of other species.* Cavity height 1–9 m, usually 3 m or less; entrance round or slightly elongated, diameter 4.6–5.0 cm. **EGGS:** 4–5; 24 x 19 mm; IP 14 days (M, F). **BEHAVIOR:** Incomplete nests are common, and some trees have multiple cavities constructed in them. Monogamous and may maintain pair bonds from one year to the next. Reliance on large standing dead trees has caused population declines.

ARIZONA WOODPECKER *Dryobates arizonae*

HABITAT: Oak or pine-oak woodlands and sycamore-walnut riparian areas. **LOCATION AND STRUCTURE:** Typically constructs nests in a dead trunk or branch of evergreen oaks and other deciduous trees. Occasionally agave stalk. Cavity height 0.6–15 m; entrance diameter 5–5.7 cm. **EGGS:** 4; 23 x 17 mm; IP 14 days (M, F); NP 24–27 days (F, M).

NORTHERN FLICKER *Colaptes auratus*

HABITAT: Variety of open woodland habitats, suburban and urban habitats. **LOCATION AND STRUCTURE:** Diversity in site selection, but often in a tree trunk or large branch of a dead tree or snag. Also uses human structures; rarely in burrows of Belted King-fisher and other bank-nesting species. Sometimes reuses existing cavities, including those created by other species. Cavity height 5–8 m (0–27 m); entrance oval or round, diameter 6.5–8.5 cm. **EGGS:** 6–8; 28 x 22 mm; IP 11–13 days (M, F). **BEHAVIOR:** Both sexes excavate.

Jonah Evans, WY

GILDED FLICKER   *Colaptes chrysoides*

HABITAT: Desert with saguaro or cardon cactus. **LOCATION AND STRUCTURE:** Most often in saguaro or cardon cactus where nest is invariably built within 3 m of top of stem. Occasionally in riparian cottonwood or willow tree. Cavity height 6 m (1–9 m); entrance oval, *wider than tall*, diameter 7 x 8.5 cm. **EGGS:** 4–5; 28 x 21 mm.

Gilded Flicker.
Thomas Machowicz, AZ

PILEATED WOODPECKER   *Dryocopus pileatus*

HABITAT: Mature forest stands with large trees. **LOCATION AND STRUCTURE:** In a large snag or decaying wood of a living tree. Entrance is taller than wide, often with a peak at top of hole. Cavity height 13–35 m (5–52 m); entrance irregular, diameter 10–12 cm x 8–9 cm. **EGGS:** 3–5; 33 x 25 mm; IP 18 days (M, F). **BEHAVIOR:** Breeding pair is resident on territory year-round. Cavity is rarely reused for a second year of nesting but may be reused as a roost. Also excavates roost cavities, often with multiple entrances to a rotten center. Nest trees tend to be larger than roost trees. Provides critical nesting habitat for large secondary cavity nesters, such as some owls and ducks.

A typical Pileated Woodpecker nest cavity high in an old-growth snag.
David Moskowitz, WA

Note the distinctive oblong shape of the cavity entrance.   David Moskowitz, MT

FALCONS and CARACARAS (Falconidae)

With the exception of caracaras, these sleek raptors *do not build nests.* Seven species breed in N. America. Most falcons are monogamous, but trios, polygyny, and extra-pair copulations are reported. Falcons can be generally divided into three categories of nest placement, though some species are quite flexible in where they build. *Cliffs* are commonly used by the largest species (Peregrine, Gyrfalcon, Prairie Falcon); *cavities* are frequently used by American Kestrels; and *nests of other bird species* are commonly used by Merlins and Aplomado Falcons, though Merlins are particularly adept at using a diversity of nesting sites.

Falconid eggs are rounded, ranging from short subelliptical to subelliptical, and are smooth and nonglossy. The base color is whitish, yellow-white, cream, or pale pink, but eggs are often so profusely marked with fine speckles and blotching that they appear deep reddish. Eggs generally hatch synchronously, but asynchronous hatching also occurs and young can vary significantly in size as they develop. The smallest young may starve to death if food is inadequate, but they usually are not killed by their larger siblings, as is typical among other raptors. Because the nests of falcons are generally hard to access, or even see clearly, their identification relies primarily on being aware of nesting pairs.

From left to right: American Kestrel, Peregrine Falcon, Crested Caracara eggs.

CRESTED CARACARA   *Caracara cheriway*

HABITAT: Variety of habitat and topography. Prefers open or semi-open grassland, brushland, and rangeland, with scattered tall vegetation. Also scrubby woodlands, arid areas with ponds, lakes and wetlands, and tree-lined arroyos and drainages. **LOCATION AND STRUCTURE:** Often in brushy areas, but also may be in trees: cypress, palms, live oaks, elms, and others that provide concealment. Also in cactus, yucca, and commonly in saguaro (AZ). Where trees aren't present, will nest on rock ledges, or occasionally on ground in good cover. Typical placement is 2.5–24 m high. Nest is well constructed, large, and bulky, made of brush, vines, briars, woody stems, and potentially other materials such as wire, blossoms, weeds, flagging. Nest structure has shallow bowl but flattens with use. **BEHAVIOR:** Both male and female collect materials and build nest in 2–4 weeks, often reusing old nest or rebuilding at same site. An early nester in FL (generally Sept. through Apr.), coinciding with dry season. **EGGS:** 2–3; 59 x 46 mm; IP ~28 days (F–M); NP 43–56 days (F–M).

Crested Caracara.
Jonah Evans, TX

TYPICALLY IN CAVITIES

AMERICAN KESTREL *Falco sparverius*

HABITAT: Prefers open habitat, sometimes sparsely treed, in country ranging from desert to mountainous areas. **LOCATION AND STRUCTURE:** An obligate secondary cavity nester, commonly using holes created by woodpeckers, often those of larger species such as Pileated Woodpecker and flickers. Also in woodpecker cavities in saguaro cactus in southwestern range. Will use crevices in rocks, natural tree cavities, buildings, and nest boxes. Rarely uses an older open nest of another species. Nest is a shallow scrape with no material added. Hole may be oriented away from prevailing weather conditions and direct midday sun. **EGGS:** 4–5; pale buffish red-white or pink; 35 x 29 mm; IP 29–30 days (mostly by F); NP 30 days. **BEHAVIOR:** Male locates potential nest sites, from which female inspects and chooses. Nestlings defecate by squirming backward and, typical of many raptors, lift tail to eject liquidy scat onto inner wall of cavity, thereby keeping it mostly off nest floor and themselves.

David Moskowitz, WA

American Kestrel eggs. Jack Bartholmai, WI

MERLIN *Falco columbarius*

HABITAT: In forested areas, riparian areas and shelterbelts in prairies, and in structures in towns. **LOCATION AND STRUCTURE:** Merlins are adaptable nesters, choosing old nests of hawks, larger corvids, and magpies, but also using structures in cities and towns. Less frequently in cliff ledges, natural tree cavities; occasionally on ground hidden in vegetation, where female scrapes a shallow bowl and may sparsely line it. **EGGS:** 5–6; 40 x 31 mm; IP 28–32 days (mostly by F); NP 25–30 days (F, M).

APLOMADO FALCON *Falco femoralis*

HABITAT: Endangered in its former habitats in s. U.S. and n. Mex.; reintroductions have been mostly unsuccessful except in s. TX. Traditionally nested in scrub deserts and grassland in low-growing trees, mesquite, Spanish bayonet, etc. **LOCATION AND STRUCTURE:** Similar to those of Merlin and other falcons that use nests of other species, including those of Common Raven and many hawks. Will also nest on platforms on power poles, and on the ground. **EGGS:** 2–3; 45 x 35 mm; IP 31–32 days (F, M); NP 28–35 days (F, M). **BEHAVIOR:** Severely affected by eggshell thinning caused by DDT, and by major loss of habit due to large-scale landscape alteration, primarily by overgrazing and agriculture.

Active Aplomado Falcon nest. The species that originally built this nest is unconfirmed, but may have been a White-tailed kite.
Mary Jo Bogatto, TX

Mary Jo Bogatto, TX

GYRFALCON *Falco rusticolus*

HABITAT: Tundra, frequently on cliffs or bluffs of coasts, rivers, offshore islands, gorges. **LOCATION AND STRUCTURE:** Commonly in cliff nests of Common Ravens, Golden Eagles, or in shallow scrape on cliff ledges. Occasionally in trees and on human structures. **EGGS:** 3–4; 59 x 46 mm.

Kate Fremlin, NU

PEREGRINE FALCON *Falco peregrinus*

HABITAT: A globally distributed falcon; nests in widely diverse habitats, from deserts to mountainous forests to wetlands. **LOCATION AND STRUCTURE:** Generally a shallow scrape on cliff ledges, often a third of the way down from top on sheer cliffs. Nests may be placed fairly close to ground to 395 m or higher; prefers ledges between 45 and 198 m, depending on location. Also uses ledges on large buildings, platforms on smokestacks and silos, bridges, etc. May use old nests (particularly where cliffs are nonexistent) of other birds, including Common Raven, cormorants, and hawks. **EGGS:** 3–4; 52 x 41 mm; IP 28–29 days (mostly by F); NP 35–42 days (F, M). **BEHAVIOR:** Single-brooded in N. America. Male makes shallow scrapes in various locations along cliff ledges, etc., from which female chooses.

Kate Fremlin, Amaroq Wildlife Services, NU

Kate Fremlin, Amaroq Wildlife Services, NU

PRAIRIE FALCON *Falco mexicanus*

HABITAT: Wide variety: arid shrub-steppe, canyons, grasslands, prairie escarpments, alpine tundra. **LOCATION AND STRUCTURE:** Primarily on cliffs, rock outcrops, and bluffs, in ledges, crevices, and cavities and potholes or shallow caves in softer sandstones, similar; may use dirt banks. Commonly uses existing nests of Common Raven and Golden Eagles in these locations (and frequently shares nesting cliffs with them). Where cliffs are nonexistent, nest location varies. Will use nests of other birds in trees, on utility poles, etc. Makes a shallow scrape in loose substrate, in a location often under a slight overhang that protects nest. Typically nests halfway up a cliff face. Nest sites are 6–120 m high, often 9–12 m. **EGGS:** 3–4; 52 x 41 mm; IP 29–31 days (F); NP 40 days (F, M). **BEHAVIOR:** Aggressively defends nesting sites. Often returns annually to same nest site, though in regions with abundant nesting sites may move location year to year.

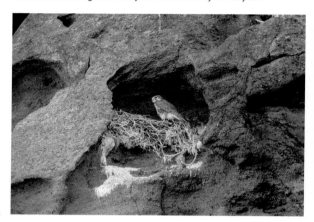

Robert C. Power, CA

TYRANT FLYCATCHERS and BECARD
(Tyrannidae, Tityridae)

These small to medium-sized passerines show a wide variety in nest placement, design, and size. Thirty-four species of tyrannids breed in N. America, and one tityrid. Accounts are grouped primarily by genus, except for Sulphur-bellied Flycatcher, and those in the section "Flycatchers with Globular, Domed, or Pendulous Nests."

Most of these species build open-cup nests, but they are often of recognizable design and placed distinctively: on or near the ground; in forks in shrubs and trees; saddled atop horizontal branches; in cavities; and on ledges or walls beneath a solid overhang. Some are attached to vertical surfaces and built with mud, others are constructed mostly of coarse twigs, and some of finer plant materials bound heavily with spider silk. Three species, all of which are found near the U.S.–Mexico border, build a domed, globular, or pendulous nest.

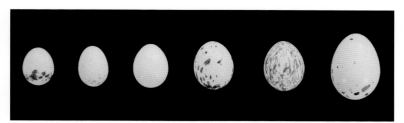

From left to right: Eastern Wood-Pewee, Willow Flycatcher, Black Phoebe, Western Kingbird, Ash-throated Flycatcher, Great Kiskadee.

Tyrant flycatchers tend toward monogamy, but in many species extra-pair copulations are possible, particularly after the first brood. Males of many tyrannids participate in or influence nest site selection, and often perform nest site selection displays (fluttering, calling, etc.) at potential nesting sites, and a female often "tests" a spot by lowering her body into it and mimicking nest-building movements. The female appears to be the sole nest builder in many species, but the male frequently accompanies her and may stand guard as she constructs. In most species the female alone incubates (but see Kingbirds), often fed by the male, and both parents feed nestlings. The altricial young hatch asynchronously over 1–3 days. True to their name, tyrant flycatchers can be exceptionally aggressive toward predators and other birds, and frequently chase conspecifics or similar species. Cowbird parasitism has occured in almost all species but is generally less frequent than in other families. Eastern Phoebes and Acadian Flycatchers are frequent hosts.

CRESTED FLYCATCHERS
and SULPHUR-BELLIED FLYCATCHER

These 5 species are obligatory cavity nesters, opportunistically using a wide variety of cavities, from woodpecker holes to natural hollows in trees to nest boxes. Cavities may be reused in successive breeding seasons. *Myiarchus* species may keep territories year-round, and are typically socially monogamous or assumed so. Sulphur-bellied Flycatchers are considered monogamous.

Nest components are similar: a foundation is built of various bulky materials such as grasses, twigs, forbs, weed stems, bark fibers, rootlets, mosses, fur (may occur in large amounts), and dung. Nests frequently contain shed snakeskin, though its use varies geographically, and similar materials such as plastic wrappers may serve as substitutes. *Myiarchus* species will also use other animal parts, and nests may contain bits of dead birds and other animals, mammal skin, etc. Nests are lined with fine plant fibers or rootlets, fur, hair, feathers. (Sulphur-bellied nests are composed of slightly different materials.) Dimensions vary greatly depending on the size of the cavity, and deeper cavities are often simply filled with bulking materials to achieve a desired base. Nests are completed mostly or entirely by the female in about 2–7 days and are best identified when active.

Eggs (usually 4–5) range in size across *Myiarchus* species from about 20 x 15 mm to 24 x 18 mm (Dusky-capped Flycatcher the smallest, Brown-crested Flycatcher the largest) and are subelliptical, short subelliptical, or elliptical. Sulphur-bellied are larger (see below.) They are creamy white to ivory and heavily marked: spotted, speckled, and blotched, often with longitudinal speckling or streaks. Incubation period is 13–17 days, NP 14–18 days.

DUSKY-CAPPED FLYCATCHER *Myiarchus tuberculifer*

HABITAT: In U.S., in low to mid-elevation oak and pine woodlands, and streamside in well-wooded riparian areas. **LOCATION:** In woodpecker hole or natural tree cavity, often high aboveground, but range is 1.2–12 m. **BEHAVIOR:** Less aggressive than other *Myiarchus* species.

Bettina Arrigoni, AZ

ASH-THROATED FLYCATCHER *Myiarchus cinerascens*

HABITAT: Arid and semiarid landscapes; desert scrub, mesquite, pinyon, yucca and cactus country, open deciduous woodland and riparian corridors. In any habitat, requires plant substrates large enough to provide suitable cavities. **LOCATION:** Primarily in natural cavities, woodpecker holes, yucca tree and columnar cactus, and in trunks and large limbs of trees and shrubs, often in dead portions but also live. Adapts readily to nest boxes and other cavities in human structures such as poles, fenceposts, pipes, etc., oriented either vertically or horizontally. Rarely in old Cactus Wren nests or behind loose bark. **BEHAVIOR:** May evict other birds from nesting cavities.

GREAT CRESTED FLYCATCHER *Myiarchus crinitus*

HABITAT: The only cavity-nesting flycatcher in e. N. America. Breeds in a wide variation of habitats: deciduous and mixed woodlands, orchards, parks, often at edges of clearings. **LOCATION:** In natural cavity (knothole, broken trunk or limb, etc.) or woodpecker hole, dead or live trees, 3–21 m high, often 1.8–6 m. In deciduous and mixed woodlands, more often in natural cavities than in woodpecker holes; in coniferous forests, more frequently in woodpecker holes. Frequently in human structures, and comfortable around human habitation; in poles, pipes, nest boxes, etc. **EGGS:** May be long elliptical.

Great Crested Flycatcher. Matt Monjello, ME

Inside the nest cavity. Note the snakeskin. Jack Bartholmai, WI

BROWN-CRESTED FLYCATCHER *Myiarchus tyrannulus*

HABITAT: Mostly in mature riparian deciduous woodlands and columnar cactus desert, mesquite. **LOCATION:** In natural cavity or woodpecker hole in trees, or woodpecker hole in saguaro cactus. More dependent on holes of larger woodpeckers than Ashthroated .

Gus Lane, AZ

SULPHUR-BELLIED FLYCATCHER *Myiodynastes luteiventris*

HABITAT: Riparian canyons in se. AZ. Commonly in sycamore, but also in oak, pine, cypress, walnut. **LOCATION:** Nests in cavities (primarily natural) in large trees. Nest is built of pine needles and/or walnut petioles. Eggs are similar in color to other crested flycatchers, but are larger and heavily marked across entire egg with blotches, spots, speckles; 26 x 19 mm. Assumed monogamous.

Bettina Arrigoni, AZ

KINGBIRDS

Nests of kingbirds vary, but there are overlapping features. With the exception of Thick-billed, Gray, and Tropical Kingbird nests, most are bulky with somewhat untidy exteriors, but contain a well-built interior cup and lining. They are typically placed along a horizontal branch, in crotches in trees, or in slanted or upright forks. Nest building appears to be a leisurely process for some species and takes at least a week. Eastern Kingbird may take 2 weeks, while Scissor-tailed Flycatcher builds in as few as 2–4 days. Eggs are similar: smooth, slightly glossy, white to creamy white or very pale pink, and heavily marked with blotches, large bold spots of dark and light browns, lilac, and blackish purples, mostly at the larger end. Sometimes wreathed, and occasionally have longitudinal streaks. Incubation period 13–16 days, NP 13–14 days. Male Eastern and Thick-billed Kingbirds are known to participate in incubation.

WESTERN KINGBIRD *Tyrannus verticalis*

HABITAT: Large variety of open-country habitats; grasslands, arid areas with scattered trees, desert scrub. Often associated with human habitation (parks, rest stops, yards); utility poles, oil derricks, fenceposts, etc. **LOCATION AND STRUCTURE:** Wide variety of sites, except on ground. Trees and shrubs seem preferable, usually on a horizontal or upward-slanting branch, centralized in canopy, often away from trunk but sometimes against it. Height varies by available vegetation; 2.4–12 m, often 4.5–9 m. Nest is a bulky cup, frequently of forb and grass stems, thin twigs, rootlets, tree barks (e.g., cottonwood), plant fibers intermixed with hair or wool, feathers, cocoons, paper, snakeskin, etc. Lined with finer, similar materials, felted into well-made soft cup *(lining of Eastern Kingbird is different, usually lacking cottony material)*. Size varies: outside diameter averages 15.2 cm; height 7.6 cm; inside diameter 7.6 cm; depth 4.5 cm. **EGGS:** 4; *slightly smaller than eggs of Eastern Kingbird.* **BEHAVIOR:** Both sexes may build in 4–8 days, or male guards. Occasionally uses old nests of the same or other species.

Western Kingbird nest tree.
David Moskowitz, WA

Western Kingbird nest on human structure. Garth Olson, OR

EASTERN KINGBIRD *Tyrannus tyrannus*

HABITAT: The most widely distributed kingbird in N. America. Open habitats along woodland edges, fields, or rangelands with scattered trees, golf courses, etc. Where possible, near water at edges of swamps, ponds, lakes, rivers, riparian belts. **LOCATION AND STRUCTURE:** Often on a horizontal limb, well away from trunk, often closer to ends of branches, 0.6–18 m up, 1.8–7.6 m common. In snags at top, or on branch crotch close to trunk. Often in a snag or live branches over water. Nest is bulky, large for size of bird, roughly but well-built exterior often with materials hanging from sides, of small twigs, roots, forbs, grasses,

Janet Bauer, WA

Mark Nyhof, BC

weed stems, mosses, twine. Lined with fine rootlets, fine grasses, occasionally plant down, hair, feathers. *Similar to nest of Cassin's Kingbird; lining of Western Kingbird nest is more cottony.* **EGGS:** 3–4; 23 x 17 mm. **BEHAVIOR:** Both sexes incubate. Extra-pair paternity is common, despite species' tendency toward social monogamy. Single-brooded. Parental care for fledged young is particularly long, 3–5 weeks.

CASSIN'S KINGBIRD *Tyrannus vociferans*

HABITAT: Nests at higher elevations than other kingbirds in N. America, though overlap occurs with Western Kingbird at lower elevations. Mountains, foothills, riparian woodlands, canyons. Usually in tall trees of various species, dependent on region (e.g., sycamore, cottonwood, pine, eucalyptus), but also in smaller trees such as pinyon and juniper. **LOCATION AND STRUCTURE:** Height from ground varies greatly with tree species, 6–12 m common, sometimes lower. Usually on a horizontal limb, toward outer end. Nest is large and bulky, *similar to those of Western and Eastern Kingbirds.* Often exterior and rim are decorated in dry blossoms, or feathers. **EGGS:** 3–4; *often less marked than other kingbird eggs;* 23 x 17 mm. **BEHAVIOR:** Female builds. Reuses old nests on occasion. Often nests near Western Kingbirds (sometimes in same tree) but rarely near conspecifics.

SCISSOR-TAILED FLYCATCHER *Tyrannus forficatus*

HABITAT: Open land with occasional trees and shrubs (honey mesquite country is common); ranches, agricultural fields, golf courses, etc. Shade trees around human habitation, greenbelts along small waterways. In shrubs, often in tallest. Also in human structures such as utility poles, towers. **LOCATION AND STRUCTURE:** Generally on a horizontal branch or fork, 2–9 m up, sometimes lower. Nest is a rough, bulky cup, thick-rimmed, of twigs, rootlets, stems, corn husk, twines. Cudweed is common in some regions, and Spanish moss where it occurs. Cup and lining are tightly woven with fine materials: rootlets, plant down, string, etc. Sometimes cigarette butts, tissue, and other debris. **EGGS:** 3–5; *similar to those of Western and Eastern Kingbirds,* but can be smaller, occasionally with few or no markings. **BEHAVIOR:** Nest is built mostly by female in 2–4 days. Second broods are uncommon.

COASTAL- OR SOUTHERN-BREEDING KINGBIRDS

Behavior, nest placement, and eggs are similar to those of other kingbirds. Nests of Tropical, Thick-billed, and Gray Kingbirds are dissimilar to those of other kingbirds.

TROPICAL KINGBIRD *Tyrannus melancholicus*

Primarily se. AZ, near water. Habitat similar to that of Eastern, Western, and Cassin's Kingbirds. Nest is unique, however, in that it is *a frail, thin, shallow bowl,* untidily built, often with materials hanging from sides. Usually 1.8–6 m aboveground, sometimes higher.

COUCH'S KINGBIRD *Tyrannus couchii*

Thorn forest, riparian habitats, brushy clearings, often close to water. *Western Kingbird generally prefers more open habitats around human settlements, agricultural areas, but nests are similar.* Often in a tree at forest edge, or in a tree set apart from others. On a horizontal branch or built on 3–4 thin branches, 5.4–9 m aboveground.

THICK-BILLED KINGBIRD *Tyrannus crassirostris*

In riparian woodlands of cottonwoods, sycamores, but also on brushy slopes. Less commonly in desert scrub. Often with water nearby. Sometimes builds in mistletoe clumps. Nest is commonly high in tall trees, 9–18 m, but can be low to ground. A *flimsy, flat cup*, often thin enough to see through, with materials projecting outward, appearing bristly. Both sexes incubate. Eggs can be slightly larger than those of other kingbirds.

GRAY KINGBIRD *Tyrannus dominicensis*

Restricted to Neotropical climates and coastal habitats. Nest is unique among kingbirds: a *large, loosely constructed shallow bowl, sometimes thin enough to see eggs from below.* Usually low, often over water, 1–3.6 m high, but can be higher.

Gray Kingbird.
Robin McLendon
Diaz, FL

PEWEES, WOOD-PEWEES, and OLIVE-SIDED FLYCATCHER

Nests of the 3 species of pewees (Including Wood-pewees) are quite similar: tidy, compact, lichen-covered cups that are camouflaged to their surroundings, and whose various materials (including grasses, weed stems, thin twigs, dry leaves, catkins, shredded bark, pine needles) are bound with spider or caterpillar silk and may look from below like a branch knot. The nest is firmly attached in place with silk, and generally covered in lichen flakes. Olive-sided Flycatcher builds a more loosely constructed, flattish or saucerlike nest. Nests are generally *saddled on horizontal limbs or forks*, and materials are often collected on the wing, hovering above the ground or at upturned root balls, weeds, etc. Females appear to be the primary builders and incubators. Pewees build in ~3–7 days or more, Olive-sided in 5–7 days. Pewees: IP 12–13 days, NP 14–18 days.

GREATER PEWEE *Contopus pertinax*

HABITAT: Pine and pine-oak woodlands; riparian zones in canyons with sycamore. **LOCATION AND STRUCTURE:** High in pines or other tall trees, saddling a generally fairly robust horizontal branch or fork. Outside diameter 10.5–13.5 cm; height 5–8 cm; inside diameter 7.6 cm; depth 2.5–5 cm. *Similar to wood-pewee nests but may appear more loosely built.* **EGGS:** 3–4; *similar to those of other pewees*; 21 x 16 mm.

WESTERN WOOD-PEWEE *Contopus sordidulus*

HABITAT: Common and widespread. Coniferous (from pinyon to sequoia), deciduous, and mixed woodlands. Along riparian areas, in orchards, human habitation. Mostly absent in dense forest. May be in drier habitats than Eastern Wood-Pewee where overlap occurs. **LOCATION AND STRUCTURE:** Nest is typically 4.5–12 m aboveground, sometimes lower, to higher than 25 m. Often placed toward outer edge of tree foliage. Nest is often *slightly larger than that of Eastern Wood-Pewee, and usually swathed heavily in silk, but lacks lichen flake decoration.* Outside diameter 8.5 cm; height 6.6 cm; inside diameter 4.8 cm; depth 3.4 cm. **EGGS:** 3; creamy white to pale yellow, with brownish spots or blotches, primarily on larger end, loosely wreathed; 18 x 14 mm. *Identical to those of Eastern Wood-Pewee.*

Sylvia M. Robertson, WA

Mark Nyhof, BC

EASTERN WOOD-PEWEE *Contopus virens*

HABITAT: A habitat generalist, using almost every type of eastern wooded habitat: deciduous, coniferous, and mixed woodlands. May avoid dense canopy and prefers drier over moist microhabitats. **LOCATION AND STRUCTURE:** In dead or live trees. Commonly at 2.4–7.6 m but can be much higher (29 m). *Nest can be slightly smaller than that of Western Wood-Pewee.* **EGGS:** 3; see Western Wood-Pewee. **BEHAVIOR:** Among the last spring migrants to return to breeding grounds from S. America.

Thomas Schultz, WI

OLIVE-SIDED FLYCATCHER *Contopus cooperi*

HABITAT: Mountainous coniferous habitats, but also in mixed woodlands. Often associated with forest edges and openings: meadows, burns, clear-cuts, lakes, beaver ponds, etc. Usually in conifers, but also in deciduous trees in some parts of range. **LOCATION AND STRUCTURE:** Saddled on a horizontal branch or branchlets, often well away from trunk, placed in upright clusters of live twigs and needles. Commonly at 2–15.4 m, but can be

Mark Peck, ON

higher than 30 m. Nest is flattish, saucer-shaped, loosely built. A platform or outer shell is made mostly of small twigs, occasionally rootlets, possibly moss and lichens (particularly *Usnea*). Lined with fine rootlets, *Usnea*, pine needles. *Somewhat similar to nest of Pine Grosbeak, but shallower.* **EGGS:** 3; creamy white or pale pink, often *strongly wreathed* at larger end with brownish, reddish blotches; short subelliptical; 22 x 16 mm; IP 16–17 days; NP 15–19 days. **BEHAVIOR:** Among the most aggressive tyrant flycatchers.

EMPIDONAX FLYCATCHERS

While these small flycatchers are very similar in appearance and notoriously difficult to distinguish visually in the field, the nests of many species are built or placed distinctly enough in varying habitats to be a reliable means to help identify the maker. In general, nests are constructed of fine, fibrous materials and appear as many shredded materials bound together. They range from a thick-walled, pucklike shape to small, compactly built cups wrapped in silk, or can be thinly walled. Nests are placed on the ground, atop larger branches, in forks or crotches, or on small ledges of broken snags, etc. Eggs vary among species in size and slightly in shape (subelliptical to short subelliptical) and typically are smooth, dull, whitish or cream colored, and typically lightly spotted, speckled, and blotched with brownish or purplish reds, mostly at the larger end. *Empidonax* are commonly single-brooded, but some species are occasionally double-brooded.

YELLOW-BELLIED FLYCATCHER *Empidonax flaviventris*

HABITAT: Northern boreal forests, boreal bogs, alder swamps. Typically deeply shaded, moist, cool, flat, and poorly drained. **LOCATION AND STRUCTURE:** *Built on or near ground*, famously well hidden, with very little to no sun exposure, in upturned roots of fallen trees, in logs, banks, in hummocks of sphagnum moss, hidden beneath herbage or at the base of ferns, and tucked in recessed hollows at the base of conifers. Nest is generally a

Mark Peck, ON

deep, well-insulated cup of mosses, weed stems, rootlets, though structural walls may vary geographically. Lined with fine rootlets, likely horsehair fungus, fine mosses, moss stems, and occasionally pine needles, fine grasses. Outside diameter 8–10 cm; height 6 cm; inside diameter 5 cm; depth 2.5 cm. **EGGS:** 3–4; white; occasionally with sparse speckling; 17 x 13 mm; IP 15 days; NP 13 days.

ACADIAN FLYCATCHER *Empidonax virescens*

HABITAT: Swampy or moist woodlands, often deciduous but will use coniferous forests, particularly in southeastern range; generally requires mature forests. Bottomlands, shaded ravines, often associated with water. **LOCATION AND STRUCTURE:** Nest is *supported hammock-like in a horizontal fork*, generally toward the end of a slim branch and never close to trunk, often near or above water. Commonly at 2.4–7.6 m. Nest is a somewhat pensile, flimsy, saucerlike cup (eggs are often visible through the thin walls and bottom) of fine, dry weed stems, plant fibers, bark shreds, fine rootlets, catkins or bud scales, small tree blossoms. Lightly lined with fine grass stems, plant down, Spanish moss. Materials are bound lightly in silk. *Long, distinctive tails, or streamers, of plant materials hang from bottom of nest,* often 30–60 cm in length. Outside diameter 8 cm; inside diameter 4 cm; depth 2 cm. **EGGS:** 3; creamy to buffy white, sparsely marked toward larger end with brownish spots; 18 x 14 mm; IP 15 days; NP 13 days.

David Pierce, MO

Matt Monjello, AL

ALDER FLYCATCHER *Empidonax alnorum*

HABITAT: Alder and Willow Flycatcher were long treated as a single species (Traill's Flycatcher). Alder Flycatcher ranges much farther north and nests in boreal bogs, swamps, wet alder

Alder Flycatcher. Don Gorney, IN

Alder Flycatcher. Chris McCreedy, AK

thickets, brushy swamps, at marsh and lake edges, damp areas, often near water. **LOCATION AND STRUCTURE:** Often *low in shrubs*, under 2 m, average 0.6 m. Nest is a somewhat loosely built, untidy cup of coarse dry grasses, bark shreds from woody plants, rootlets, twiglets, and *small amounts* of cattail down or other silky material (see Willow Flycatcher). Lined with fine wiry grass, hair, and occasionally a few feathers. Often has long streamers. Generally in an upright fork, built around supports. Nest is described as *similar to Song Sparrow and Indigo Bunting nests.* Outside diameter 8 cm; height 7 cm; inside diameter 5 cm; depth 3.5 cm **EGGS:** 3–4; lightly marked; 18 x 13 mm; IP 13–15 days; NP 12–15 days.

WILLOW FLYCATCHER *Empidonax traillii*

HABITAT: Tends toward dry, shrubby thickets, frequently in willow stands, but also in moist, mixed deciduous woodlands. **LOCATION AND STRUCTURE:** Placement *similar to that of Alder Flycatcher*, but nest may be more loosely built when not in fork. Dry, bleached grasses and bark can give nest a silvery or gray appearance. Feathers are often worked into rim. Large use of plant down

Bill Summerour, CO

Bill Summerour, CO

lends a much more *cottony appearance*, and nest may have streamers but *less often than Alder Flycatcher nest does*. Nest is *more compactly and neatly built than Alder Flycatcher's* and described as *similar to those of American Goldfinch and Yellow Warbler*. **EGGS:** *Similar to Alder Flycatcher's*, often with brown spots on larger end. **BEHAVIOR:** In arid reaches of its range, depends heavily on riparian habitat along riverways. Human development, overgrazing, river water use, and river alteration have caused sharp population decreases continent-wide.

LEAST FLYCATCHER   *Empidonax minimus*

HABITAT: The smallest and most common *Empidonax, often in drier woodland habitats than many others*. Open deciduous and mixed woodlands, orchards, shade trees in city parks and settlements, edges of lakes or along streams. In openings but also in forests with dense canopies. Often in a deciduous tree, sometimes in a conifer. **LOCATION AND STRUCTURE:** In a vertical crotch, fastened to horizontal limbs, near a trunk in fork, or more rarely toward ends of limbs in smaller forks, often in lower to middle canopy. Commonly at 1.5–6 m, but possible at 0.6–18 m. Nest is a neat, deep, compact cup with a thickish wall and tidy rim. Built of bark strips, grasses, plant down, fine fibers, and shredded plant stems, bound with spider silk. Lined with fine grasses, plant down, hair. *Resembles nests of American Redstart, Yellow Warbler, and possibly Blue-gray Gnatcatcher.* Outside diameter 6.4–7.6 cm; height 4.5–6.4 cm; inside diameter 4.5–5 cm; depth 3–3.8 cm. **EGGS:** 4; *unmarked*; 16 x 13 mm; IP 14–16 days; NP 13–16 days. **BEHAVIOR:** Both sexes appear to choose nest site; nest is built by female in 6–8 days. Known to steal materials from other birds' nests. Despite it small size, is highly aggressive toward other birds. Interestingly, numerous territories are set in close proximity.

Thomas Schultz, WI

HAMMOND'S FLYCATCHER *Empidonax hammondii*

HABITAT: Typically at higher elevations, mountainous. Coniferous or mixed forests, often in a conifer but also in aspen, maple, birch, cottonwood, etc. **LOCATION AND STRUCTURE:** *Nest and location are often more similar to that of pewees than of other* Empidonax, *and nest is generally much higher aboveground than Dusky Flycatcher's.* Saddled on a large horizontal limb, typically well away from trunk, at 1.8–18 m but averaging 7.6 m. Nest is compact, thick-walled, made of plant fibers, bark strips, rootlets, grasses and stems, silk cocoons; lined with grasses, feathers, sometimes hair. Outside diameter 8.5 cm; height 6 cm; inside diameter 5.5 cm; depth 3 cm. **EGGS:** 4; 17 x 13 mm, creamy white or white, usually unmarked; IP 15 days; NP 17–18 days.

DUSKY FLYCATCHER *Empidonax oberholseri*

HABITAT: Mountain slopes, foothills, scrub. Open, brushy lands, scarcely treed, often in patches of shrub. **LOCATION AND STRUCTURE:** Low to ground, typically in deciduous shrubs or trees, at 1.2–2 m, sometimes higher. Nest, usually woven into an upright crotch, is a firm, compact, tidily built cup of shredded grasses, forb stems, and plant fibers, lined with fine bark fibers, hair, feathers. Sometimes made just of weed stems, and feathers may be few to none. Bound with spider silk. Outside diameter 7 cm; height 7 cm; inside diameter 5 cm; depth 3.5 cm. **EGGS:** 3–4; creamy white or white, usually unmarked; 17 x 13 mm; IP 15–16 days; NP 18 days. **BEHAVIOR:** This and other small temperate-zone *Empidonax* may be particularly susceptible to severe weather, causing population-scale mortality or nest failure.

Mark Nyhof, BC

Libby Megna, WY

GRAY FLYCATCHER *Empidonax wrightii*

HABITAT: Sagebrush and other overgrown desert brush; semiarid habitats, including shrub-steppe and very open coniferous woodlands of pinyon-juniper or pine. **LOCATION AND STRUCTURE:** Commonly in sagebrush or similar, or along horizontal branches, twigs, or on a branch against trunk of pine, juniper. Often well hidden and blending into surroundings. Nest is untidy, made of bark strips (sage, juniper, etc.), dead weeds, pine needles possible, plant fibers, spider silk; lined with wool, hair, small feathers. *Nests tend to have a shredded, fibrous appearance, while those of Hammond's and Dusky often are more compact and are bound in spider silk. Can be slightly larger than Dusky Flycatcher nest, with a shallower cup.* **EGGS:** 3–4; 18 x 13 mm. Creamy white, unmarked; IP 14 days; NP 16 days.

Janet Bauer, WA

PACIFIC-SLOPE FLYCATCHER *Empidonax difficilis*

CORDILLERAN FLYCATCHER *Empidonax occidentalis*

Sibling species primarily distributed in western coastal mountain ranges (Pacific-slope) and Rocky Mts. (Cordilleran). Both species use a wide variety of nesting locations but are commonly associated with cool, shaded areas along waterways, and share similar nests, eggs, and placement. **HABITAT:** Moist deciduous, mixed, and pure coniferous forests (mostly coastal), steep and wooded ravines, canyon bottoms. Often associated with streams (commonly over them), shade. Cordilleran Flycatcher is often in cooler, more arid, and denser coniferous forests but also mixed forests. **LOCATION AND STRUCTURE:** Often near ground, but to 9 m. Nest is typically supported from below and behind, in upturned tree roots, in shrub or medium-sized tree, behind loose bark, in cavities or crevices in cut banks and trees, or in human structures (covered trailhead sign, etc.). Nest is made of green and dry

Pacific-slope Flycatcher. Both this species and Cordilleran Flycatcher commonly build in structures such as porches, trailhead signs, and the like.
David Moskowitz, WA

Pacific-slope Flycatcher. Mark Nyhof, BC

Cordilleran Flycatcher. Casey McFarland, NM

Cordilleran Flycatcher. Matt Monjello, AZ

mosses, lichens, inner bark strips, bark shreds, leaves, stems, small annuals. Spider silk binds mosses and lichens, other cup wall materials, and anchors nest base in place. Often includes artificial debris when built near human habitation. Lined with fine materials such as bark strips, grasses, stringy lichens. Outside diameter 11 cm; height 5 cm; inside diameter 5.7 cm; depth 4.5 cm. **EGGS:** 4; 17–18 mm x 13 mm (Cordilleran Flycatcher the larger); IP 14–15 days; NP 15–18 days. **BEHAVIOR:** Pacific-slope Flycatcher may use nests of other birds.

BUFF-BREASTED FLYCATCHER *Empidonax fulvifrons*

HABITAT: Localized. Found in mountain canyons, often near riparian areas, commonly in open pine or pine-oak forests with grassy, sparse understory. **LOCATION AND STRUCTURE:** Nest is saddled atop a horizontal branch or placed directly against trunk on limb, usually with overhead cover, ~7 m aboveground. Nest is well built and attached with spider silk. *Often oval in shape*, of spider silk, rootlets, leaves, lichen, with an outer layer of feathers, bark bits, leaves, lichen; somewhat similar in design to Eastern Wood-pewee but slightly smaller.

PHOEBES and VERMILION FLYCATCHER

Phoebes prefer nest sites sheltered by a substantial overhang, and nests are generally placed on a ledge or attached to a vertical wall. They are quite different than the nest of Vermilion Flycatcher (see species account). Phoebe ranges and numbers have increased as human infrastructures such as buildings and bridges have become ubiquitous across the N. American landscape. Phoebes are typically monogamous, and all 3 species tend to reuse old nests, their own and those of other species. Nests may vary in shape depending on where they are built or affixed, but the cup generally remains perfectly circular. Nest dimensions for all 3 species can be similar, but because of reuse, size can increase dramatically. Single-use nests on occasion can also be surprisingly large (Eastern Phoebe). General dimensions for phoebe nests: outside diameter 11.4–12.7 cm; height 5.7–10.2 cm; inside diameter 6.4–7 cm; depth 3.2–4.5 cm. Say's and Black Phoebe males visit new and old nest sites during pair formation, accompanied by females, and perform nest site-showing displays. Females are likely the primary nest builders and are the sole incubators. Nests are built in 5–7 days, or as long as 3 weeks, and are often finished 10–14 days before laying starts. Eggs number 4–5 and are subelliptical to short subelliptical, generally white, smooth with little to no gloss, usually unmarked, and measure 19–20 mm x 14–15 mm (Say's Phoebe largest). All 3 species are typically double-brooded. Eastern Phoebe makes 2 types of nests: one adhered to the side of a wall like that of Black Phoebe, and one placed atop a ledge like that of Say's Phoebe. Say's Phoebe nests typically lack mud.

BLACK PHOEBE *Sayornis nigricans*

HABITAT: Invariably near or associated with water, permanent or semipermanent; mud is a requirement for nest building. Desert riparian areas along creeks and riverbanks, coastal cliffs, lake and pond borders, frequently around human habitations, nesting in structures. **LOCATION AND STRUCTURE:** Typically adhered to the side of a wall, cliff face, often beneath overhang, with lip of nest often within 2.5–7 cm of ceiling. Nest is a half-hemispherical shape, plastered to vertical surface with mud; the base and lower wall of cup are a mixture of mud and dry grasses, rootlets, and other vegetation, and sometimes hair. Upper wall (half to third) is often woven tightly of plant fibers,

Black Phoebe.
Kim A. Cabrera, CA

containing less mud than the supportive base. Lined with wool, hair, fine plant fibers, fine rootlets, sometimes feathers. **EGGS:** IP 15–18 days; NP 21 days.

EASTERN PHOEBE *Sayornis phoebe*

HABITAT: Wide variety; site and available nesting locations seem to drive primary preference. Farms and other human structures (culverts, bridges, etc.), open woodlands, occasionally on cliffs. **LOCATION AND STRUCTURE:** On shelf or ledgelike projections: any dry, suitable surface, sometimes surprisingly narrow, often with overhead protection. Also plastered to vertical surface with mud, like nest of Black Phoebe. Nest is built of mud pellets, *invariably with green moss,* grasses, forbs, lined with fine grasses, rootlets, and hair. Despite mud use, nests may be located relatively far from water sources. Nests placed atop ledges are rounder and have less mud. **EGGS:** IP 14–16 days; NP 15–17 days. **BEHAVIOR:** Among the earliest migratory species to return north in the spring.

Casey McFarland, CT

Matt Monjello, ME

SAY'S PHOEBE *Sayornis saya*

HABITAT: Among the widest-ranging flycatchers, breeding from cen. Mex. to n. AK. Not as closely associated with water as Black and Eastern Phoebes, and generally avoids dense forest, watercourses, agricultural lands. Open, semiarid land-scapes: badlands, canyons, desert borders, foothills, sagebrush flats. **LOCATION AND STRUCTURE:** Requires sheltered ledges or recesses; nest is not adhered to vertical surface. Protected ledges on cliffs, arroyo banks, caves, old mine shafts, etc. In many cases seems to prefer sites in human structures such as abandoned buildings. Nest is a flattish, bulky cup of small stones (in base), weed stems, dry grasses, moss, plant fibers, sage bloom (regionally), forbs, cocoons and spider silk, hair. *Mud is infrequent.* Lined with wool, hair, human-made materials, some-times feathers. **EGGS:** IP 12–14 days; NP 12–14 days. **BEHAVIOR:** Male cares for fledged young of first brood while female renests. One nest was used for 5 consecutive years.

Say's Phoebe. Pat Leigh Photos, WA

Say's Phoebe. Mark Nyhof, BC

An old Say's Phoebe nest tucked inside a small alcove in a desert wash wall. Chris Byrd, AZ

VERMILION FLYCATCHER *Pyrocephalus rubinus*

HABITAT: Desert scrub, mesquite bosque, sometimes willow thickets, desert riparian woodlands (often in groves near or along water). Often with water nearby. **LOCATION AND STRUCTURE:** Set deeply into nook of a fork or crotch of horizontal branch so that the nest wall *may be noticeable from only one side*, if at all. Commonly at 1.8–6 m, occasionally much higher. Nest is a flat-tish, well-built cup of fine twigs, weed stems, rootlets, dead leaves, bark strips, spider cocoons and silk, lined to varying degrees with plants, feathers, hair, fur. *A heavy use of fine twigs can make nest look quite different from those of other tyrannids.*

Vermilion Flycatcher. Scott Olmstead, AZ   *Vermilion Flycatcher.* Charles Robertson, TX

Nest cup is very shallow. Outside diameter 6.4–7.6 cm; height 5 cm; inside diameter 4.5–5 cm; depth 2.5 cm. **EGGS:** 3; white or creamy white, no gloss, heavily marked toward larger end with spots and blotches of browns, lavenders; short subelliptical to short oval; 17 x 13 mm; IP 14 days; NP 14–16 days. **BEHAVIOR:** Stunningly colored male performs spectacular courtship flight displays. Usually single-brooded.

FLYCATCHERS with GLOBULAR, DOMED, or PENDULOUS NESTS (U.S.–Mexico Border)

ROSE-THROATED BECARD *Pachyramphus aglaiae*

This species in not a tyrant flycatcher, but rather the single species of Becard (Tityridae) in N. America, where it breeds in deciduous, riparian woodlands. Nest is a pendent structure, suspended from tips of tree branches, 4.5–18 m aboveground.

Scott Olmstead, AZ

Globular nest is woven of and into twigs with bark strips, grasses, thin wiry vines, plant stems, branching lichens, spider cocoons and silk, long pine needles, plant down, other similar items. Lined with soft materials. Side entrance, directed downward, is sometimes concealed. Walls 3.8–6 cm thick. Nest is large: Outside diameter 22–31 cm; height (length) 30–36 cm. Both sexes build. May continue to build after incubation begins.

NORTHERN BEARDLESS-TYRANNULET
Camptostoma imberbe

Uncommon, primarily in AZ, in deciduous riparian woodlands, often near water, in cottonwoods and willow; mesquite thickets. Nest is small (outside diameter 10 cm), *globular or domed,* and thick-walled, usually far out on horizontal limbs, or in hanging branches, 1.8–15 m aboveground. Entrance is on side near top. Frequently concealed in mistletoe clumps or in tent caterpillar webs.

GREAT KISKADEE *Pitangus sulphuratus*

Borders of lakes, ponds, rivers, swamps, or brush savanna or woodlands typically near water. Nest is often in a major fork in larger branches of tree, 3–9 m high, sometimes higher. Nest is a football-shaped globular structure, built mainly of grasses but also of moss, thin twigs, vine tendrils, perhaps rags, paper, sometimes plastic strips. Shallow-cupped, lined with fine material. **EGGS:** 3-4; larger than those of kingbirds, but similarly marked. 28 x 21 mm. Both sexes build, probably in 3–10 days. Highly aggressive, harassing raptors, snakes, and other animals.

David Pierce, TX

SHRIKES (Laniidae)

These predatory, primarily monogamous songbirds generally nest in a low, small tree or large shrub with dense foliage. Nests are typically bulky, well-built cups consisting of an outer shell of long twigs, roots, similar, and a *thick-walled, well-insulated inner cup* of soft materials. Proportions of nest components vary with weather conditions; shrikes will adjust nest cup thickness, for example, in response to insulative requirements. In an effort to attract females, males create an impressive cache site of prey at the beginning of the breeding season. Both sexes build the nest, or both gather materials and the female primarily builds, in 6–11 days. Eggs are oval to subelliptical, smooth and nonglossy, dull white to grayish buff, and marked with speckles, spots, and blotches of browns, grays, purples (mostly on larger end, occasionally on smaller end); 24–26 mm x 19 mm (Northern the larger). Incubation period is 14–16 days (F, fed by male), NP 17–21 days (F, M). Nest reuse (relining or building atop) occurs among Loggerhead Shrikes but is uncommon for Northern Shrike, though both species will reuse materials from previous nests.

LOGGERHEAD SHRIKE *Lanius ludovicianus*

HABITAT: Open country, often with short vegetation: pasture-lands, old orchards, agricultural fields, roadside thickets, areas of scattered trees and bushes. **LOCATION AND STRUCTURE:** Often in a tree or large shrub with dense foliage; thorny species preferred. May be in mistletoe or, when trees or shrubs are lacking, in brush piles, tumbleweeds, etc. Well concealed out on a fork in branch, or in crotch, commonly between 0.9–3 m, height dependent on size of trees/shrubs. Sometimes uses nests of thrashers. Outside diameter 15–25 cm; height 11.5 cm; inside diameter 7.6 cm; depth 6.4 cm. *Nest is similar to that of LeConte's Thrasher but smaller.*

Sandra Young, NM

Sandra Young, NM

NORTHERN SHRIKE *Lanius excubitor*

HABITAT: Taiga, tundra, open landscapes with suitable, scattered trees, shrubs. Scrublands. Open deciduous and coniferous woodlands; avoids dense coniferous forest; found at clearings, openings, lakes, muskegs, etc. **LOCATION AND STRUCTURE:** Often in a bush in dense stands, thickets, low in bushes and small trees, but also higher in larger trees, deciduous and coniferous. Commonly at 0.6–4.9 m, occasionally much higher. Nest is large for bird's size, often with insulative materials worked into lattice of outer twig shell, a compact middle layer of grasses and moss, and a deep, well-insulated cup (which may conceal all but the tail tip of incubating female) of rootlets, hair, fur, feathers. Can be similar in size or considerably larger than Loggerhead Shrike nests. Largest nests: outside diameter 40 cm; height 25 cm; inside diameter 12 cm; depth 15 cm. **EGGS:** 5–7; *more heavily marked than Loggerhead Shrike's.*

Chris Smith, AK

VIREOS (Vireonidae)

Twelve species of these small insectivorous songbirds breed in N. America. Yellow-green Vireo *(Vireo flavoviridis)*, a rare breeder in extreme s. TX, is not included below. Typical nests are easily recognizable, pensile open cups constructed from various materials, including plant fibers, grasses, bark strips, lichen, moss, spider egg cases, cocoons, and leaves. Materials are bound with spider and caterpillar silk, and the nest is suspended by its rim to a horizontal twig fork. Nests are lined with fine grasses, animal hair, plant fibers, rootlets, feathers, bark fibers, and similar materials.

Vireos can be loosely divided into 2 groups based on breeding-habitat preference. *Thicket or shrubland nesters* nest in shrubby growth and low trees. *Woodland nesters* are often more arboreal in deciduous and mixed forests, and generally nest higher up. Where species overlap in range, their differences in habitat preference, as well as foraging behavior, may minimize competition.

Vireos are generally monogamous, and pair bonds may last for 1 season or several, depending on the species. They are single- or double-brooded, with clutch sizes of 3–5. Eggs are smooth, generally nonglossy, subelliptical to oval, and white. Depending on the species, they may be unmarked or lightly speckled and spotted with browns, blacks, or reddish browns, mostly at the larger end. Both sexes incubate, for 12–17 days, varying by species. Altricial young fledge in 10–12 days but may continue to be fed by parents up to 30 days or more. Vireos are frequent hosts to parasitic cowbirds, and some species have seen significant population declines as a result.

THICKET OR SHRUBLAND NESTERS

BLACK-CAPPED VIREO *Vireo atricapilla*

HABITAT: Areas dominated by dense, deciduous scrub (especially oak) with openings and gaps between low thickets and brushy clusters: dry eroded hillsides, rocky ravines, regenerating burned areas, similar. **LOCATION AND STRUCTURE:** Usually 0.5–2 m up and concealed from top and sides. Pensile nest is placed on bottom half of shrub (often oak), attached to a horizontal fork in inner portions of a bush or interior opening in a cluster of bushes. Sometimes places nest on outer margins of shrub, rarely in a small tree or vine tangle. *Indistinguishable from nest of Bell's Vireo.* Outside diameter 5.5–5.9 cm; height 5.6–6.2 cm; depth 3.7–3.9 cm. **EGGS:** 3–4; white, unmarked; 18 x 13 mm; IP 14–17 days (F, M); NP 9–12 days (F, M). **BEHAVIOR:** Occasionally polygynous and polyandrous. *May occupy more open areas than White-eyed Vireo where ranges overlap.* Unmated males build small foundational platforms in nesting sites, or "bachelor pads," which a female may choose to build nest on.

WHITE-EYED VIREO *Vireo griseus*

HABITAT: Various areas with deciduous scrubby growth, including riparian thickets, reclaimed strip mines, wooded edges, shrubby abandoned pastures; second-growth woodlands; also mangroves (coastal FL). *Where range overlaps with Bell's Vireo, typically favors habitats in later stages of development.* **LOCATION AND STRUCTURE:** Nest is often located close to edge of territory. Usually well concealed by dense surrounding vegetation. In a shrub, sapling, or other substrate, 0.3–1.8 m up. Pensile nest is attached to a twig fork, *narrowing at bottom.* Made of leaves, soft wood, bark shreds, plant fibers, bits of wasp nest, lichen, moss, bound with silk. Lined with fine grasses, hair, rootlets. Outside diameter 6.3–8.9 cm; height 6.9–14.6 cm; inside diameter 4.4–6.4 cm; depth 3.8–7.3 cm. **EGGS:** 4; white, sparingly marked; 19 x 14 mm; IP 13–15 days (F, M); NP 9–11 days (F, M). **BEHAVIOR:** Seasonally monogamous. Female chooses nest site, both sexes build, female lines nest. Construction takes 3–5 days.

White eyed Vireo. Casey McFarland, AL   *White eyed Vireo.* Eric C. Soehren, AL

BELL'S VIREO   *Vireo bellii*

HABITAT: Commonly near water in areas with thick, low shrubby growth: riparian scrub, shrubby fields, scrub oak, wooded edges, early or intermediate-stage successional woodlands, coastal chaparral, scrubby ravines and gullies, mesquite brushlands. Typically avoids areas of intensive farming, prairies, or open scrubby deserts. **LOCATION AND STRUCTURE:** In a small tree or shrub, occasionally herbage. Commonly within 1 m of ground, sometimes higher. *Usually located in small openings of vegetation, near outer margins of plant.* Nest is a pensile cup with rim firmly attached to a twig fork. Made of stems, grasses, small dead leaves, paper, bark strips, plant down, feathers. Spider silk and cocoons are woven into nest and outer layer, sometimes plastic or debris. Lined with fine grass stems. *Indistinguishable from nest of Black-capped Vireo.* Outside diameter 7 cm; height 7–9.9 cm; inside diameter 4.5 cm; depth 4–5 cm. **EGGS:** 4; white, sparingly marked or occasionally unmarked; 17 x 13 mm; IP 14 days (F, M); NP 10–12 days (F, M). **BEHAVIOR:** Generally monogamous. Both sexes build, sometimes just female, usually in 4–6 days. Population of endangered Least Bell's Vireo (CA and Baja CA) continues to decline owing to riparian habitat loss and cowbird parasitism.

Matt Monjello, AZ   Heidi Blankenship, AZ

GRAY VIREO *Vireo vicinior*

HABITAT: Dry shrubby montane slopes, scrubland, pinyon-juniper, oak-juniper scrub and similar associations, dwarf conifer woods. From 914 to 2,380 m elevation. *Where ranges overlap, observed occupying lower elevations than Plumbeous and Hutton's Vireos and higher elevations than Bell's Vireo.* **LOCATION AND STRUCTURE:** Typically low (1.3–3.4 m up) in a short shrub or tree, commonly on west- or north-facing side of substrate, concealed by surrounding vegetation. Nest is a pensile cup suspended from a fork, made of bark shreds, plant fiber, leaves, spider silk and cocoons, outside layer occasionally decorated with sagebrush or other whole leaves. Sometimes materials are loosely woven. Lined with fine grass. Outside diameter 5–8.5 cm; height 4.5–7 cm; inside diameter 3.5–7 cm; depth 2.8–4.8 cm. **EGGS:** 3–4; white, sparingly marked; 18 x 13 mm; IP 12–14 days (F, M); NP 13–14 days (F, M). **BEHAVIOR:** Generally monogamous. Both sexes build.

Sandra Young, NM

Sandra Young, NM

WOODLAND NESTERS

BLACK-WHISKERED VIREO *Vireo altiloquus*

HABITAT: Coastal FL mangrove swamps; subtropical hardwood slopes. **LOCATION AND STRUCTURE:** Nest is 1.5–4.6 m aboveground in a mangrove or deciduous tree or shrub, on horizontal branch. Nest is a deep, pensile cup with typical vireo nest material, occasionally with seaweed or other debris. Outside diameter 6–8.8 cm; height 5.3–7 cm; inside diameter 3.8–6.5 cm; depth 3.6–4.9 cm. **EGGS:** 3; white, sparingly marked; 21 x 15 mm; IP 14 days (F); NP likely similar to that of other vireos (F, M). **BEHAVIOR:** Socially monogamous. Female builds in ~6 days.

YELLOW-THROATED VIREO *Viero flavifrons*

HABITAT: Mature deciduous and mixed deciduous-coniferous woodlands. Typically along edges, including streams, lakes, wetlands, rivers, roadsides. Also in parks, orchards, and wooded residential areas that contain tall deciduous trees. Occasionally in interior forest openings. Avoids coniferous woodlands. From sea level to 1,100 m elevation but typically below 900 m (South) or 600 m (Northeast). *More common along edges and in more open habitats than Red-eyed Vireo. More common in forests lacking understory than Blue-headed Vireo.* **LOCATION AND STRUCTURE:** Suspended in a fork of branch, usually located close to trunk in upper portion of a large deciduous tree (often pine in South) and 6–15 m up, sometimes 1–24 m. Nest is a deep, pensile, rounded cup of grasses, bark strips, plant down, hornet paper, etc., lined with fine grasses, rootlets, pine needles, bark strips. Decorated with moss, lichens, spider egg cases, birch bark. Outside diameter 6–8.5 cm; height 5.5–7 cm; inside diameter 4–7 cm; depth 4–5 cm. **EGGS:** 4; white or pinkish white, sparingly marked; 21 x 15 mm; IP 13 days (F, M); NP 13 days (F, M). **BEHAVIOR:** Monogamous. Both sexes build in 8 days.

Laurie Pocher, LP3
Photography, ME

WARBLING VIREO *Vireo gilvus*

HABITAT: Prefers mature, mixed-deciduous forests with tall trees, moderately open canopy, with or without thick undergrowth. Often near water (streams, lakes, wetlands) but may be found in drier uplands. Also parks, gardens, brushy fence lines, deciduous groves in pine woods, deciduous grow-back in logged areas. Infrequently in conifer woodlands. From sea level to 3,200 m elevation. **LOCATION AND STRUCTURE:** On horizontal fork

Warbling Vireo. Mary Kiesau,
Mountain Kind Photography, WA

Warbling Vireo. Mark Nyhof, BC

of a tree or shrub, usually well away from trunk or main stem, 1–37 m aboveground. Nest is a pensile suspended cup of grasses, plant fibers, leaves, bark strips, plant down, rootlets, hair, fine twigs, lichens, rarely feathers, bound and fastened with spider silk. Lined with fine grass stems, pine needles, leaves, fibers, rootlets. Outside diameter 6.5–8 cm; height 5–7 cm; inside diameter 4.5–5 cm; depth 3.5–4.5 cm. **EGGS:** 3–4; white, sparingly marked; 19 x 14 mm. **BEHAVIOR:** Monogamous. Female constructs most of nest, usually within 7 days.

HUTTON'S VIREO *Vireo huttoni*

HABITAT: Prefers habitats with medium to heavy canopy cover and understory: oak and mixed oak-coniferous woodlands, especially habitats with evergreen oaks. Also tall chaparral or pine woodland, temperate coastal coniferous and mixed forests, in riparian woodlands of montane canyons. **LOCATION AND STRUCTURE:** Commonly built near but not on edge of forest clearings or openings. Occasionally over streams. Also along roadsides or in cultivated areas. In a deciduous tree or conifer, usually well concealed, anchored to a horizontal fork near branch tips; occasion-

Matt Monjello, AZ

ally suspended from foliage; infrequently in mistletoe. Commonly 0.9–13.7 m aboveground. Nest is a deep, round, pensile cup of *Usnea* and other lichens, mosses, plant or tree down, other typical vireo nest material, bound with spider silk. Lined with fine grasses, occasionally hair. Outside diameter 7.6–8.3 cm; height 7 cm; inside diameter 4.8–6 cm; depth 4–4.5 cm. **EGGS:** 4; white, sparingly marked; 18 x 13 mm; IP 14–16 days (F, M); NP 14 days (F, M). **BEHAVIOR:** Monogamous. Both sexes build in 3–10 days.

PHILADELPHIA VIREO   *Vireo philadelphicus*

HABITAT: Prefers poplars, willow, ash, and alder. Often along edges of young to second-growth deciduous and mixed woodlands. Also regenerating burns and cut-over areas, aspen groves, cherry coppices, willow and alder riparian thickets. Will occupy mixed coniferous and mature forests that contain preferred tree species. Usually absent in residential areas. **LOCATION AND STRUCTURE:** In a deciduous tree, close to trunk and often high up in canopy, 7.5–20 m aboveground. Sometimes much lower (2.5 m) when near water. Nest is a pensile cup, suspended from a horizontal twig fork. Made of grass blades, birch bark strips, willow seed tufts, feathers, twine, *Usnea*. Lined with pine needles, grass. Outside diameter 6.5–8 cm; height 5.5–10 cm; inside diameter 4.5–5 cm; depth 3.5–6 cm. **EGGS:** 4; white, sparingly marked; 19 x 14 mm; IP 11–13 days (F, M); NP 14 days (F, M). **BEHAVIOR:** Monogamous. Female typically builds in 6 days.

RED-EYED VIREO   *Vireo olivaceus*

HABITAT: Deciduous and mixed deciduous-coniferous forests with an understory of shrubs and saplings. Also occupies alder thickets and aspen stands (northern population), deciduous-dominated streams of southeastern pine forests. Occasionally near human habitations in areas that contain large trees. More abundant in interiors than along forest edges. **LOCATION AND STRUCTURE:** Usually in a deciduous tree or shrub and 2.5–4.3 m aboveground, sometimes 0.4–19 m. Nest is a round, pensile cup suspended in a twig fork. Made of grasses, bark strips, rootlets, vine tendrils, wasp-nest paper, spider egg cases, pine needles, lichen. Lined with grasses, pine needles, plant fibers, occasionally hair. Outside diameter 6.8–7.6 cm; height 5.4–8 cm; inside

Matt Monjello, NH   Richard Besser, ON

diameter 5.4–5.9 cm; depth 3–5.4 cm. **EGGS:** 3; white, sparingly marked; 20 x 14 mm; IP 11–14 days (F); NP 10–12 days (F, M). **BEHAVIOR:** Monogamous. Female builds in 4–5 days.

BLUE-HEADED VIREO *Vireo solitarius*

PLUMBEOUS VIREO *Vireo plumbeous*

CASSIN'S VIREO *Vireo cassinii*

HABITAT: Blue-headed Vireo: Typically in coniferous and mixed coniferous-deciduous woods, preferring large tracts of intermediate to mature forests with heavy canopy cover, moderate shrub or sapling understory, near openings of wetlands and other water. Plumbeous Vireo: Occupies dry montane coniferous or mixed coniferous-deciduous forests, mountainous riparian woodlands, oak shrublands. Commonly at 920–2,500 m elevation, *often higher than Gray Vireo.* Cassin's Vireo: Prefers dry, open habitats: oak woodlands, coniferous or mixed coniferous-deciduous forest, deciduous riparian woods. *More common in deciduous habitats than Blue-headed and Plumbeous.* **LOCATION AND STRUCTURE:** Typical vireo nest but may appear more loosely woven and bulkier than those of other species. Suspended from a horizontal fork by nest rim. Typical vireo nest materials, including spider egg sacks, cocoons, wasp nest paper, lined with grasses and rootlets. Plumbeous often includes catkins, sometimes plastic. Blue-headed commonly adds birch bark, only occasionally lichen. **EGGS:** 4; white, sparsely marked; 20 x 14 mm; IP 13–14 days (F, M); NP 13–14 days (F, M).

Plumbeous Vireo. Casey McFarland, NM

Plumbeous Vireo. Casey McFarland, NM

Cassin's Vireo.
David Moskowitz, ID

JAYS, CROWS, RAVENS, and ALLIES (Corvidae)

Medium-sized to large perching birds, corvids nest in trees and bushes, on cliff ledges, in artificial structures, and among rocks on the ground. Sixteen species breed commonly in N. America. Two additional species, Island Scrub-Jay and Brown Jay, are extremely localized; Brown Jay *(Psilorhinus morio)* is not included below.

Most corvids build attractive, well-constructed, open-cup nests with sticks and twigs, with some species including mud in the outer structural layer and in the base. They are commonly lined with a distinct, thick inner cup of soft and fine materials. Magpie nests differ from those of other corvids in being *globular, with a domed canopy* of sticks covering the inner bowl. Both sexes of corvids build, but females generally do most of the work. Some species have helpers that contribute to nest construction and other parental duties.

Corvids are mostly single-brooded, although some species may renest following a failed first attempt. They lay 3–8 short subelliptical to long subelliptical smooth and slightly glossy eggs. Eggs are generally bluish green with variable markings of green, gray, brown, and olive. Females are typically the sole incubators (some incubation by males is reported in Steller's Jay and possibly among crows and ravens). Clark's Nutcracker is unique in that males have an equally well developed brood patch and share in incubating. Depending on the species, young hatch in 16–21 days. They are cared for by both parents, and in some cooperative-breeding species, helpers aid in rearing young.

Left to right: Woodhouse's Scrub-Jay, Yellow-billed Magpie, American Crow, Common Raven.

JAYS

Jay nests are smaller than those of ravens and crows, and vary widely in design from a simple structure of twigs and rootlets to large, robust cups laden with mud.

CANADA JAY *Perisoreus canadensis*

HABITAT: Coniferous and mixed coniferous-deciduous forests, often with spruce trees. Also in subalpine fir-hemlock-cedar forests. **LOCATION AND STRUCTURE:** In a dense conifer, often next to trunk, sometimes exposed on a horizontal branch; 1.2–9 m up. *Often on northern edge of forest opening.* Nest is bulky and deeply cupped, made of twigs (often spruce, tamarack), caterpillar

Canada Jay. George Peck, ON

Canada Jay. Brett Forsyth
Photography, www.brettforsyth.com, ON

cocoons, lichens, bark strips. Lined with feathers, fur, and down. *Nest bowl becomes shallower as nestlings develop.* Before hatching: outside diameter 14–16 cm; height 10–15 cm; inside diameter 6.7–8.8 cm; depth 4.9–6.3. **EGGS:** 3–4; light greenish white or gray, finely speckled, spotted, and blotched with brown, dark olive, reddish brown; 29 x 21 mm; IP 18–22 days (F); NP 22–24 (F, M). **BEHAVIOR:** Assumed to be monogamous. Single-brooded. Stays in territory all year and begins breeding in late winter.

PINYON JAY *Gymnorhinus cyanocephalus*

HABITAT: Commonly on hilly terrain in pinyon-juniper woodlands, but also occupies sagebrush, scrublands, chaparral, ponderosa or Jeffrey pine forest, pine-oak woodland. Tends toward mid-elevation habitat. **LOCATION AND STRUCTURE:** Often in a juniper or pine, 0.9–6 m aboveground and up to 35 m high in taller pine forests. Nest is a bulky cup with a foundation of twigs and an inner cup of coarse grasses, shredded bark. Lined with finer

Elisabeth Ammon,
Great Basin Bird
Observatory, NV

bark shreds, feathers, hair, fine rootlets, plant fibers. Outside diameter ~28 cm; height 17 cm; inside diameter 10 cm; depth 8 cm. **EGGS:** 3–4; bluish white to light blue, blotched and speckled with dark brown to reddish brown, often concentrated at larger end; 29 x 22 mm; IP 16–17 days (F); NP 21 days (F, M). **BEHAVIOR:** Monogamous. Breeds in colonies of ~30 up to 300 individuals, often starting in late winter. Young may form creches. Helpers, typically male offspring from previous year, may assist with tending nestlings but do not participate in nest construction or incubation.

STELLER'S JAY *Cyanocitta stelleri*

HABITAT: Coniferous and various mixed coniferous-deciduous forest types, neighborhoods, parks, and orchards. Common along edges of fragmented and altered habitats. From 1,000 to 3,500 m elevation in interior range and from sea level to tree line along West Coast. **LOCATION AND STRUCTURE:** Nest is in a tree, often a conifer, occasionally a deciduous tree or shrub. Sometimes on a building. Commonly near the trunk in top portion of tree, on horizontal branches. Usually 3–5 m aboveground, but may be ground level or 30 m up. Nest is a bulky cup of twigs, dry leaves, moss, paper, sometimes filled with mud. Lined with rootlets, pine needles, hair, grasses. Outside diameter 25–43 cm; height 15–18 cm; inside diameter 11–13 cm; depth 6–9 cm. **EGGS:** 4–5; bluish green, irregularly marked with fine brown, olive spots; 31 x 22 mm; IP 16 days (F); NP 16 days (F, M). **BEHAVIOR:** Socially monogamous. Quiet and secretive near nest unless flushed. Juveniles may stay with parents as a family group into fall or winter.

Ian C. Tait, CA

Quinn Bailey, WA

BLUE JAY *Cyanocitta cristata*

HABITAT: Wide variety of deciduous, coniferous, mixed wood-lands. Also in wooded residential areas, parklands. Often found along woodland edges, wooded riparian buffers, power-line cuts. Less abundant in dense forests. **LOCATION AND STRUCTURE:** In a coniferous or deciduous tree. Nest is placed in vertical crotch of trunk or on outer fork of a large horizontal branch, 1–30 m high. Occasionally in ivy growing on tree, rarely on building or in shrub. Nest is often a conical, bulky cup of twigs, bark strips, moss, grass, leaves, paper, other debris, sometimes with mud included. Lined with rootlets, occasionally leaf mold. Outside diameter 17–21 cm; height 10–12 cm; inside diameter 8.5–10.5 cm; depth 6 cm. **EGGS:** 4–5; color variable, in shades of blue, greenish, buff, olive, light brown, marked with shades of brown, red, gray, usually concentrated on larger end; 28 x 20 mm; IP 17–18 days (F); NP 17–21 days (F, M). **BEHAVIOR:** Socially monogamous. Often very secretive around nest site.

Matt Monjello, ME

David Pierce, MO

FLORIDA SCRUB-JAY *Aphelocoma coerulescens*

HABITAT: Oak scrub habitat with sparse ground cover, sandy soil, and very few pines. **LOCATION AND STRUCTURE:** Near center of dense-sand live oak or other shrub, often with vine as cover, occasionally in citrus groves adjacent to oak scrub. Usually 0.5–2.5 m aboveground, occasionally higher. Nest is a bulky cup with outer layer of coarse twigs, inner cup of smaller twigs and root-

Gerrit Vyn, FL

David L. Sherer, FL

lets, lined with palmetto fibers or fine rootlets. **EGGS:** 3–4; pea green to pale glaucous green, irregularly spotted and blotched with rusty reds; IP usually 18 days (F); NP 18 days (F, M). **BEHAVIOR:** Monogamous, rarely polygynous. Cooperative breeders with 1 breeding pair and up to 6 helpers.

ISLAND SCRUB-JAY *Aphelocoma insularis*

Restricted to Santa Cruz I. off coast of s. CA, in chaparral, scrub oak, and woodlands. Nest and eggs similar to those of other scrub-jays.

CALIFORNIA SCRUB-JAY *Aphelocoma californica*

HABITAT: Scrub oak, chaparral, and oak-pine woodlands. Also riparian areas, gardens, orchards, residential areas, parks, golf courses. Unusual habitats include pure mountain mahogany, montane pine-spruce, mangrove swamps, and stands of single-leaf pinyon pine. **LOCATION AND STRUCTURE:** Well concealed in a tree, shrub, or vine. Common substrates include oak, sumac, bay, elderberry, poison oak. Placement is highly variable, usually 2–4 m aboveground, occasionally up to 15 m. *Nest is similar to that of Woodhouse's Scrub-Jay.* **EGGS:** Similar to those of Woodhouse's Scrub-Jay; 28 x 20 mm; IP 17–18 days (F); NP 18–22 days (F, M). **BEHAVIOR:** Socially monogamous. Pairs typically stay in territory all year.

Damon Calderwood, OR

WOODHOUSE'S SCRUB-JAY *Aphelocoma woodhouseii*

HABITAT: Various pinyon, oak, and juniper forest types. Less commonly, and in certain parts of range, in riparian woodlands, cactus forests, scrublands, orchards and gardens, scrubby edge of tropical forest. *Typically in lower-elevation and more arid and open habitats than Pinyon and Steller's Jays.* **LOCATION AND STRUCTURE:** Well concealed in a tree or shrub. May be in a branch or trunk fork, on a terminal or horizontal branch. Usually low, 1.2–2.6 m up. Nest is an open basket-shaped cup with outer layer of twigs, inner cup of rootlets or finer twigs, weed stems. Lined with fine

*Woodhouse's
Scrub-jay.*
Casey McFarland,
NM

rootlets and hair. In NM nests, inner cup and lining was commonly made only of fine rootlets. Outside diameter 10 cm; height 10 cm; inside diameter 9.5 cm; depth 6 cm. **EGGS:** 3–5; bluish green, blue, pea green, light green, spotted, speckled, sometimes blotched with olive or brown; sometimes eggs are light gray or green with reddish marks; 28 x 20 mm; IP 17–18 days (F); NP 18–22 days (F, M). **BEHAVIOR:** Assumed to be socially monogamous.

MEXICAN JAY *Aphelocoma wollweberi*

HABITAT: Oak, pine, and juniper woodlands, and oak-dominated riparian areas. **LOCATION AND STRUCTURE:** Close to main trunk or in fork of a horizontal branch of tree; commonly oak, juniper, or pine. Usually well concealed. Height variable, often in top half of tree; documented at 3.7–24.2 m up. Nest is an open cup of sticks and twigs. Lined with fine rootlets, plant fibers (yucca), possibly hair. **EGGS:** 4; pale greenish blue to pale green, speckled and blotched with light browns, frequently unmarked in AZ subspecies, occasionally in TX subspecies; IP 18 days (F); NP 25–28 days (F, M). **BEHAVIOR:** Commonly monogamous, but polygynandry occurs. Stays in flocks and defends permanent group territory. One territory may hold multiple active nests, and nests may contain eggs fertilized by different males.

GREEN JAY *Cyanocorax yncas*

HABITAT: Mainly in open woodlands and in dense thickets with mesquite, acacia, and ebony. Also citrus orchards. **LOCATION AND STRUCTURE:** Well hidden on horizontal branches in a dense thicket, tree, or shrub, 0.9–7.6 m aboveground. Nest is a loosely woven cup of sticks and thorny twigs so thinly constructed that eggs are often visible from below. Lined with fine rootlets, vine stems and fibers, moss, dry grasses, sometimes leaves. **EGGS:** 4; off-white or light greenish white, spotted with shades of brown, gray, and lavender, usually concentrated at larger end; 27 x 20 mm; IP 17–18 days (F, M); NP 16–18 days (F, M). **BEHAVIOR:** Monogamous. Permanent resident of territory. Offspring of previous year aid in territorial defense during egg and nestling stages, but then leave flock.

CLARK'S NUTCRACKER *Nucifraga columbiana*

HABITAT: Open to semi-open montane pine forests, including ponderosa pine, pinyon/juniper, Jeffery pine, Dougals-fir, white-bark pine, limber pine, subalpine fir, Engelmann spruce. Also mixed-coniferous subalpine forests. **LOCATION AND STRUCTURE:** 1.8–24 m up in live, dead, or dying conifers. A large, deep nest with woven platform of twigs, outer cup of wood pulp, bark fibers, and a mud or soil layer in base of cup. Lined with dried grasses, bark strips, sometimes moss, hair. Outside diameter 20–33 cm; height 10–23 cm; inside diameter 10.2–11.4 cm; depth 6.4–8.9 cm. **EGGS:** 3; pale green, grayish green, or greenish white, sparsely speckled and spotted with brown, olive, gray; 32 x 23 mm; IP 16–18 days (F, M); NP ~20 days. (F, M). **BEHAVIOR:** Apparently monogamous. Will re-nest up to twice after failed attempts. Breeding behavior may begin in late winter, but common flurry of nesting activity begins in March. Nesting sites are located around previous year's seed stores.

Gerrit Vyn, WY

Taza Schaming, WY

MAGPIES

Magpies build large, recognizable nests, usually identifiable from afar. Their large *domed or globular structures* are more often mistaken for nests of woodrats or squirrels than those of other birds. The nest is typically a large, loose, *spherical mass of sticks and twigs with a prominent canopy* that houses an inner bowl of mud or dung. The cup is lined with rootlets, thin grass and weed stems, and possibly hair. Nests may be anchored to supporting substrate with mud. Eggs are subelliptical or occasionally long oval (Black-billed Magpie), greenish gray or olive brown, and profusely spotted and speckled (blotching possible) with dark browns, sometimes concentrated at larger end.

BLACK-BILLED MAGPIE *Pica hudsonia*

HABITAT: Riparian groves in open brush-covered country: rangeland, cultivated fields, meadows, open woodlands and suburbs. Avoids large tracts of contiguous forests and very arid habitats.
LOCATION AND STRUCTURE: Nest location is variable. Usually on

Black-billed Magpie. Note the spherical appearance. David Moskowitz, WA

This Black-billed Magpie nest entrance reveals the well-built inner cup.
Neal Wight, CO

Black-billed Magpie, close-up. Note the heavy use of mud.
David Moskowitz, WA

branches of a tall thorny bush, sometimes in trees and on power lines. Usually 4.6–9 m aboveground, with range of 1.5–18 m. Size varies significantly. Outside diameter 50 cm; height commonly ~50–75 cm; inside diameter 17 cm; depth 9 cm. **EGGS:** 4–9; IP 16–21 days (F); NP usually ~27 days (F, M). **BEHAVIOR:** Monogamous. Often in small colonies. May reuse nests, with new nests being built on top of old ones.

YELLOW-BILLED MAGPIE *Pica nuttalli*

HABITAT: Open oak savanna in valleys and foothills, in streamside groves near open areas, including cultivated sites. **LOCATION AND STRUCTURE:** Usually placed high up (average 14 m) and in outer margin of branch in tall tree (especially oak, sycamore, cottonwood), possibly as predator deterrence. Often built in mistletoe clumps. *Larger than nest of Black-billed Magpie, with outside diameter of ~1 m.* **EGGS:** 6–7; 33 x 23 mm; IP 16–18 days (F). **BEHAVIOR:** Socially monogamous. Nests in small colonies, but infrequently in same tree. Young may form creches days after leaving nest.

CROWS

Crows make large, attractive nests with well-made inner cups with soft linings. Nest is a large basket-shaped cup of dead sticks, twigs, weed stems, bark strips, and mud or soils; lined with finer material, including plant fibers, bark fibers (commonly cedar), grasses, mosses, pine needles, hair, rootlets, debris. Both sexes build in 9–14 days, sometimes longer. Crows are monogamous, with long-term pair bonds, or presumed so (Fish Crow); extra-pair copulations are possible.

AMERICAN CROW *Corvus brachyrhynchos*

HABITAT: Broad range of both natural and altered habitats. Prime sites include edge habitats along woodlands, rivers, creeks and marshes, farmlands, agricultural fields, pastures, city parks and neighborhoods, golf courses, campgrounds, garbage dumps, clearings, and orchards. Typically absent in large tracts of forest and areas that lack suitable perches. **LOCATION AND STRUCTURE:** In coniferous or deciduous trees and shrubs, generally located

Matt Monjello, ME

Matt Monjello, ME

near upper portion of substrate, close to the trunk, in a crotch or on horizontal branch, 3–21 m up; rarely on ground or artificial structure. Nest may include more mud than Fish Crow nest. Outside diameter 35.6–76.2 cm; height 10–38 cm; inside diameter 16–36 cm; depth 10.2–20.3 cm. **EGGS:** 4–5; light bluish green or greenish olive, irregularly blotched with browns and grays; markings may be sparse or profuse, usually concentrated at larger end; 41 x 29 mm; IP 16–18 days (F); NP 20–36 days (F, M). **BEHAVIOR:** May have additional helpers.

NORTHWESTERN CROW *Corvus caurinus*

HABITAT: Often near coast: wooded shorelines, estuaries, bays, rivers, ponds, offshore islands, human settlements and campgrounds; occasionally inland along fields. From sea level to 1,700 m elevation. Avoids dense forests. **LOCATION AND STRUCTURE:** Usually in a fork of a tree, shrub, or blackberry thicket, sometimes on ground near trunk of woody substrate, in grass or weeds sheltered by cliffs, under fallen tree. Ground nests are built similarly but contain few sticks and twigs in outer layer. Outside diameter 24–45 cm; height 16–36 cm; inside diameter 14–20 cm; depth 7–11 cm. **EGGS:** 4; *similar to those of other crows*; 40 x 28 mm; IP 17–19 days (F); NP 29–35 days (F, M). **BEHAVIOR:** Nests in isolated pairs or in small groups. Helpers may assist in territory defense.

Bob Armstrong, AK

R. Wayne Campbell, BC

FISH CROW *Corvus ossifragus*

HABITAT: Coastal habitats: beaches, marshes, tidal rivers, estuaries, lakes. Also inland along agricultural fields, in urban and suburban settlements, city parks, golf courses. Population is increasing and breeding range is expanding north and inland. **LOCATION AND STRUCTURE:** Usually in a conifer, also in deciduous trees, palms, and mangroves. May nest in shrub. Nest is commonly in top portion of substrate, 1.5–24.4 m aboveground. Nest is a large, bulky open cup of sticks and twigs, with inner cup of mud and vegetation. Lined with bark strips, moss, plant fibers, hair, pine needles, paper, etc. **EGGS:** 4–5; *similar to those of Amer-*

ican Crow but smaller; 37 x 27 mm; IP 16–19 days (F); NP 32–40 days (F, M). **BEHAVIOR:** May nest in small groups and in heron rookeries.

RAVENS

Ravens build the largest cup nests of all N. American passerines. Both species produce a large nest of sticks and twigs, often with an inner cup and lining of a variety of materials, including fine twigs, fur, bark shreds, grasses, leaves, mud, wool, metal wires, debris, rags, feathers, rope, bark, plant fibers, moss, paper. Both species may reuse nests.

COMMON RAVEN *Corvus corax*

HABITAT: Wide variety: seacoasts, islands, prairies, deserts; boreal coniferous and deciduous woodlands, mountains, tundra; rural, urban, and suburban areas. *Generally in denser woodlands than Chihuahuan Raven.* **LOCATION AND STRUCTURE:** Variable. Can be found nesting on cliffs, quarries, trees, and artificial structures such as buildings, bridges, telephone poles. Cliff nests are commonly on ledges and under an overhang. Tree nests (often in conifers) are usually located in a crotch and close to canopy. Sticks in outer wall ~0.9 m long x 3–25 mm wide. Outside diameter 0.4–1.5 m; height 20–61 cm; inside diameter 22–30 cm; depth 13–15 cm. **EGGS:** 4–6; greenish, olive, light blue to blue, variably marked with blotches, spots, streaks, scrawls of dark brown, olive brown, dull brown, occasionally without markings; 50 x 33 mm; IP 20–25 days (F); NP 35–42 days (F, M). **BEHAVIOR:** Assumed to be monogamous.

Casey McFarland, ID

This nest is heavily lined with elk hair.
Casey McFarland, WY

CHIHUAHUAN RAVEN *Corvus cryptoleucus*

HABITAT: Dry grasslands with scattered shrub and trees such as mesquite, yucca, acacia. Also scrublands dominated by mesquite, creosote bush, or shinnery oak, in open and hilly pinyon-juniper forests, riparian woodland edges, short-grass prairies. **LOCATION AND STRUCTURE:** In a tree (commonly mesquite), large yucca, shrub, or artificial structures such as buildings, utility poles, windmills. Tree nests are often 3–5 m aboveground; may be much higher on artificial sites. Outer wall is made of thin sticks, thorny twigs, often includes metal wires. Outside diameter 31–62 cm; height 30–36 cm; inside diameter 15–21 cm; depth 13–15 cm. **EGGS:** 5–7; *similar to those of Common Raven*; 44 x 30 mm; IP 18–22 days (F, M); NP 30 days (F, M). **BEHAVIOR:** Assumed to be socially monogamous. Occasionally nests in loose colonies.

LARKS (Alaudidae)

HORNED LARK   *Eremophila alpestris*
HABITAT: Wide ranging. Grassland and other open country from desert to tundra, from sea level to 3,962 m elevation. Does well in heavily grazed areas. **LOCATION AND STRUCTURE:** Prefers bare ground, typically with a protective clump of grass or similar structure on at least one side of nest. Where alpine habitat is shared with American Pipit, *Horned Lark often uses drier sites.* Commonly on north or northeastern side of vegetative clump. Loose thick cup, often shallow, is built into a natural or excavated depression. Outer layer is constructed of grasses and other vegetation, lined with plant down, fur, feathers, and fine plant material. Often collects items, "pavings," such as dirt clods, pebbles, or dung, and places them over excavated soil around nest. Outside diameter 8–10 cm; inside diameter 7 cm; depth 4.5 cm. **EGGS:** 4; cream colored with heavy spotting; ovate; 22 x16 mm; IP 11days (F); NP 9–12 days. **BEHAVIOR:** Monogamous.

Note the dirt-clod pavings.
R. Adam Martin, WA

Note the pebble pavings.
R. Adam Martin, WA

SWALLOWS (Hirundinidae)

These well-known aerial specialists are common in open areas (often with nearby water) with a rich supply of flying insects, including rivers, ponds, woodlands, farms, residential areas, and grasslands. The 8 species that breed regularly in N. America can be divided into 3 nesting strategies: *cavity and hole nesters* (some species also use burrows), *burrow nesters or excavators*, and *mud-nest builders*. Swallows nest abundantly in natural sites but frequently use a wide variety of human-made structures, such as rafters, eaves, ledges and walls beneath bridges, crevices in buildings, drainpipes, nest boxes, and similar settings.

All species are monogamous, and pairs often occupy the same nesting site year after year. The degree of coloniality during the breeding season varies among species; they may be highly gregarious, with breeding pairs numbering from hundreds to thousands, or nest in single pairs. Swallows may have up to 2 broods per season and, depending on species, usually lay 4–7 (4–5 is common) subelliptical to oval white eggs. Eggs of mud nesters may be finely speckled and spotted with varying shades of reddish or purple-brown. The extent of parental involvement in nest construction and incubation differs among species, but males help feed young.

CAVITY AND HOLE NESTERS

PURPLE MARTIN *Progne subis*

HABITAT: In open areas, often near water. Eastern populations formerly occupied woodpecker cavities in dead trees along wooded edges and areas adjacent to water, but prime habitat is now mostly near human settlements where birdhouses are provided. Western populations nest in natural crevices or cavities in montane and low-elevation habitats, including clearings and burned areas, aspen groves bordering wetlands, and saguaro deserts; rarely occupy birdhouses. **LOCATION AND STRUCTURE:** Artificial nests commonly include birdhouses, hollowed-out

Purple Martin nesting in an old woodpecker cavity.
David Moskowitz, MT

gourds; occasionally crevices in human-made structures. Natural sites include woodpecker holes in snags or on main stems of saguaro, crevices in rocky cliffs, and cavities in live trees. Nest is made of twigs, grasses, leaves, feathers, other plant material. Mud is added to base of cavity entrance. May add green leaves to inner cup through incubation. **EGGS:** 24 x 17 mm; IP 15–18 days (F); NP 27–36 days (F, M). **BEHAVIOR:** Western populations are largely solitary; eastern populations nest in colonies of varying size depending on site. Both sexes build.

TREE SWALLOW *Tachycineta bicolor*

HABITAT: Variety of open and semiopen sites, including fields, marshes, ponds, wooded swamps, and shorelines. Generally prefers areas close to water, but may breed a distance away if nesting sites and flying insects are present. **LOCATION AND STRUCTURE:** Favors open areas more than edges. Uses dead tree cavities, sapsucker holes in live trees, nest boxes, hollowed-out stumps; occasionally in rock and building crevices, fenceposts, old Cliff Swallow nests, or ground cavities. Nest is a cup of mostly dry grasses, occasionally pine needles, moss, rootlets, other material. Lined with feathers from other birds (quill ends positioned in cup so tips curl up and over eggs); material depth 2–8 cm. Inside diameter ~5 cm; depth 3 cm. **EGGS:** 17 x 14 mm; IP 11–20 days (F); NP 15–25 days (F, M). **BEHAVIOR:** Female mostly builds in a few days to 2 weeks. *Prefers south- or east-facing cavities early in season.* Nests both solitarily and semicolonially. May have helper.

Matt Monjello, NH

A full clutch. Note the heavy use of feathers in the lining, common among many swallow species. Mark Nyhof, BC

Note how feathers are placed so they over the nestlings in the nest. Matt Monjello, NH

VIOLET-GREEN SWALLOW *Tachycineta thalassina*

HABITAT: In open deciduous, coniferous, and mixed forests, but also in forested canyons, around human habitation and grasslands. From sea level to 3,500 m elevation, most commonly 1,500–3,000 m. Favors montane habitats in southern range, lower elevations in north. *Often in more open areas than Tree Swallow where range overlaps.* **LOCATION AND STRUCTURE:** In natural tree cavities and woodpecker holes, cavities in cliffs, tall cacti and dirt banks, building crevices, nest boxes; old Cliff or Bank Swallow nests. *Nest is similar to Tree Swallow's.* **EGGS:** 18 x 13 mm; IP 13–15 days (F); NP 23–24 days (F, M). **BEHAVIOR:** Both sexes build, mostly female in a few days. May nest solitarily, but common in small colonies.

BURROW NESTERS OR EXCAVATORS

NORTHERN ROUGH-WINGED SWALLOW
Stelgidopteryx serripennis

HABITAT: Across a wide range of open country. Often close to water, including exposed banks along streams and rivers, ravines, railroad banks, highway cuts, gravel pits; coastal cliffs. Also nests along dry gullies in arid regions. **LOCATION AND STRUCTURE:** Usually a burrow in a steep alluvial bank. May excavate, but commonly uses vacated burrows of other species (kingfishers, Bank Swallows, ground squirrels). Also occupies cavities in buildings, bridges, culverts, wharves, drainpipes, etc.; rarely tree holes. Nest is often well inside burrow, occasionally close to entrance or on ledge. Nest is a bulky foundation of twigs, weed stems, bark shreds, roots, other plant material. Nest cup is lined with fine grass, rootlets, green vegetation, occasionally manure, rarely feathers. **EGGS:** 6–7 (larger clutch than is typical for other swallows); 18 x 13 mm; IP 15–16 days (F); NP 17–21 days (F, M). **BEHAVIOR:** Nest is mostly built by female in 3–7 days (up to 20). Foundation material too bulky to fit into burrow is dropped at entrance. Nests solitarily or in small colonies. When nesting near Bank Swallows, often found at colony edges. May build atop old nests, including those of other species.

David Moskowitz, WA

BANK SWALLOW *Riparia riparia*

HABITAT: Common in steep, loose sediment banks along open, low-lying rivers, streams, lakes, wetlands, coasts; also in artificial sites, sand and gravel pits, road cuts. From sea level to 2,100 m elevation. **LOCATION AND STRUCTURE:** An excavated burrow and chamber, generally located near top of bank. Nest is typically a flattish structure of straw, grass stalks, leaves, rootlets, twigs; lined with feathers after eggs are laid. **EGGS:** 18 x 13 mm; IP 13–16 days (F, M); NP 18–24 days (F, M). **BEHAVIOR:** Dense colonies up to 1,500 pairs. Will excavate new tunnels each year; sometimes reuses existing burrows in good condition. Male does majority of excavation, female primarily building after cavity is completed. Excavation takes 4–5 days (up to 14), nest 1–3 days. Entire colony may synchronize breeding.

David Moskowitz, WA

Doug Backlund, SK

MUD-NEST BUILDERS

CLIFF SWALLOW *Petrochelidon pyrrhonota*

HABITAT: Once limited to western montane areas with steep cliff walls and overhangs, now widespread where it has taken advantage of artificial nest sites. Usually close to open areas near water with mud source, but may occur a good distance away from mud. **LOCATION AND STRUCTURE:** On cliff face, or along entrance in caves; under bridges; building and dam overhangs; culverts lacking vegetation. *Retort nests* are plastered at corner of vertical wall and overhang, but swallows will also build on other nests in colony. Completed nests are gourd-shaped. Made of mud pellets, with a globular chamber that narrows outward, forming a small, igloolike entrance tunnel, with opening projecting downward. Tunnel construction continues throughout season, sometimes becoming a long tube. Chamber is sparsely lined with grasses, dry stems. Feathers found in lining are from another bird previously occupying nest. Sometimes uses Bank Swallow burrows, crannies in cliffs. **EGGS:** 20 x 14 mm; IP 10–19 days (F, M); NP 20–26 days (F, M). **BEHAVIOR:** Dense colonies. Both sexes build. Mud gathering is generally synchronized. Commonly reuses nests, especially in smaller colonies. Will use nests of other swallows, Say's phoebe.

A collection of Cliff Swallow nests under a bridge.
David Moskowitz, ID

CAVE SWALLOW *Petrochelidon fulva*

HABITAT: Once limited to natural caves or sinkholes, now found under bridges, in culverts, buildings and similar structures. **LOCATION AND STRUCTURE:** In caves and sinkholes, nests are often found in dim lighting along crevices, in pockets. Under bridges and culverts, preferred sites have similar lighting conditions and are placed close to ceiling on vertical walls or atop ledges. Builds 2 distinct nest structures, varying by site: in natural sites, nests are commonly half-cup-shaped with a stout rim, made of mud pellets, guano. Used year after year, they become quite tall and large as material is added. Artificial-site nests are similarly shaped, and mud walls often extend over the cup, nearly enclosing it. Lined with dry grass, plant fiber, down, feathers, rarely debris. Occasionally unlined. **EGGS:** 20 x 14 mm; IP 15–18 days (F, M); NP 20–22 days (F, M). **BEHAVIOR:** Colonial, but nests are usually less clustered than those of Cliff Swallows. Both sexes build.

Jason Kleinert, TX

BARN SWALLOW *Hirundo rustica*

HABITAT: Wide variety. Common in open terrain near water with a mud source: farmlands, rural and suburban areas, marshes, lakes, along highways, etc. Sea level to 3,000 m elevation. **LOCATION AND STRUCTURE:** Former sites, now rare, were in caves, cliffs, and tree hollows. Now common on human-made structures that provide a vertical wall or overhang, with or without some underlying support, including barn rafters, open buildings, culverts, bridges, docks. Nest is an open, shallow cup of mud pellets, plastered to beams, eaves, or vertical wall, mixed with grass stems, similar. Lined with finer stems, horsehair, often large amounts of feathers. When built on support, nests may be fully cup-shaped. When attached to walls, nests are semicircular. In rock cavities, nests lack mud. **EGGS:** 20 x 14 mm; IP 13–15 days (F, M); NP 15–27 days (F, M). **BEHAVIOR:** Breeds singly or in small colonies. Both sexes build. May reuse and repair nest each season and for consecutive broods, and other swallows, phoebes, and House Wrens will use these nests.

Barn Swallow eggs in the nest, lined with feathers. Casey McFarland, OR

Barn Swallow nest built on a platform with no support against the back.
Casey McFarland, OR

Common appearance of a Barn Swallow nest attached to a vertical surface.
Bettina Arrigoni, TX

CHICKADEES and TITMICE (Paridae)

Chickadees and titmice are *cavity nesters*. Chickadees will readily excavate their own nest in soft rotting wood or may acquire an existing natural cavity or an abandoned woodpecker cavity. Titmice use only existing cavities, though they may refurbish or improve a natural cavity. The female incubates and is fed by the male, and both sexes feed and tend the young.

CHICKADEES

Chickadees are gregarious and well-known birds of forest, garden, and park. One or more of these 6 species can be found in every forested environment in N. America, and many of them readily use human-made resources of feeders and nest boxes. Chickadees most often nest in the soft wood of rotting trees and limbs in which they are able to excavate their own cavities or expand on naturally existing ones. They tend to carry excavated material away from the nest, as opposed to letting it fall to the ground directly below. They will use existing cavities if available, as well as nest boxes. Some species are more likely to use a nest box if it has been filled with shavings or sawdust, giving the birds something to excavate. When the cavity is in a tree limb rather than trunk, the entrance is often on the underside. Inside the cavity, nests are typically constructed with a base of coarse plant material, including moss, bark, wood chips, and an inner lining of mammal fur of various sorts. Entrance size varies depending on whether the chickadee excavated the cavity or is using a preexisting cavity, but it is typically just large enough to allow entrance.

Habitat can help distinguish species in the absence of the bird, but otherwise chickadee nest cavities are very hard to distinguish from each other. Species are divided below into two categories: *Deciduous- and Mixed-Forest Chickadees*, and *Coniferous-Forest Chickadees*. Black-capped Chickadees are the most abundant and widespread, and where they overlap with a variety of other species that prefer coniferous forest, they typically nest in deciduous stands.

Chickadees are primarily monogamous, and pair bonds can last for several years. In some species, flocks maintain territories throughout the year, and it is the dominant males and their mates that breed in that territory.

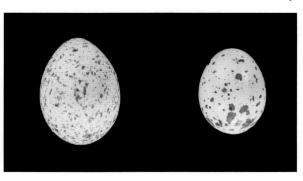

White-breasted Nuthatch (right) and Black-capped Chickadee (left). The eggs of some species in these families can look similar.

Males and females both participate in excavation, but females alone typically construct the inner nest. Second broods of the season are usually laid in a new nest cavity, but cavities may be reused again in subsequent years. Nest cavity and eggs of all species are very similar. Cavity diameter 3.6–5.5 cm; cavity height typically less than 4 m (up to 26 m reported). Most species lay 6–8 eggs (as few as 4 and as many as 12 are possible) that are rounded ovate (the small end can be quite pointed), commonly white with reddish dots concentrated at the larger end or distributed throughout. Eggs are typically ~15–16 mm x 12 mm. Incubation period varies by species, 11–15 days (F); NP 17–21 days.

DECIDUOUS- AND MIXED-FOREST CHICKADEES

CAROLINA CHICKADEE *Poecile carolinensis*

HABITAT: Variety of deciduous woodlands, sometimes pine or mixed pine-hardwood forests. Also swamps, savanna, and wooded urban and suburban landscapes. **LOCATION AND STRUCTURE:** Cavity is excavated in rotten wood or fencepost; natural cavity or nest box is often near forest edge. **EGGS:** 6. **BEHAVIOR:** Does not seem to have preference for nest boxes filled with shavings over those that are empty.

Matt Monjello, NC

Bill Summerour, AL

BLACK-CAPPED CHICKADEE *Poecile atricapillus*

HABITAT: Deciduous and mixed deciduous-coniferous forests. Also parks, suburban areas. **LOCATION AND STRUCTURE:** Excavates cavities in rotten wood of birch, aspen, or other softwood, or uses existing cavity in a variety of tree species. Will use nest boxes. **EGGS:** 6–8. **BEHAVIOR:** Prefers nest boxes filled with shavings, which it "excavates" prior to nesting.

Black-capped Chickadee eggs in the fine nest lining that is constructed inside the nest cavity. Jack Barthomai, WI

A Black-capped Chickadee peering out from a nest constructed in a natural cavity. Matt Monjello, AZ

CONIFEROUS-FOREST CHICKADEES

MOUNTAIN CHICKADEE *Poecile gambeli*

HABITAT: Montane coniferous forest, especially pine, spruce-fir, and pinyon-juniper. Also in mixed coniferous-deciduous and riparian woodlands, where it still is more associated with coniferous species. **LOCATION AND STRUCTURE:** Likely does little excavation; instead mostly uses preexisting cavities or those excavated by other species. Readily uses nest boxes. Wide variety of sizes for entrance hole but typically just large enough for bird to enter. Uses a significant amount of fur, including a cap or plug that covers eggs when they are not being incubated. **EGGS:** 6–12; *usually white, unmarked*, but can have markings similar to those of other species. Can be slightly larger than those of other species.

MEXICAN CHICKADEE *Poecile sclateri*

HABITAT: Montane coniferous forest in extremely limited range in U.S. **LOCATION AND STRUCTURE:** Nest similar to that of other chickadees, though cavities on underside of limbs are particularly common. Will use nest boxes. **EGGS:** 5–6.

CHESTNUT-BACKED CHICKADEE *Poecile rufescens*

HABITAT: Primarily associated with moist coniferous forests of Pacific Northwest coast and inland Northwest. Also urban and suburban settings, and expansion into forested habitats of Sierra Nevada. **LOCATION AND STRUCTURE:** If excavating cavity, exclusively in very rotten, punky wood. May be in a snag, stump, fence-post, similar. Uses existing cavities created by other species as well as nest boxes. May tend toward areas of dense rather than open vegetation. Nest inside cavity often includes bark strips, especially incense cedar, as well as notable quantities of fur. A thick covering of fur and bark is created that is used to cover eggs when both adults are away from nest. **EGGS:** 6–7.

Mark Nyhof, BC

BOREAL CHICKADEE *Poecile hudsonicus*

HABITAT: Boreal spruce and fir forest, at higher elevation in mountain regions where it overlaps with Mountain Chickadee. **LOCATION AND STRUCTURE:** Excavates cavity in snag, decayed

Mark Peck, ON

stump, or living tree with significant heart rot. Will also use cavities created by other species and natural cavities. Will use nest boxes, especially if filled with wood chips. Usually a side entrance, as is typical for chickadees, but about a third of nests have a *top entrance* in a stump or broken top of a tree. Interior nest is similar to that of Chestnut-backed and Mountain Chickadees, though some are constructed almost exclusively of fur. **EGGS:** 4–9; can be slightly larger than those of other species.

TITMICE

Small songbirds, titmice nest in natural cavities or abandoned cavities created by other birds, as well as in nest boxes. They may do some minor expansion of preexisting cavities in very rotten wood. The nest cavity may be 1–28 m off the ground but typically is below 7 m. Within the cavity, the titmouse builds a cup nest with bark fibers, dry grasses, fur, feathers, snakeskins, and the like. The female alone constructs the interior. Titmice may reuse favored nest cavities in subsequent years. They are monogamous or socially monogamous (varying by species). Females do not abandon the nest easily at signs of danger. There is little geographic overlap among species, but cavity entrances are unremarkable for distinguishing from other cavity nesters. Eggs are subelliptical or rounded ovate (the small end can be quite pointed). Incubation period is 12–16 days and NP ~15–21 days, varying by species.

BRIDLED TITMOUSE *Baeolophus wollweberi*

HABITAT: Oak or oak-pine-juniper woodlands and riparian woods in mountains of sw. U.S. into Mex. *Where it overlaps with Juniper Titmouse, Bridled Titmouse is more likely to be in dense oak scrub.*
LOCATION AND STRUCTURE: Primarily uses natural cavities, especially rotted-out knotholes. Also readily uses nest boxes. Grass and plant leaves are prevalent in nest structure, also cottonwood down and silk from spiders and caterpillars. Cavity height 0.5–10 m. **EGGS:** 5–7; white, *unmarked*; 16 x 13 mm. **BEHAVIOR:** Only member of this family in which nest helpers seem to be frequent, assisting with feeding young and defending territory.

OAK TITMOUSE *Baeolophus inornatus*

HABITAT: Oak or oak-pine woodlands. Less commonly in pine or juniper woodlands without oak trees. **LOCATION AND STRUCTURE:** Primarily in natural cavities, to a lesser extent in abandoned woodpecker holes. May partially excavate existing cavity in very soft wood. Readily uses nest boxes, other artificial sites; occasionally in bank burrow of other species.

Bringing nesting material to a natural cavity. Alan Vernon, CA

Lining includes grass, moss, fur, feathers. **EGGS:** 6–7; white, sometimes with fine markings, *similar to those of Juniper Titmouse*; 18 x 14 mm. **BEHAVIOR:** Pairs mate for life and defend permanent territories. Nest may be reused from year to year by same or different pair.

JUNIPER TITMOUSE *Baeolophus ridgwayi*

HABITAT: Arid juniper and pinyon-juniper woodlands, *at higher elevations than Oak Titmouse where they overlap*. **LOCATION AND STRUCTURE:** Often in natural crevice of a twisted juniper tree. Also in old woodpecker holes or in nest boxes. Base of nest contains more shredded bark and hair than does nest of Oak Titmouse. Cavity height 2 m (3–4 m). **EGGS:** 4–7; *see Oak Titmouse*. **BEHAVIOR:** Incubating female is extremely reluctant to leave nest when disturbed, and will often remain inside cavity, hissing.

TUFTED TITMOUSE *Baeolophus bicolor*

HABITAT: Deciduous, sometimes mixed, forest, usually with dense canopy and a variety of tree species. **LOCATION AND STRUCTURE:** Usually in a cavity created by other species, but occasionally uses nest boxes. Does not excavate. Nest of vegetation also usually includes hair, often collected directly from live animals. Cavity height 11 m (1–28 m). **EGGS:** 5–6; cream colored with reddish spots evenly distributed; 18 x 14 mm. **BEHAVIOR:** Pairs typically stay on territory year-round. Sometimes 1 young from previous year will remain with adults to help raise next year's brood.

BLACK-CRESTED TITMOUSE *Baeolophus atricristatus*

HABITAT: Oak, mesquite, and riparian woodlands. Also orchards and suburban areas. **LOCATION AND STRUCTURE:** Appears to prefer nest sites in areas free of undergrowth. Primarily in old woodpecker cavities, but will also use nest boxes and crotches of trees. As with other titmice, nest consists of vegetation and other soft material, especially hair. When using nest box, often packs empty space between nest cup and wall of box with moss. Cavity height 1–7 m. **EGGS:** 6; white with evenly distributed speckles; 17 x 14 mm.

Mikael Behrens, TX

VERDIN (Remizidae)

VERDIN *Auriparus flaviceps*

HABITAT: Desert scrub with thorny brush, riparian washes in desert. **LOCATION AND STRUCTURE:** Nest is built in periphery of shrub or small tree, often on edge of thicker vegetation or along edge of open wash. Nests are often quite visible. Easily identifiable, globular enclosed nest, either spherical or oblong. Entrance is ~2.5 cm in diameter and located on side of nest toward bottom, with projecting tunnel or platform of twigs. Outer layer, attached to substrate with spider webs, is constructed of interwoven thorny twigs and is rough in appearance. Inside is a more compact layer of leaves, stems, and other softer materials. Final inner lining of fur, plant down, and feathers of various birds often almost completely fills interior of nest and serves as insulation from heat and cold. Outside diameter 14–20 cm; height 2 m (0.8–4 m). **EGGS:** 4 (3–6); pale green or blue-green with fine speckling concentrated on larger end; elongate to elliptical; 15 x 11 mm; IP 14–18 days (F); NP ~21 days (F, M). **BEHAVIOR:** Multiple nests of same family group may be built close together. Verdins also construct and maintain smaller similar structures for roosting throughout year. Assumed to be monogamous. Male leads construction of outer layer and structure of nest. Female constructs inner lining alone, with materials often delivered by male.

Matt Monjello, AZ

BUSHTIT (Aegithalidae)

BUSHTIT *Psaltriparus minimus*

HABITAT: Open deciduous and mixed woodlands, chaparral- and shrub-dominated landscapes. Common in suburbs and parks. **LOCATION AND STRUCTURE:** Multiple adults help construct an easily recognizable, fully enclosed, hanging, pendulous nest with a side entrance near top. Abundant spider silk is used, adding strength and stretch to structure, which includes a variety of soft plant materials such as moss, lichens, leaves. Nest is often well camouflaged, suspended from branches in a shrub or tree. Often well concealed, but not always. Height above ground: 1–30 m; length 15–30 cm; outside diameter (of bowl) 2.5–10 cm. **EGGS:** 6 (4–10); white; subelliptical; 14 x 10 mm; IP 12–13 days (F, M); NP 14–15 days (F, M). **BEHAVIOR:** Territories belong to flocks, not pairs, and multiple active nests are possible at one time within territory. Breeding pair often has additional helpers assisting with nest construction, during incubation, and with nestling rearing. Helpers are often unpaired adult males, though they may be juveniles or females. When nest failure occurs, a pair may attempt to take over another active nest, and if unsuccessful may instead join in using the nest along with the original pair. Polygyny, polyandry, and polygynandry are likely common. Multiple females may lay eggs in same nest, resulting in clutch sizes of up to 15. During breeding season multiple adult birds may sleep in nest.

David Moskowitz, WA

NUTHATCHES (Sittidae)

These 4 species of small songbirds are common breeders in mature woodlands across much of N. America and are found in boreal, coniferous, deciduous, and mixed coniferous-deciduous forests. They are generally year-round residents throughout most of their range, but northernmost populations of White-breasted Nuthatches migrate south in winter, and Red-breasted Nuthatches demonstrate irruptive or irregular movements, leaving their common habitats during years when food becomes scarce.

All species are *secondary cavity nesters*, primarily taking up residence in existing woodpecker holes, but will also use natural cavities which they may enlarge and modify. With the exception of White-breasted Nuthatch, species may excavate new nesting holes themselves. A cavity entrance may take from a week to 2 months to complete, and its shape may vary from round to oblong and ragged. When building a nest, nuthatches fill the cavity floor with a foundation of material such as grasses, bark strips, rootlets, pine needles, lumps of earth, and dry leaves. They then add a soft inner cup of finer material, including fur, feathers, wool, cocoons, plant down, hair, bark shreds, grasses, rootlets, weed stems, and occasionally snakeskin. One of the unusual features of nuthatch nests is the application of sticky resin (Red-breasted) and crushed smelly insects (White-breasted) around the inside and outside of the cavity rim. This behavior may aid in nest defense by deterring predators and competitors.

Nuthatches are monogamous, and pair bonds may last from 1 season to several years or the life of the pair. Cooperative breeding has been observed in both Pygmy and Brown-headed Nuthatches, whereby helpers—generally first-year males that appear to be closely related to the parents—assist in territorial defense, nest excavation and construction, sanitation, and rearing of young. Except in White-breasted Nuthatch, males appear to select the nest site. Nuthatches generally lay 5–9 subelliptical to short elliptical white eggs, speckled and blotched with various colors from red, reddish brown to purple, purplish red, purplish brown. See Chickadees, p. 295. Females alone incubate, for 12–17 days. Young fledge in 2–3 weeks.

RED-BREASTED NUTHATCH *Sitta canadensis*

HABITAT: Common in mature mixed-coniferous forests, especially habitats with a variety of species (spruce, fir, hemlock, larch, pine, cedar); also in mixed coniferous-deciduous woodlands (especially eastern population). Rarely in pure stands of pine, hemlock, or aspen. *Generally replaces White-breasted Nuthatch in boreal forest zone.* **LOCATION AND STRUCTURE:** Uses both dead and dying coniferous and deciduous trees. Excavates new cavity each year but may use existing holes, rarely nest boxes. Commonly 3–4.5 m aboveground, sometimes 1.5–12 m up. Cavity height 1.4 cm; diameter 2–9.5 cm. *Sticky resin is applied around entrance to rim. As it's removed from cavity, sawdust often accumulates at base of tree.* **EGGS:** 5–8; 15 x 22 mm; IP 12–13 days (F); NP 18–21 days (F, M). **BEHAVIOR:** Both sexes excavate. Male often begins cavity, female performs majority of excavation.

*Red-breasted
nuthatch.*
David Moskowitz, UT

WHITE-BREASTED NUTHATCH *Sita carolinensis*

HABITAT: Common in mature deciduous forests. Also mixed deciduous-coniferous woodlands; orchards, neighborhoods and parks with large trees; arid pine and Douglas-fir forests of Northwest and Rocky Mts. Rarely in boreal coniferous forests. **LOCATION AND STRUCTURE:** Prefers woodland edges, often occupying areas along water, roads, and clearings. Nest is generally in a natural cavity or old woodpecker hole in both hardwoods and conifers. May excavate and enlarge existing cavity; rarely uses nest boxes. Commonly 5–20 m aboveground. Cavity diameter 1.8–15 cm (usually 3.8–5.7 cm). **EGGS:** 5–9; 19 x 14 mm. **BEHAV-**

Matt Monjello, ME

Note the modified entrance excavated by the White-breasted Nuthatches using this crack for a nest. Matt Monjello, AZ

White-breasted nuthatch full clutch, laid in a deep mat of fur. Casey McFarland, NM

IOR: Female builds. Often reuses same cavity from previous year. May apply noxious-smelling insects to inside and outside of cavity. Pairs become quiet when nesting begins.

PYGMY NUTHATCH *Sitta pygmaea*

HABITAT: Ponderosa and other long-needle pine forests. Especially abundant in late-successional, undisturbed woodlands with plenty of dead and dying trees; also mixed pine-deciduous and pine-coniferous forests. **LOCATION AND STRUCTURE:** Often oriented south or east and away from prevailing winds on dead or dying long-needled pine, especially ponderosa; occasionally in fir or deciduous tree (aspen, cottonwood, maple, oak). Cavity is generally on trunk and may be positioned near a branch. Will excavate new cavity but may use natural cavities and woodpecker holes or nest boxes when available, 1–21 m aboveground. Cavity entrance height 2.75–5 cm; width 2.5–4.5 cm. *Nesting material may be stuffed in cracks along cavity wall to weatherproof nest.* **EGGS:** 5–9; 15 x 12 mm; IP 12–17 days (typically 14–15) (F); NP 14–22 days (F, M). **BEHAVIOR:** Male appears to select nest site, both sexes excavate. Uses cavities year-round for roosting. During breeding season, a pair and a helper may roost together. In winter, family groups and large flocks may share a cavity.

David Moskowitz, WA

BROWN-HEADED NUTHATCH *Sitta pusilla*

HABITAT: Southeastern pine forests: open mixed pine-deciduous, young pine woodlands, pine and grasslands bordering cypress swamps. Most abundant in mature woodlands of loblolly, slash, longleaf, and shortleaf pine with open understory and plentiful snags. Also in residential areas and parks. **LOCATION AND STRUCTURE:** Commonly in a decayed tree, usually pine, occasionally deciduous. Also in dead portions of live pine or cypress; wood piles, artificial structures (nest boxes, fenceposts, etc.). Cavity entrance height 5 cm; width 3.8 cm. *Nest material often includes pine seed wings.* **EGGS:** 3–7; 15 x 12 mm; IP 14 days (F); NP 18–19 days (F, M). **BEHAVIOR:** Male selects nest site, both sexes excavate, sometimes with assistance from helpers. May begin multiple cavities before choosing one.

Eric C. Soehren, AL

TREECREEPERS (Certhidae)

BROWN CREEPER *Certhia americana*

HABITAT: Woodlands throughout N. America. Often found in mature coniferous, deciduous, and mixed woodlands where large living trees, snags, and dying trees are present. **LOCATION AND STRUCTURE:** Favors dense tree groves. Commonly built *between a loose slab of bark* and the trunk on a dead or dying tree, occasionally in dead sections of living trees. Usually concealed, though sometimes nest material may be sticking out from behind loose bark. Usually 1.5–4.6 m aboveground. Nest is a *crescent-shaped outer cup*, consisting of twigs, roots, and bark strips, bound together and fastened to underside of bark with silk, egg cases, and cocoons. Inner nest cup of fine bark shreds, wood fibers, grasses, moss, lichens, hair, spider egg cases. **EGGS:** 5–6; white, speckled with pink or reddish brown, concentrated toward larger end, occasionally unmarked; smooth and nonglossy; subelliptical; 15 x 12 mm; IP 13–17 days (F); NP 14–20 days (F, M). **BEHAVIOR:** Pairs may inspect multiple nest sites before one is chosen. Generally singled-brooded. Female is primary builder (6–7 days) during first nesting attempt, though both male and female take part in later nests. Male may vocalize at nest site while female gathers material and builds.

Brown Creeper nests are often crescent-shaped, conforming to the space between the trunk and loosened bark.
Matt Monjello, AZ

Mark Nyhof, BC

WRENS (Troglodytidae)

These small to medium-sized songbirds occupy a wide range of habitats, including arid deserts, rainforests, and deciduous and coniferous woods. Ten species breed in N. America and are found in brackish estuaries, along interior lakes and waterways, in dense and open woodlands, near human habitations, and at varying elevations from seaside thickets to montane habitats. Wrens use an impressive variety of nest sites, ranging from cavities in trees, rock fissures, and earthen banks to cattail thickets, dense tree branches, shrubby vegetation, and cacti. They also use many artificial sites, including buildings, farm machinery, flowerpots, and similar settings. They show an equal diversity in nest design; nests may be *loosely built open cups or domed or fully enclosed oblong or globular structures*. Nest site and design vary not only within the family but in many cases within a species. In most

Left to right:
Marsh Wren,
Bewick's Wren,
House Wren.

species both sexes build, and females are often responsible for adding the final lining.

Although the majority of wrens are predominately monogamous, there are some notable exceptions. Some species demonstrate varying degrees of extra-pair copulation, and polygyny occurs among House and Marsh Wrens and others. Wrens are highly territorial, defending breeding areas from their own and other species. Many species commonly destroy eggs of competitors.

Wrens generally have 2 broods per year, sometimes 3, with clutch size typically 2–7 (rarely 10). Eggs are subelliptical to oval, smooth and slightly glossy, varying from white with little to no speckling of reds and browns, to pinkish or brown with varying spotting and mottling. Incubation period is 12–18 days; altricial young are tended by both parents and fledge the nest in 12–21 days.

ROCK WREN *Salpinctes obsoletus*

HABITAT: Variety of arid and semiarid habitats: coastal and montane cliffs, steep gullies, dry washes, rocky canyons, outcroppings, escarpments, scree and talus slopes, road cuts, from sea level to over 3,600 m elevation. **LOCATION AND STRUCTURE:** In bare or sparsely vegetated substrate: in cavity on rocky slope under overhang, cliff crevice, in a hole or under a large boulder, cranny of building; occasionally an animal burrow along a bank. Foundation of nest and entrance are often paved with small stones, sometimes bone and other material; may be built up enough to partially shield entrance. Nest is a loosely constructed cup of fine grasses, weeds, bark strips, moss, hair. Lined with hair, wool, rootlets, spider silk. Outside diameter 9 cm; height 3.4 cm; inside diameter 6.7 cm; depth 2.9 cm. **EGGS:** 5–6, sometimes 4–10; white, sparsely speckled and spotted with reddish browns, browns, or purple; smooth, slightly glossy; subelliptical to short subelliptical; 19 x 15 mm. **BEHAVIOR:** May reuse nests in subsequent years, rarely during same season.

Casey McFarland, ID

CANYON WREN *Catherpes mexicanus*

HABITAT: Sheer-walled canyons, cliffs, outcroppings, and boulder fields with or without a permanent water source. Rarely near human settlements. **LOCATION AND STRUCTURE:** Commonly under an overhang on ledge, in crevice on cliffs and canyon walls, crannies among rocky outcroppings, boulder fields; occasionally in building crevices. Nest is a cup with a foundation of twigs and coarse material, including moss, grasses, dead leaves, bits of wood, occasionally debris. Lined with softer material such as wool, plant down, feathers, cocoons, spider webs, etc. Outside diameter 10.2–20.3 cm; height 9.5 cm; inside diameter 5.6 cm; depth 1.9–5 cm, average 3.9. **EGGS:** 5–6; white, with sparse markings sometimes concentrated at larger end; can be nonglossy; 18 x 13 mm. **BEHAVIOR:** Nest is occasionally reused in same season or in subsequent years.

David Moskowitz,
WA

HOUSE WREN *Troglodytes aedon*

HABITAT: Broad variety, with an understory of shrubs and thickets interspersed with cavities: open deciduous, mixed deciduous-coniferous, and mixed coniferous woodlands, forested edges, wooded wetlands and openings, riparian groves; parks, gardens, treed suburbs, and farms. Sea level to 3,000 m elevation. Usually absent in mature, dense coniferous forests. **LOCATION AND STRUCTURE:** Readily, even preferentially, uses nest boxes and other artificial substrates. Will also nest in preexisting tree hollow or cavity; occasionally uses other substrates, including rock crevices, bank burrows, nests of other species, etc. Typically avoids heavily vegetated sites with limited view of surroundings. Nest is a platform of twigs with a soft inner cup of plant fibers, grasses, rootlets, hair, bark strips, feathers, and various other material, including snakeskin, rubbish, metal. In cavities with larger entrances, materials may be built up to effectively constrict opening to a desired entry size, possibly to maintain temperatures or block predation and nest parasitism. **EGGS:** 6–7; white to pinkish white or grayish, thickly speckled with fine

House wren removing a fecal sac from nest. Matt Monjello, AZ

House wren nest cup and eggs.
Mark Nyhof, BC

This House Wren nest entrance has been made smaller with a collection of twigs.
David Moskowitz, WA

purplish red or purplish brown dots often concentrated on larger end; 16 x 13 mm. **BEHAVIOR:** Male builds dummy nests of twigs in available cavities in territory, attracting 1 or more mates. Female chooses nest and completes construction.

PACIFIC WREN *Troglodytes pacificus*

WINTER WREN *Troglodytes hiemalis*

HABITAT: Most abundant in interior of mature coniferous, deciduous, and mixed coniferous-deciduous forests, occasionally along shrubby forest openings and edges, to 3,720 m elevation. Less common in treeless areas in north of range, riparian areas in dry habitats, rocky shorelines of Pacific Coast. Commonly near streams, wetlands, and lakes in areas with large trees, thick understory, fallen and standing dead wood, moss, and ferns. **LOCATION AND STRUCTURE:** Cavity sites include upturned tree roots, decayed trees, stumps or logs, old woodpecker holes, and rocky crevices; in hanging moss hummock under overhanging bank or building, in crotch of shrub, occasionally suspended from conifer branches. Ground level to 7 m. Non-cavity nests *typically are domed with entrance on side.* Made of moss, grasses, bark, fine twigs, rootlets, lined with animal hair, feathers. Conifer twigs are interwoven around entrance hole. Outside diameter

Pacific Wren nest built into the upturned rootwad of a fallen tree. David Moskowitz, WA

Winter Wren nest in the upturned rootwad of a fallen tree. Marcus Reynerson, VT

Pacific Wren nest. David Moskowitz, WA

Interior of Winter Wren nest.
Marcus Reynerson, VT

8.5–14 cm; height 6.5–8 cm; inside diameter 4–4.5 cm; depth 3.5–5.5 cm. **EGGS:** 5–7; white, unmarked or finely speckled with reddish brown at larger end; 18 x 13 mm. **BEHAVIOR:** Male builds majority of nest but does not line it. May build numerous dummy nests in territory. Female selects unlined nest, adds lining just before egg laying.

SEDGE WREN *Cistothorus platensis*

HABITAT: Common in moist meadows within tall sedge or grassy areas; in higher and drier portions of freshwater and coastal marshes, bogs, and other water; in hayfields, regenerating cultivated fields. *Generally avoids deep water, preferring drier sites than Marsh Wren;* open areas lacking dense vegetative growth. **LOCATION AND STRUCTURE:** Well hidden and very close to ground

Sedge Wren.
Joe Kosack, PA

or over water, 10–100 cm up. Nest is a globular ball built among dense vegetation with entrance hole on side, lined with fine grasses, feathers, hair, plant down. Occasionally builds in low shrub. Outside diameter 8–13 cm; height 8–13 cm; entrance 1.5–2.5 cm. **EGGS:** 6–7; *white, unmarked.* **BEHAVIOR:** Males often polygynous, females may be polyandrous. Male builds multiple nests, female selects nest to line.

MARSH WREN *Cistothorus palustris*

HABITAT: Stands of cattails, bulrush, sedges, phragmites of large freshwater marshes; in tall grass and rushes in salt or brackish marshes and along tidal waters; rarely in mangroves. **LOCATION AND STRUCTURE:** Usually 0.3–1 m above water, sometimes up to 1.8 m. Oblong structure is lashed to standing stems of emergent vegetation (cattail, bulrush, cordgrass) with entrance on the side and toward top. Woven of wet cattail, sedges, grasses, lined with fine plant material, cattail down, feathers, rootlets, leaves. See

Casey McFarland,
WA

Sedge Wren. Outside diameter 13 cm; height 18 cm. **EGGS:** 4–6, sometimes 3–10; *pale brown to dark brown* and evenly speckled with dark brown spots or blotches; wreathed or capped at larger end. **BEHAVIOR:** Males typically polygynous, females socially monogamous. Male builds multiple dummy nests; female selects from these nests and adds lining. Construction takes 5–8 days.

CAROLINA WREN *Thryothorus ludovicianus*

HABITAT: Variety, with moderate to thick woody understory, including deciduous and mixed woodlands, dense brushy ravines, thickets, riparian woodlands, oak and pine-oak scrub, wooded swamps, farmlands, suburban areas, and gardens. More common in wetter forests and along wooded riverine embankments than in swamps or drier uplands. **LOCATION AND STRUCTURE:** Usually in a tree hollow, stump, or woodpecker hole, up to 3 m high. Also among upturned roots, forks and branches of trees, brush piles, on ground beneath thick vegetation (especially early season nests), building crevices, nest boxes, and artificial structures (bridges, etc.). Nest is generally a bulky, open oblong dome with substantial side entrance. Often has a slight perch at entrance. Made of bark strips, dried grass, leaves, weed stems, moss, rootlets. Lined with fine grasses, rootlets, hair, feather, snakeskin, bits of trash. Early nests are more substantial than later ones. **EGGS:** 4–6; white, cream, or pale pink, finely speckled with reddish brown, often wreathed at larger end. **BEHAVIOR:** Mates for life.

Matt Monjello, AL

BEWICK'S WREN *Thyromanes bewickii*

HABITAT: Often in areas with dense brushy undergrowth and open woodlands. Western populations in a broad variety of habitats, including chaparral slopes, riparian groves, oak, mixed evergreen, pinyon-juniper-chaparral woodlands, montane conifers, desert washes, brushy hillsides, suburbs. Avoids closed canopy forests and open areas lacking scrubby growth. Eastern populations (uncommon) are usually along brushy and wooded

Bewick's Wren nest in tree cavity.
Scott Olmstead, AZ

Bewick's Wren nest cup and eggs.
Mark Nyhof, BC

edges adjacent to farm buildings. **LOCATION AND STRUCTURE:** Natural and artificial cavities: woodpecker holes; tree cavities; brush piles; crannies in rocks, walls, or buildings; nest boxes; fenceposts. Ground level to 10 m up. Nest is a bulky cup, *sometimes domed.* Made of varied material, including twigs, grasses, moss, leaves, rootlets, spider egg cases. Lining is a soft cup of feathers, wool, moss, plant down, hair, sometimes snakeskin. One nest in WA: outside diameter 12 cm; height 10 cm; inside diameter 6 cm; depth 3 cm. **EGGS:** 5–7; white, finely speckled with irregular reddish brown and purple spots, concentrated at larger end. **BEHAVIOR:** Male may build unfinished dummy nests.

CACTUS WREN *Campylorhynchus brunneicapillus*

HABITAT: Variety of arid habitats dominated by cactus, thorny scrub and shrubs. Also in brushland near human habitations, in native coastal scrub mixed with mature cacti (s. CA). **LOCATION AND STRUCTURE:** From 1 to 6 m (usually 1–3 m) aboveground in a variety of substrates: cholla, giant saguaro, prickly pear, catclaw, mesquite, yucca, palo verde; occasionally in mistletoe, salt cedar, desert willow, bird boxes. Nest is a large, bulky, *horizon-*

David Moskowitz, AZ

tally situated globular structure with a tube-shaped entrance. Made of coarse grasses, fine twigs, plant stems, lined with fine grasses, feathers, plant down. May include debris and trash. Chamber diameter 18 cm; passageway entrance ~7 cm above base of chamber; entrance diameter 9–15 cm; overall length from chamber wall to entrance up to 30.5 cm. **EGGS:** 4–5; white to pale pink, heavily speckled or spotted with brown; 24 x 17 mm. **BEHAVIOR:** Construction takes 1–6 days. New nests may be built for first or subsequent broods, but previous nests may also be repaired and reused, depending on geography. Nests may be used as winter roosts.

GNATCATCHERS (Polioptilidae)

These very small songbirds build beautiful and readily recognizable nests. Four species breed in N. America, though Black-capped Gnatcatcher (*Polioptila nigriceps*), rare in se. AZ, is not covered below. Gnatcatchers are monogamous, but whether they maintain their pair bond for more than a season is unclear. Males typically defend the territory (though females give chase to other females) and advertise with song from perches and while foraging. Nests are compact, sturdily built, and *appear somewhat like a larger version of a hummingbird nest*; they are made of fine materials such as plant fibers, bark strips, and grasses, and lined with plant down, hair, fibers, and possibly feathers. The nest is typically bound and fastened in place with spider silk, and in Blue-gray Gnatcatcher it is *decorated with lichen flakes*. Dimensions vary slightly across species but are similar. Outside diameter 5–6.4 cm; height 5.7–6.4 cm; inside diameter 3.2–3.8 cm; depth 3.2–5 cm.

Gnatcathers are often double-brooded, mostly in separate nests, but materials from the first nest may be used to build the second. First nest is built in 1–2 weeks, sometimes longer, and precedes egg laying by 10–14 days. Both sexes build, though the male may be the dominant builder. There are commonly 4 eggs per clutch, 14 mm x 11–12 mm. Eggs are generally similar across species (*Black-tailed Gnatcatcher's can be more lightly marked, but can also be similar to California Gnatcatcher's*): short elliptical to short subelliptical, smooth with little or no gloss, pale to pale blue, lightly marked with fine spots, speckles, and blotches of reddish browns, purples. Both sexes incubate, though males do not have a brood patch, and eggs hatch in 14 days. Both parents care for young. Despite their tiny nest size, gnatcatchers are frequent hosts to cowbirds.

BLUE-GRAY GNATCATCHER *Polioptila caerulea*

HABITAT: The only migratory *Polioptila*, with the largest and northernmost range. Wide variety of habitats; often in moist deciduous woods, but also in oak woodlands-chaparral and pinyon-juniper in Southwest, pine-oak in Southeast. Streamside thickets. Riparian areas are used particularly in northern range. Prefers habitat edges. **LOCATION AND STRUCTURE:** Nest frequently saddles a small horizontal or slanted branch, *most often in a live deciduous tree*, often well out from trunk (more than halfway to

Blue-gray Gnatcatcher. Matt Monjello, NC   Matt Monjello, NC

branch tips). Typically uses a twig, small branch, *or knot* to build nest against for support. Height variable, 0.9–24 m, often more than halfway up nesting tree. Nest is deep-cupped, *often decorated extensively with lichen flakes*, possibly similar material.

CALIFORNIA GNATCATCHER *Polioptila californica*

HABITAT: Limited range, in coastal sage scrub, brushlands, occasionally in chaparral or riparian areas adjacent to aforementioned habitats. **LOCATION AND STRUCTURE:** In a low tree or bush (0.6–0.9 m), fastened securely to several twigs. Nest is *more cone-shaped* than those of other gnatcatchers, *and lacks lichen covering.* May make as many as 10 nesting attempts in a season.

BLACK-TAILED GNATCATCHER *Polioptila melanura*

HABITAT: Confined mostly to arid regions; saguaro forests and other cactus country, creosote, mesquite, catclaw, etc. Thickets in desert washes, arroyos. **LOCATION AND STRUCTURE:** Usually 0.6–1.2 m, but to 2.7 m, sometimes higher. Nest is often built into a fork or crotch of more than 2 small branches, usually in a dense shrub or tree, well shaded from above and sides. Deep cup narrows slightly at rim and *lacks lichen covering.*

Casey McFarland, AZ

DIPPERS (Cinclidae)

AMERICAN DIPPER *Cinclus medicanus*

HABITAT: Mountain and coastal streams. Scattered range in desert Southwest, limited to areas with permanent water source. Usually along clear, fast-moving streams, up to 15 m wide, with stony bottoms and little aquatic vegetation. **LOCATION AND STRUCTURE:** Found overlooking fast running water on a ledge that provides protection from flooding, weather, and predation: cliff ledges, rock surfaces and crevices, midstream boulders, rock piles. Often under bridges. Occasionally behind waterfalls or in tree roots, rotted tree stumps, dirt banks, culverts. Frequently close enough to water that nest remains damp. Nest is a dense, *large globular structure with side entrance*. Often built *mostly of moss*, or grass and leaves where moss is not available. Inner cup is made of similar materials, sometimes bark shreds. When built under overhang, can be an open cup. Dome diameter 30.5 cm;

Look closely for the gaping yellow mouths of begging nestlings, just an arm's reach from the rushing water below. Nest placement by default must initiate young to an aquatic life.
Bob Armstrong, AK

American Dipper using a natural rock crevice for a nest. Matt Monjello, CA

One dipper nest constructed atop an older one under a bridge over a mountain stream. David Moskowitz, BC

American Dipper
nest with young.
Bob Armstrong, AK

entrance width 7.5 cm; entrance height 5–10 cm; inner cup diameter 5–7.5 cm. **EGGS:** 4–5; smooth, glossy, white, and unmarked; subelliptical to oval; 26 x 18 mm; IP 16 days (F); NP 18–25 days (F, M). **BEHAVIOR:** Generally uses same habitat year-round. Built by both sexes. In areas with limited nest sites, pairs may build near old nest, or directly on top.

KINGLETS (Regulidae)

GOLDEN-CROWNED KINGLET *Regulus satrapa*

HABITAT: Mature boreal and montane coniferous forests, especially spruce and fir. Also riparian softwoods in Southwest. **LOCATION AND STRUCTURE:** Typically built high in conifer, sometimes close to trunk but often near end of a branch. Beneath overhanging foliage, making viewing from above or nest level difficult or impossible. Nest is a *deep, softball-sized, dense cup, suspended from up to 4 twigs*; rarely resting on top of branches. *Cup is small for size of nest.* Outer layer is constructed of *moss*, twigs, lichens, pine needles, and other coarse material. Also used throughout are spider silk, fine strips of bark, insect cocoons, hair, and plant down. Lined primarily with feathers. Nest is 2–20 m aboveground (most often at high end of range). Outside diameter 7.5 cm; height 7.7 cm; inside diameter 40 cm; depth 4 cm. **EGGS:** 5–10; white with light speckling; elliptical oval; 13 x 10 mm; IP 14–15 days (F). **BEHAVIOR:** Monogamous. Both male and female build, may build multiple dummy nests that are smaller, less tidy. Two broods are common. Female begins a second nest while her first brood is still in nest; she begins to lay the day after first brood fledges. Male then tends to first set of fledglings, which are mostly independent by the time second brood hatches.

RUBY-CROWNED KINGLET *Regulus calendula*

HABITAT: Boreal and montane coniferous or mixed coniferous-hardwood forests. **LOCATION AND STRUCTURE:** Secluded and rarely observed. Built close to top of a conifer and usually close to trunk, though sometimes out on a limb. Nest is usually sus-

Ruby-crowned Kinglet. George Peck, ON

Ruby-crowned Kinglet.
Katherine Hocker, AK

pended from branches, sometimes resting on top, but always under overhanging branches that visually obscure it. *Globe-shaped cup nest has top sometimes very constricted so as to almost fully enclose interior, and sometimes fully open*; this may be a regional variation. Nest stretches to accommodate growing brood. *Built of moss*, spiderwebs, bark strips, twigs, rootlets, etc. Inside is composed of fine plant material, feathers, and fur. Nest is 0–12 m, up to 27 m, aboveground. Outside diameter 6–10 cm; height 7–15 cm; inside diameter 3–5 cm; depth 3–5 cm. **EGGS:** 5–11; white with speckles encircling larger end; ovate; 14 x 11 mm; IP 12–14 days (F). **BEHAVIOR:** Monogamous. Female constructs entire nest.

LEAF WARBLERS (Phylloscopidae)

ARCTIC WARBLER *Phylloscopus borealis*
HABITAT: Primarily in shrubby willow thickets along streams or in spruce forest; also in shrubby birch. **LOCATION AND STRUCTURE:** On ground, tucked into tussock or vegetation. Nest is a readily identifiable *domed structure with side entrance*. Built of grasses, mosses, leaves, lined with fine grasses and ungulate hair. Outside diameter 11–13 cm; height 7–13 cm; inside diameter 5.7–7 cm; depth 5 cm. **EGGS:** 5–7; white, finely speckled with reds, browns, particularly at larger end; short subelliptical; 16 x 13 mm; IP ~13 days (F); NP ~12 days (F, M). **BEHAVIOR:** Single-brooded.

OLD WORLD FLYCATCHERS (Muscicapidae)

BLUETHROAT *Luscinia svecica*

HABITAT: Scrubby thickets of willow, birch, shrubs in foothills, some tundra; thickets along streams, rivers, lakes. **LOCATION AND STRUCTURE:** On ground, usually in hummock, tussock, steep bank, protected from above by vegetation. Also at base of shrubs. **EGGS:** 5–7; pale green or greenish blue or blue, but so finely speckled throughout as to have a reddish brown or rusty tint; subelliptical; 19 x 14 mm; IP 14–15 days (F); NP 14 days (F, M).

Rick Bowers,
Bowersphoto.com,
AK

NORTHERN WHEATEAR *Oenanthe oenanthe*

HABITAT: Arctic nester. Mountain areas, with rocky, barren hills or hilltops; tundra with rocky outcroppings. **LOCATION AND STRUCTURE:** In open habitat, *on ground in crevice beneath stones or loose rock, or in crevice in rock wall.* Nest is a large, bulky, loose

*This nest was
hidden beneath a
large, flat rock.*
Mark Peck, AK

pile of dry ferns, heather, mosses, other plants. Inner cup is distinct from foundation, well made of similar finer materials, and lined with mosses, wool, plant fibers, feathers, leaves. **EGGS:** 5–6; pale blue, unmarked; subelliptical; 21 x 16 mm; IP 12–15 days (F, occasionally M); NP 15 days (F, M). **BEHAVIOR:** Female builds, male accompanies. Mostly socially monogamous, but returns to same nesting site each year, where it may re-pair with prior mate. Winters in sub-Saharan Africa.

WRENTIT (Sylviidae)

WRENTIT *Chamaea fasciata*

HABITAT: Coastal scrub, coastal and montane chaparral, but also in other habitats, including variety of woodlands with dense understory. Parks, gardens, blackberry thickets. **LOCATION AND STRUCTURE:** Usually low, 0.3–1.2 m aboveground, typically in crotch of 2 or more branches. Nest is a compact cup mostly of fine bark strips from shrubs and plants, but also other fibers, fine stems, grasses, with spider silk used for foundation and to bind materials. Lined with fine stems, fine grasses, fibers, sometimes hair. Materials are often collected within a few meters of nest. Outside diameter 9 cm; height 7 cm; inside diameter 6.5 cm; depth 4.8 cm. **EGGS:** 4; little to no gloss, greenish blue, no markings; subelliptical; 18 x 14 mm; IP 15–16 days (F, M); NP 15–16 days (F, M). **BEHAVIOR:** Both sexes build (~6–7 days), incubate, and care for young. Monogamous, forming lifelong bonds, often early in life. Long-lived, and pairs have been reported to maintain their small home ranges for more than a decade. Dispersing young don't wander far, establishing territories nearby.

David L. Sherer, CA

David L. Sherer, CA

THRUSHES (Turdidae)

Thirteen species of these small to medium-sized woodland songbirds nest in N. America. Thrushes are found in a wide variety of habitats from sea level to high mountains, including deciduous, coniferous, and mixed woodlands, prairies, and high desert. They are also common nesters in suburbs, parks, farmlands, and similar areas. Nests are typically a large, open cup, lined with grasses, twigs, weeds, moss, and depending on the species, may be reinforced with mud or leaf mold. They are built in forks and on horizontal branches of trees, saplings, and shrubs, and on the ground. All 3 species of bluebirds nest in cavities.

The female usually builds the nest, but males may assist in construction and often stay close during the building process. Both parents feed the altricial young, and generally have 2 broods per year. Thrushes usually lay 2–4 (sometimes 1–8) subelliptical to oval, smooth and glossy, light blue or greenish blue eggs. Depending on the species, eggs are either unmarked or speckled, blotched, and spotted with reddish browns.

BLUEBIRDS

Bluebirds are generally socially monogamous (polygyny and polyandry do occur), and *mostly nest in natural tree hollows, woodpecker holes, and nest boxes*. Nests are loose cups that include dry grasses, pine needles, weed stems, and fine twigs, and are lined with fine grasses, bark fibers, and occasionally hair and feathers. The female builds in a few days, or in ~ 1 week. All species can be double-brooded.

Western Bluebird in a woodpecker cavity.
Matt Monjello, AZ

Mountain Bluebird in a natural rock cavity. Gerrit Vyn, WY

Eastern Bluebird eggs in a nest box.
Bill Summerour, AL

Western Bluebird in a nest box.
Matt Monjello, CA

EASTERN BLUEBIRD *Sialia sialis*

HABITAT: Variety of open habitats with scattered trees and sparse understory, including orchards, farms, parks, golf courses, agricultural fields, pasture; open pine woodlands, clear-cuts, burned-over areas, roadsides. **LOCATION:** Rarely will build an open-cup nest outside a cavity, in a burn scar, or on a large limb. **EGGS:** 3–7; IP 11–19 days (F); NP 17–21 days (F, M). **BEHAVIOR:** Rarely in socially polygynous and polyandrous associations. May build nests in multiple cavities before choosing one. Will reuse nests and roost in nest boxes during winter.

WESTERN BLUEBIRD *Sialia mexicana*

HABITAT: Common in open coniferous, mixed, and deciduous woodlands. Favors clearings with open canopy and dead standing trees, such as wooded edges, burned areas, moderately harvested forests, riparian woodland, farmland. Less abundant in large open areas than other bluebirds. **LOCATION:** Sometimes between trunk and bark; rarely in building crevice and swallow nest. **EGGS:** 5; IP 12–18 days (F); NP 18–25 days (F, M). **BEHAVIOR:** Pairs in some populations may have helpers at nest, usually adult male relatives.

MOUNTAIN BLUEBIRD   *Sialia currucoides*

HABITAT: Prefers more open country than other bluebirds, with some trees, short grasses, and scattered shrubs, including mountain meadows, alpine tundra bordering tree line, sagebrush flats, lowland savanna; newly burned or cut-over areas, roadsides. **LOCATION:** Occasionally nests in hole in dirt bank, or crannies in cliffs and buildings. **EGGS:** 5–6; IP 13 days (F); NP 18–21 days (F, M). **BEHAVIOR:** Cavities are often reused, both during and outside breeding season.

TOWNSEND'S SOLITARE   *Myadestes townsendi*

HABITAT: Montane coniferous forests to high elevations, usually in fairly open forest with scant understory. **LOCATION AND STRUCTURE:** *In a cavity or hollow under overhanging shelter,* such as a rock, log, shrub, tree, dirt bank cavity of road cut, among tree roots, at base of a tree or decayed stump; infrequently aboveground in a stump or live tree cavity, no more than 3.5 m high. Nest is a cup of pine needles, weed stems, dry grasses, rootlets, moss, lined with bark strips, grass stems, similar material. Structure and size vary with placement. Nests built on level ground generally are simple cups built into a shallow depression. Nests built into slopes are larger with a foundation of twigs and other plant material that may extend downward from cavity, creating a tail of material up to 30 cm long. Outside diameter varies, 20 cm at base; inside diameter ~7 cm; depth ~5 cm. **EGGS:** 3–4; color variable, white, gray blue, greenish blue, or pinkish white and heavily spotted or blotched with purple, brown, or red, uniformly or concentrated at larger end; IP 11–13 days (F); NP 10–14 days (F, M). **BEHAVIOR:** Socially monogamous. Female builds in 2–14 days.

Sarah Hegg,
National Park
Service, WY

VEERY *Catharus fuscescens*

HABITAT: Typically in moist, young deciduous woodlands, often near water in areas with a dense understory, including wetlands, streamside thickets, damp ravines, early successional forests; disturbed habitats. Occasionally in mature wet woodlands and mixed deciduous-coniferous forest. *Where range overlaps with Hermit Thrush, Veery nests are usually in damper, younger woods.* **LOCATION AND STRUCTURE:** Built on or close to ground, usually well concealed by dense vegetation. Ground nests are placed next to cover, including fallen tree trunks, branches, at base of shrubs, saplings, trees, or stumps; in grass tuft; under debris or brush or on moss hummock. Elevated nests are usually in a shrub, sapling, stump, low tree fork, or small tree stub with new shoot growth. Nest is a platform of dead leaves with firm inner cup of weed stems, grass, bark fiber, twigs, leaf mold. Lined with rootlets, leaf mold, bark fibers, pine needles. Outside diameter 8–15 cm; height 8.8–14.3 cm; inside diameter 6–7.5 cm; depth 3.3 6 cm. **EGGS:** 4; pale blue or greenish blue, unmarked; 22 x 17 mm; IP 10–14 days (F); NP 10–12 days (F, M). **BEHAVIOR:** Polygynandrous. Female builds in 6–10 days. Male may feed young at multiple nests. Upon spring arrival, male delays commencement of song; uses calls to set up territorial boundaries. Male begins full song after female begins incubating.

Matt Monjello, CT

Matt Monjello, ME

GRAY-CHEEKED THRUSH *Catharus minumus*

HABITAT: To tundra or tree line in dense, scrubby undergrowth in coniferous or dwarf coniferous forest or in dense thickets of willow, alder, etc. Also in thick undergrowth of cottonwood floodplains. **LOCATION AND STRUCTURE:** Usually in fork of a small shrub, occasionally on horizontal branches of a conifer, on fallen tree trunks or stumps, or on ground or building ledge. Usually within 2.5 m of ground, sometimes to 6 m. Nest is a dense, deep cup of small twigs, grass stems, horsetails, sedges, rootlets, with inner cup of wet mosses, detritus, sometimes mud. Lined with fine grasses. Outside diameter 11–13 cm; height 7–10 cm; inside diameter 6–7 cm; depth 5 cm. **EGGS:** 4; pale greenish blue, spotted, speckled, and blotched with reddish brown or light brown, concentrated at larger end or evenly distributed; 23 x 17 mm; IP 12–13 days (F); NP likely ~11–13 days (F, M). **BEHAVIOR:** Presumed to be monogamous.

BICKNELL'S THRUSH *Catharsus bicknelli*

HABITAT: Limited to fragmented range in ne. U.S. and e. Canada. Can be found in fir-dominated montane forests above 1,000 m elevation. Favors disturbed areas with dead but standing conifers and dense successional coniferous growth such as logged clearings, cut trails; exposed slopes and ridgelines. Sometimes lower elevations in second-growth coniferous or mixed coniferous-deciduous woodlands. **LOCATION AND STRUCTURE:** Usually along edge of a clearing in dense stands of small to medium-sized fir or spruce. Nest is generally placed where 1–4 horizontal branches meet the trunk of small tree, sometimes farther out on horizontal branches of larger tree. Often in balsam fir, but occasionally in spruce, birch, or in dead fir, usually 0.8–3.2 m aboveground (to 10 m). A bulky outer cup is made mostly of twigs, moss, sometimes with grasses, sedges, flower or fern stalks, dried leaves, bark strips, and lichen added to structure, with inner cup of leaf mold and other decaying vegetation. Lined with rhizomorphs, sometimes fine grass, sedge stems. Outside diameter 10.3–14 cm; height 7–9.6 cm; inside diameter 5.8–8.7 cm; depth 3.8–6.5 cm. **EGGS:** 3–4; bluish green, speckled or blotched with variable amounts of light brown, generally on larger end but may occur over entire egg; 22 x 17 mm; IP 9–14 days (F); NP 9–13 days (F, M). **BEHAVIOR:** Sometimes polygynandrous. Males are not territorial on breeding grounds; often sing within same area. Sire and feed young in more than 1 nest. Female builds in 2–11 days.

Kent McFarland,
KPMcFarland.com,
VT

SWAINSON'S THRUSH *Catharus ustulatus*

HABITAT: Northern and montane coniferous, mixed coniferous, and mixed coniferous-deciduous forests with dense shrubby understory; also along deciduous riparian forests, thickets, and scrub. From sea level to 2,600 m elevation. **LOCATION AND STRUCTURE:** Usually in moist, shaded areas in understory. Often in dense deciduous shrubs or conifer saplings; occasionally on horizontal branches of large trees and away from trunk. Usually within 3 m of ground, sometimes higher. Nest is a small, compact cup of grasses, moss, fine twigs, bark shreds, sometimes mud. Lined with leaf mold, rootlets, rhizomorphs, lichens, or moss. Nesting material varies by region, changing outer appearance: some nests are made predominantly of grasses or moss or twigs. Outside diameter 9.4–15.4 cm; height 3.8–10.5 cm; inside diameter 5–8.5 cm; depth 2.5–6 m. **EGGS:** 4; blue to greenish blue, speckled or spotted with reddish or brown markings that may cover entire egg or be concentrated on larger end; 22 x 17 mm; IP 10–14 days (F); NP 10–14 days (F, M). **BEHAVIOR:** Presumed to be monogamous.

David Moskowitz, WA

Kim Shelton, WA

HERMIT THRUSH *Catharus guttatus*

HABITAT: Variety of forests: boreal, coniferous, deciduous, or mixed woodlands, in both moist and dry habitats. Often near wetland edges, mountain streams and meadows, small clearings, windthrows, etc.; occasionally in more open areas such as fields, golf courses. **LOCATION AND STRUCTURE:** Generally concealed by surrounding vegetation. Western populations: Typically build elevated nests in small conifers, deciduous trees, shrubs and bushes, 0.5–3 m high. Eastern populations: Usually build ground nests under a tree, sapling, shrub, or herbaceous plant; occasionally elevated in shrubs and trees. Nest is a compact bulky cup of grasses, leaves, bark strips, moss, bits of wood, twigs, weeds, rootlets, lichen, rarely mud (eastern population). Lined with pine needles, plant fibers, fine grass, rootlets, moss, and

Hermit Thrush nest. Placed near the base in a small conifer. David Moskowitz, BC

Hermit Thrush nest. Placed just above head height in the crotch of a small deciduous tree. Casey McFarland, NM

A Hermit Thrush ground nest tucked below a thorny shrub. Matt Monjello, ME

willow catkins. Outside diameter 10–15 cm; height 5–8 cm; inside diameter 6–7 cm; depth 3.5–5.5 cm. **EGGS:** 4; light blue or greenish blue, unmarked; 25 x 19 mm; IP 11–13 days (F); NP usually 12 days, sometimes 10–15 days (F, M). **BEHAVIOR:** Presumed to be monogamous. Female builds in 7–10 days.

WOOD THRUSH *Hylocichla mustelina*

HABITAT: Moist, shaded deciduous and mixed woodlands. Prefers larger tracts of forest than broken ones, but will breed in smaller lots, wooded parks, and residential sites. Common in habitats with a diverse deciduous component of tall trees, fairly dense subcanopy, moderate understory, and open forest floor with decomposing litter. **LOCATION AND STRUCTURE:** In saplings, shrubs, and trees (especially oaks, hemlock, beech). Usually shaded in a crotch or fork of horizontal branches, away from trunk, occasionally adjacent to it, usually 2–4 m aboveground, sometimes much higher. Nest is a large outer cup of dead grass, weed stems, leaves, sometimes paper or plastic. An inner cup of

Wood Thrush. Matt Monjello, CT   *Wood Thrush.* Matt Monjello, CT

mud is added and *lined with rootlets*, dead leaves. Outside diameter 10–14 cm; height 5–14.6 cm; inside diameter 7–8.5 cm; depth 3.2–5 cm. **EGGS:** 3–4; pale greenish blue, unmarked; 25 x 19 mm; IP 12–13 days (F); NP 12–15 days (F, M). **BEHAVIOR:** Socially monogamous. Sometimes double-brooded. Most nest material is gathered close to site, but mud may be gathered off territory. Female mainly builds, in 3–6 days.

CLAY-COLORED THRUSH *Turdus grayi*
HABITAT: Expanding in southernmost TX along Lower Rio Grande Valley, often in densely vegetated parks and riparian edge habitat. **LOCATION AND STRUCTURE:** Typically well concealed in a tree or small shrub, 1.5–3.5 m high. Nest is a bulky cup of grass, moss, feathers, leaves, with *inner cup of mud*. Lined with rootlets, fibers, grasses. **EGGS:** 2–4; pale blue, occasionally with reddish brown spots; 29 x 21 mm.

AMERICAN ROBIN *Turdus migratorius*
HABITAT: Broad range, with mixture of shrubs and trees, often in residential and urban areas. Parks, farms, yards, roadside and agricultural edges, forest, open woodlands, riparian habitats, clearings, brushy fields. **LOCATION AND STRUCTURE:** Typically in

The tidy construction of this nest completely conceals the use of mud in the inner cup.
Casey McFarland, WA

An American
Robin ground
nest, more
common in areas
where trees or
shrubs are sparse
or nonexistent.
Chris Helzer, NE

fork or on horizontal branch of a tree or shrub, often under heavy foliage. Also nests on ground and in thickets (western prairies), on cliff ledges, human-made structures, tree stumps, posts. Usually 1–7.6 m high. Nest has a bulky outer layer of dead grass, twigs, weeds, grass stems, occasionally human-made materials, rootlets, moss, and feathers. An *inner cup of mud is added* and lined with *fine, dry grasses* (which may be removed from old nests to reveal the hardened inner layer). Pacific Northwest populations often have a platform of twigs supporting nest. Outside diameter 8–20 cm; height 4–27 cm; inside diameter 3–12.7 cm; depth 3–13 cm. **EGGS:** 4; light blue, unmarked; 28 x 20 mm; IP 12–14 days (F); NP 14–16 days (F, M). **BEHAVIOR:** Socially monogamous. Double- or treble-brooded. Female selects nest site and builds; male may bring nesting material. Typically builds new nest for each brood.

VARIED THRUSH *Ixoreus naevius*

HABITAT: Most strongly associated with mature wet coniferous and mixed forests with deep shade. In drier portions of range, found near forest pond and streams. Also possible in alder and cascara grow-back in logged areas. **LOCATION AND STRUCTURE:** Usually on branches near trunk of a small conifer, but also on distal ends of branches on larger conifers; also in deciduous

David Moskowitz, WA

Mark Nyhof, BC

trees, shrubs, vines, on buildings, or on ground. Generally 2–4 m high, but up to 18 m. Nest is a bulky cup with outer layer of loosely woven twigs, leaves, lichens, bark fibers, with moss hanging over rim and outer wall. Inner cup is made of rotted wood, moss, leaf mold or mud possible, wet grasses. Lined with fine grasses, moss, dead leaves. Outside diameter 17–22 cm; height 9–12 cm; inside diameter 9–11 cm; depth 6 cm. **EGGS:** 3–4; light blue, sometimes scantly marked with small brown spots; IP 12 days (F); NP 13–15 days (F, M). **BEHAVIOR:** Presumed to be monogamous. Female builds. New nests may be built near previous year's nest, occasionally directly atop it.

MOCKINGBIRDS and THRASHERS (Mimidae)

Mimids are small to medium-sized songbirds, known for their melodious songs and mimicking ability. They live in a wide variety of habitats, from lush, moist forests to extreme desert. Ten mimids breed in N. America, eight of them thrashers. Nests typically are bulky and coarse, and in many cases appear as loosely constructed baskets of projecting twigs. In most species the outer layer is platformlike, built of twigs and sticks, and lined to various degrees with finer materials. Three or more distinct layers may be visible. Nests are generally low to the ground and well concealed or in protected locations, such as cholla cactus. Mimids are primarily monogamous. Nests are built by both sexes (though more by female in Gray Catbird), and both sexes care for young; except in Northern Mockingbird and Gray Catbird, both parents incubate. Thrasher young typically leave the nest with weak flying ability, relying on their ability as capable runners. Some mimids, particularly the western thrashers and Northern Mockingbird, overlap in range, but generally species occupy distinct habitats. The nests of many species can appear similar; it's important to note which species occur in particular regions and cross reference the accounts. Eggs are smooth, glossy, bluish or greenish, and when marked are speckled or spotted with reddish brown. Many mimids are often-double brooded; exceptions are noted below.

Left to right: Gray Catbird, Northern Mockingbird, Brown Thrasher, LeConte's Thrasher.

GRAY CATBIRD *Dumetella carolinensis*

HABITAT: Dense, low thickets, brushy areas, and woodland understory. Frequently in edge habitat: along marshes or waterways, hedgerows, woodland edges, roadsides, paths, small roads, and trails, etc. **LOCATION AND STRUCTURE:** In dense thickets, vine tangles, saplings, low trees, usually deep within branches and leaves; 0.9–3 m aboveground, averaging 1.5 m, rarely higher. Nest is quite bulky, thick, deeply cupped, typically consisting of 3 layers: a coarse outer structure of twigs (platformlike), weed stems, vines, grasses, dead leaves, and sometimes human-made materials; a middle layer of similar materials more tightly woven; and an inner lining of materials, usually fine rootlets or tendrils, but may contain fine grasses, bark shreds, skeletonized leaves, hair, and plant down. Outside diameter 14 cm; height 8.9 cm; inside diameter 8.3 cm; depth 4.5 cm. **EGGS:** 4; blue to deep greenish blue, *unmarked*, darker than those of eastern thrushes; subelliptical; 23 x 17 mm; IP 12–13 days (F); NP ~11 days (F, M). **BEHAVIOR:** Two broods, sometimes 3. Female builds (5–6 days), though male may collect materials and contribute somewhat to construction. Recognizes and ejects cowbird eggs, a learned behavior, and has been reported to destroy eggs of other passerines. Male, commonly a loud songster, sings quietly when near nest.

Casey McFarland, AL

Matt Monjello, AL

CURVE-BILLED THRASHER *Toxostoma curvirostre*

HABITAT: Found at higher elevations than other thrashers (to about 2,100 m). Open and semiopen arid country. Sparse scrub and brushlands; creosote habitat with cholla cactus; chaparral with prickly pear; grasslands with cholla; pinyon, juniper, and oak habitats (NM). **LOCATION AND STRUCTURE:** Most commonly in cholla cactus, but also in mesquite, yucca, mistletoe, small trees, 1–1.5 m up, sometimes slightly higher. Nest is made of twigs, sometimes flowering weeds, often thorny, well built into a deep cup (tail tip may be only visible part of incubating bird). Lined with long grasses (collected by male, shaped by female), sometimes hair. Green vegetation or flowers may be put in nest prior to first egg. Outside diameter 23 cm; height 17 cm; inside diameter 11 cm; depth 10 cm. **EGGS:** 3; pale blue to bluish green, finely speckled, markings may be very subtle; subelliptical to oval; 29 x 20

Curve-billed Thrasher. Casey McFarland, NM

Curve-billed Thrasher. Bill Summerour, AZ

Curve-billed Thrasher. Casey McFarland, NM

mm; IP 13 days; NP 14–18 days. **BEHAVIOR:** Both sexes build, sometimes fairly quickly (3 days) but up to 1 month. Climate differences throughout range lead to wide variation in nesting times, from Jan. to July. A territory is usually maintained from year to year, with pairs often nesting close to prior season's nest, sometimes in same cholla. May build on old nests, or old Cactus Wren nests.

BROWN THRASHER *Toxostoma rufum*

HABITAT: Low, dense thicket, underbrush; woodland edges, shelterbelts, hedgerows, brushy fields; in Great Plains in understory of riparian cottonwood stands, wooded draws. **LOCATION AND STRUCTURE:** In a dense shrub or small tree, vine, especially

Matt Monjello, AL

Casey McFarland, AL

thorny species, usually below 2.5 m, sometimes on ground. Nest often has 3 or 4 layers: an outside basket of twigs; a layer of leaves, inner bark, vines; another layer of rootlets, fine twigs, and stems; then a lining of fine rootlets or grasses. Outside diameter 19 cm (may be up to 30 cm); height 9–15 cm; inside diameter 9–10 cm; depth 5–6 cm. **EGGS:** 4–5; white to pale blue or greenish but heavily covered in fine reddish brown speckles; subelliptical to long elliptical, long oval; 26 x 19 mm; IP 11–14 days; NP 9–12 days. **BEHAVIOR:** Generally single-brooded. Both sexes build in 5–7 days. Among the largest passerine hosts of parasitic cowbirds, but frequently rejects eggs.

LONG-BILLED THRASHER *Toxostoma longirostre*

HABITAT: Open country, riparian woodlands. **LOCATION AND STRUCTURE:** *Low in dense thickets, similar to nests of Brown Thrasher, Curve-billed Thrasher, and Northern Mockingbird.* **EGGS:** 4; *similar to Brown Thrasher's.* **BEHAVIOR:** Considered to be similar to Brown Thrasher. Much of preferred habitat has been cleared, and numbers are dramatically reduced.

BENDIRE'S THRASHER *Toxostoma bendirei*

HABITAT: Open shrubland, grasslands, woodlands, desert areas of cacti, Joshua trees, yucca, etc; *habitat typically less sparse, harsh, than that of LeConte's Thrasher.* Foothills. Sagebrush-juniper at higher elevations. **LOCATION AND STRUCTURE:** Variety of sites, including cacti, shrubs, trees (cholla, mesquite, yucca, juniper, etc.). Nest is similar to that of other desert thrashers, but often more *compactly built, smaller, and of finer materials.* Nest wall may be thin. **EGGS:** 3–4; gray green to greenish white, irregularly marked with brown spots, blotches; subelliptical to short subelliptical; 26 x 19 mm; NP 12 days.

CALIFORNIA THRASHER *Toxostoma redivivum*

HABITAT: Coastal and inland chaparral; scrub, riparian-woodland thickets; in lower-elevation range of pine-oak woodlands. **LOCATION AND STRUCTURE:** Typically well hidden in a low shrub or tree, often 1–2 m aboveground. *Similar to nest of Northern Mockingbird but larger.* Platform and sidewalls are made of rigid twigs, the largest at bottom, decreasing in diameter toward cup. Cup is fairly shallow. Outside diameter 20–30 cm (to 45 cm); height 12–21 cm; inside diameter 10.8 cm; depth 4.5 cm. **EGGS:** 3; pale blue, covered with fine reddish brown specks; subelliptical to long elliptical; 30 x 31 mm; IP 14 days (F, M); NP 12–14 days (F, M). **BEHAVIOR:** The largest thrasher, with a long breeding season (Nov.–July). Female responds to male with song. Rarely flies to or from nest, rather coming and going by foot on ground.

LECONTE'S THRASHER *Toxostoma lecontei*

HABITAT: A xerophile, LeConte's Thrasher needn't drink water. Hot, barren desert, sparsely vegetated: flats, alluvial fans, dunes, low hills, typically with sandy or silty clay substrates. **LOCATION AND STRUCTURE:** Usually in cholla cactus, but also in dense, thorny shrubs 0.5–2.5 m up. Also palo verde, sage, saltbrush, etc. Sites are often shaded from above; may be placed against arroyo walls in overhanging bushes. Nest has 3 layers: an outer shell of twigs; a middle layer of thinner twigs, rootlets, grasses, possibly human-made materials; *and a thick, species-definitive inner lining of fuzzy plant seeds, leaves, flowers,* rarely feathers or human-made items. *Within LeConte's habitat, no other species builds a nest of similar dimensions with such a thickly padded, well-insulated third lining layer.* Outside diameter 14–28 cm; height 9–23 cm; inside diameter 8–9 cm; depth 5–7 cm. *Similar to Log-gerhead Shrike's nest but larger.* **EGGS:** 3; pale greenish blue, finely marked, *(usually with fewer markings than on Curve-billed or California Thrasher eggs),* often more on larger end; subellipti-cal; 28 x 20 mm; IP 14–20 days; NP 14–17 days. **BEHAVIOR:** Both sexes build in 3–12 days, 3–5 days common. Pairs may form any time of year or remain together year-round. Nonmigratory. Unusually, singing begins in fall, peaking in midwinter. Nesting and laying begin as early as Feb.

CRISSAL THRASHER *Toxostoma crissale*

HABITAT: Low to relatively high-elevation desert: desert edges, brushy riparian scrub belts, brushy plains, canyons, large arroyos, ridge bases, mesquite thickets. At higher elevations (mesas, foothills, etc.) in scrub mixed with pinyon, juniper, and oak. **LOCATION AND STRUCTURE:** Often placed in central portions of densest shrubs in habitat. Nest is similar in size and appear-ance to that of other desert thrashers, but can be notably smaller and of finer materials (see Bendire's Thrasher). Dimensions are similar to those of other thrashers, or smaller. Outside diameter 18–28 cm (average 20.5 cm); height 10–15 cm; inside diameter 8 cm; depth 4 cm. **EGGS:** 2–3; pale blue or bluish green, like those of a robin, and *the only unmarked eggs among the thrashers*; 27 x 19 mm; IP 14 days (F, M); NP 11–12 days (F, M). **BEHAVIOR:** Known for its keen watchfulness, expert covertness, and keeping cover between itself and anyone approaching. Both parents leave and approach nest undetected.

Damon Calderwood,
AZ

SAGE THRASHER *Oreoscoptes montanus*

HABITAT: Large expanses of semiarid shrub-steppe, largely in regions dominated by sagebrush, but also in greasewood, saltbrush, and bitterbrush habitats, or similar. **LOCATION AND STRUCTURE:** In a bush, often built in a fork in densest portion of plant, sometimes beneath constructed canopy of twigs or old nest; 0.5–1 m up. Also commonly on ground at base of a dense bush. Bulky nest is made of coarse twigs and plant stems, bark shreds, lined with grasses, fine rootlets, hair or fur. Outside diameter 18–20 cm; height ~10 cm; inside diameter 9–11 cm; depth 5 cm, but to 8 cm in later-season nests. **EGGS:** 4–5; deep greenish blue (sometimes lighter), *strongly marked* with dark brown spots *(larger markings than on other thrashers' eggs)*, blotches, wreathed; subelliptical; 28 x 15 mm; IP 15 days; NP 11–14 days. **BEHAVIOR:** The smallest thrasher (more closely related to mockingbirds than to true thrashers). Known to reject cowbird eggs. Populations have been severely reduced owing to loss of sagebrush habitats.

Gerrit Vyn, WY

Sanders, WY

Rohan Kensey, MT

NORTHERN MOCKINGBIRD *Mimus polyglottos*

HABITAT: Open habitats: scrubby woodlands of scattered trees and bushes; gardens, parks, shade trees in cities and suburban areas, particularly in hedges, shrubs near bare areas; desert scrub. **LOCATION AND STRUCTURE:** Usually built in dense shrubbery or trees, but sometimes on beams, rafters in buildings, posts. Also in cactus (cholla, prickly pear), vines, sagebrush. Usually 1–3 m aboveground, but may be lower or occasionally considerably higher. Nest is a bulky, loose cup of dead twigs, sometimes with other coarse materials such as weed stems, with inner layer of dry leaves, grasses, stems, moss, and potentially a variety of trash or human-made materials. Lined with rootlets, grasses, occasionally plant down, hair, bits of trash. *May look similar to nests of several thrasher species where ranges overlap.* Outside diameter 17.8 cm; height 11.4 cm; inside diameter 7.9 cm; depth 6 cm. **EGGS:** 3–5; smooth, glossy, blue to blue-green, heavily marked with brownish speckles, spots, blotches; subelliptical to elliptical; 25 x 18 mm; IP 11–14 days (F); NP 12–14 days (F, M). **BEHAVIOR:** Two or 3 broods, usually building new nest for each; pairs may build as many as 6 nests per season.

Matt Monjello, NC

STARLINGS (Sturnidae)

EUROPEAN STARLING *Sturnus vulgaris*

HABITAT: Wide variety of habitats with nesting cavities, water, and open areas for foraging. Close association with settlements: cites, suburbs, parks, agricultural fields, etc. Avoids large tracts of forest, desert, and arid scrublands. **LOCATION AND STRUCTURE:** Almost any cavity, including woodpecker holes, natural tree cavities, holes in buildings, cliffs, nest boxes, burrows. May nest in dense vegetation on trees or on ground. Commonly 3–7.6 m aboveground, sometimes up to 18 m. Nest is a mass of grasses, weed stems, pine needles, leaves, fresh vegetation, trash. Lining is variable, including grasses, bark, feathers, leaves, and moss.

*European
Starling.*
Steve Smith, WA

Nest has slight depression near back of chamber for eggs. Inside diameter 7–8 cm; depth 5–8 cm. **EGGS:** 3–6; smooth, slightly glossy, pale blue or greenish white, unmarked; subelliptical; IP 12 days (F, M); NP 21–23 days (F, M). **BEHAVIOR:** Secondary cavity nester. Male selects nest site, begins construction. Female builds majority of nest in 1–3 days. Highly aggressive, evicting primary nesters and defending sites from variety of species, including other starlings, woodpeckers, ducks, bluebirds, flycatchers, swallows, screech-owls, kestrels. Single- or double-brooded. Intraspecific brood parasitism is common.

WAXWINGS (Bombycillidae)

These medium-sized songbirds (2 species) are among the latest breeders in N. America; nesting is apparently timed with the ripening of summer fruits. Both species prefer open woodlands at edges or clearings, often near water (to feed on aquatic insects) and in areas with fruiting plants. Neither nests in deep forest interior. Courtship involves variations of hopping toward and away from one another while perched atop a branch, often touching bills when close; hovering; fruit offerings. Both sexes search for nests sites, and the female tests each site with nest-making movements. Nest is a *thick-walled, bulky cup* of various materials often collected near nest site, and may be adorned with dangling grasses, catkins, etc. It is usually built on a horizontal branch or branches, in ~3–6 days. Both sexes collect material, but the female is the primary builder. Eggs are subelliptical to oval, with little or no gloss, and are pale gray or pale bluish gray, sparsely marked with small irregular dark brown spots or speckles, light brown or gray blotches; 22–25 mm x 16–17 mm (Bohemian the larger).

BOHEMIAN WAXWING *Bombycilla garrulus*

HABITAT: Open boreal forest of conifer, birch. **LOCATION AND STRUCTURE:** Nest is on a horizontal limb, often near trunk, usually in a conifer. Height depends on trees, often 1–6 m but can be much higher. Structural wall is made primarily of small twigs, then a layer of coarse grasses, moss. Lined with finer grasses,

Bohemian Waxwing.
Mark Peck, ON

moss, cocoons, down. Often camouflaged with moss, lichen, dangling materials. Nest can be slightly *larger* than Cedar Waxwing's, and eggs *larger*. Single-brooded.

CEDAR WAXWING *Bombycilla cedrorum*

HABITAT: Wide variety of open woodlands, orchards, gardens, shade trees. Riparian habitats are especially important in arid regions. **LOCATION AND STRUCTURE:** Usually in fork in a horizontal branch, well out from trunk in trees, sometimes in a vertical fork in bushes; 1–15 m aboveground, often ~2–7 m. Nest is made of plant stems, twigs, grasses, plant down, mosses, string, dead leaves, bark shreds, roots, ferns, etc. Lined frequently with moss, rootlets, grasses, fibers, pine needles, spider silk, etc.

Matt Monjello, ME

Cedar Waxwing.
Matt Monjello, ME

Outside diameter 11.4–15.2 cm; height 8.9–11.4 cm; inside diameter 6.4–7.6 cm; depth 5 cm. **EGGS:** Slightly *smaller* than Bohemian Waxwing's; IP 12–14 days (F, fed by M); NP 15–17 days (F, M). **BEHAVIOR:** Double-brooded.

SILKY FLYCATCHERS (Ptiligonatidae)

PHAINOPEPLA *Phainopepla nitens*

HABITAT: Desert scrublands of mesquite, palo verde, acacia. Often associated with desert mistletoe. Also in oak and sycamore woodlands and canyons. Breeds earlier in year in hotter deserts, later in other habitats. **LOCATION AND STRUCTURE:** In fork of a vertical or horizontal branch of tree, shrub, or in clump of mistletoe; 2–5 m aboveground, sometimes higher, often 2–3.5 m. Nest is a *notably shallow*, neat, compact, thick-walled, bulky cup, small for size of bird. Made of stems, small twigs, small leaves, plant fibers, oak blossoms, plant down, and bound with silk. Thinly lined with wool or hair, down. Nest may be grayish in color. Out-

Daniel Baldassarre, CA

Daniel Baldassarre, CA

side diameter 9.5–11.4 cm; height 3.8–6.4 cm; inside diameter 6.4 cm; depth 2.5–3.8 cm. **EGGS:** 2–3; smooth, some gloss, pale grayish white, covered heavily with speckles, spots of black or shades of violet, sometimes wreathed; subelliptical to short subelliptical; 22 x 16 mm. **BEHAVIOR:** Male builds, sometimes starting before he has a mate. Female may help complete nest.

OLIVE WARBLER (Peucedramidae)

OLIVE WARBLER *Peucedramus taeniatus*

HABITAT: Mountainous pine forests (largely ponderosa, but also fir) to 3,760 m elevation; pine-oak forest. **LOCATION AND STRUCTURE:** High up in conifers, 9–21 m, far from trunk in branches, often in terminal needles; very difficult to find. Nest is compactly built of lichens (including hairlike species), plant down, mosses, rootlets, moss, catkins, cone bud scales *(material composition may look like that of Bushtit nest, and nest can appear similar to that of a large version of a Blue-gray Gnatcatcher nest)*, etc., bound with spider webs; outside may be decorated. Lined with fine rootlets, plant down. **EGGS:** 3–4; smooth, glossy, pale gray or bluish white, thickly marked with spots, dots, blotches of grays, olive browns, gray browns, charcoal, especially toward larger end; subelliptical to short subelliptical; 17 x 13 mm. **BEHAVIOR:** Female builds, with male accompanying.

OLD WORLD SPARROWS (Passeridae)

HOUSE SPARROW *Passer domesticus*

HABITAT: Highly successful introduced species. Limited to areas in close association with human habitation, including farmlands and cultivated areas, suburban and urban environments; never in natural habitats that have not been altered by human activity.

Alan Vernon, CA

A House Sparrow nest built in the open, less common than cavity nests.
Rick Wright, NE

LOCATION AND STRUCTURE: Often in a cavity or crevice of buildings and wide variety of other artificial structures. Also on ivy-covered walls and trees or in tree cavities and built on horizontal branches; sometimes uses nests of other birds, such as swallows and robins. Structure varies with placement. When placed in open, nests are globular in shape with a side entrance. In cavities or crevices, nests can be simple cups or large masses of material that fill the space. Made of dried grasses, weed stems, paper, string, and other refuse; lined with feathers. **EGGS:** 3–6; smooth and glossy, white, greenish white, or bluish white, marked with brown, gray, and purple spots, often concentrated toward larger end; 1 egg in clutch is usually less profusely marked; subelliptical; 23 x 16 mm; IP 10–14 days (F, M); NP 10–14 days (F, M). **BEHAVIOR:** Commonly in small colonies. Both sexes build. Nests may be reused for both breeding and as roosts in nonbreeding season.

EURASIAN TREE SPARROW   *Passer montanus*

Limited but expanding range, occurring mostly in bordering portions of MO, IL, and IA. Nests in natural cavities in trees, or in cavities or holes in buildings, fenceposts, etc.

WAGTAILS and PIPITS (Motacillidae)

These small to medium-small songbirds are well adapted to nesting in harsh environs and climatic extremes. Five species nest in N. America. They are typically socially monogamous, and all build *ground nests* in varying locations and habitats. Males typically feed incubating females, though wagtail males may assist in incubation. Females are typically the primary builders. Nests comprise a variety of materials, including dry, coarse grasses; stems; small twigs; bark shreds; lichens; and mosses. They are lined with fine grasses, rootlets, mammal hair or fur, and feathers. Mud and other wetted vegetation are sometimes used to bind dry materials.

Sprague's Pipit nests are often domed, like those of meadowlarks. Eggs are typically subelliptical to ovate. Coloration helps them blend with the ground: shades of gray, olive, or bluish white, finely or heavily spotted with browns and grays. Wagtail eggs measure 18–19 mm x 14 mm, pipit eggs 21 x 15 mm. Incubation period varies across species, 12–14 days; NP 10–16 days.

Two species, both scarce breeders in w. AK, are not included below: Red-throated Pipit *(Anthus cervinus)* and White Wagtail *(Motacilla alba).*

EASTERN YELLOW WAGTAIL   *Motacilla tschutschensis*

HABITAT: Open, shrubby zones in tundra or coastal tundra and similar, particularly at edges where shrubs meet other ground cover. Willow along waterways near coast, grassy tussocks, vegetated hummocks, cut banks of creeks and rivers, roads, and two-track dirt roads; also near industrial cuts (mining operations). **LOCATION AND STRUCTURE:** Built in hollow on ground; in mosses, sedges, or under edges of tussocks or hummocks, or tucked in roots of low-growing shrubbery, or in cavity often created by female. May have substantial, low cover from above from surrounding substrates. Nest is a thick cup of grasses, plant stems, bark shreds, dead leaves, lichens, moss. Lined with fine grasses and may contain large quantities of feathers (ptarmigan) or mammal hair, typically caribou, Arctic fox. Outside diameter 10–11.5 cm; height 6.3–7.6 cm; inside diameter 5–5.8 cm; depth 4.3–4.6 cm. **EGGS:** 5–6; *can be quite dark in color, olive brown.* **BEHAVIOR:** Snow cover and warm-up flood cycles influence the start of nesting. Territorial behavior may be put on hiatus, with birds forming flocks, if weather turns cold.

Mark Peck, AK

*Eastern Yellow
Wagtail.*
Mark Peck, AK

AMERICAN PIPIT *Anthus rubescens*

HABITAT: Arctic and alpine tundra; grassy and rocky slopes; high-elevation subalpine meadows. **LOCATION AND STRUCTURE:** On ground tucked in hollow among carpet of low-growing plants in meadows; in tussocks, along cut banks. Almost always with overhead shelter. Nest is in an excavated shallow scrape; *a separate ring of grasses or stems is first added, leaving floor bare*; a second layer is added across entire nest hollow, and then lined with finer materials, including grasses, plant fibers, sometimes hair or wool; feathers are present in some AK nests. In shared alpine habitat, *Horned Larks often use drier sites.* Outside diameter ~9–16 cm (dependent on hollow); inside diameter 6.4–7.9

Libby Megna, WY

Libby Megna, WY

cm; depth 3.8–5.7 cm. **EGGS:** 4–6; grayish or dull bluish white, but can be *heavily marked* (more so than paler eggs of Horned Lark), making them look dark brown. **BEHAVIOR:** See Sprague's Pipit.

SPRAGUE'S PIPIT *Anthus spragueii*

HABITAT: Endemic to N. American grasslands. Grazed (but not overgrazed), healthy native mixed prairie, short-grass prairie. **LOCATION AND STRUCTURE:** On ground, usually in a natural depression (hollow, cattle hoofprint, etc.), though female may excavate depression), usually in dense vegetation (dead grasses, herbage). Nest is made of coarse and fine dry grasses or stems; lined with fine grasses, sometimes moss, hair. Adjacent grasses are often *pulled over and interwoven above to form dome* (though nests may also be mostly exposed from above) *with a small side entrance hole.* Nest dimensions are similar to those of American Pipit. **EGGS:** 4–5; grayish or buff, blotched and speckled evenly. **BEHAVIOR:** Male has a remarkable flight display, flying and singing at great heights; he may continue for 30 minutes or more (to a few hours) before returning. American Pipit has similarly displays, though from lower heights and for shorter periods. Sprague's Pipit populations have declined severely as habitat has disappeared.

John Pulliam, MT

FINCHES and ALLIES (Fringillidae)

Nests of most of these 17 N. American species are constructed in trees, shrubs, and tall vegetation. Rosy-finches are the exception, with nests hidden in crevices in craggy, rocky areas. Fringillids build cup-shaped nests that range from very tidy and compactly built to a loose assemblage of twigs and grasses; there can also be quite a bit of variation of appearance within some species. Often parents do not remove nestlings' droppings, and they can build up around the edge of the nest. Females typically construct the nest and are the sole incubators. Males may feed females on the nest. Many species are associated with coniferous trees, and several that travel in flocks much of the year nest in loose association with one another.

EVENING and PINE GROSBEAKS

Evening Grosbeaks nest higher in trees than Pine Grosbeaks. Eggs of Pine are more elongated.

EVENING GROSBEAK *Coccothraustes vespertinus*

HABITAT: Open coniferous or mixed coniferous-deciduous forest. Less common in deciduous woodlands, parks, etc. **LOCATION AND STRUCTURE:** Typically in a conifer, rarely in a hardwood tree in East and oak or willow in Southwest. Nest is a loose and fairly flat cup, flimsy in appearance, with eggs often visible through nest from below. Structure is formed of small twigs and rootlets; lined with grasses, pine needles, and other fine vegetation. Height is 5–35 m, typically just over halfway up tree, sometimes toward top. Outside diameter 12–14 cm; height 13 cm; inside diameter 8.5 cm; depth 3 cm. **EGGS:** 3–4; light blue or green with dark splotches on larger end; subelliptical; 23 x 17 mm; IP 12–14 days (F); NP 13–14 days (F, M). **BEHAVIOR:** Socially monogamous. General nesting location may be associated with spruce budworm outbreaks.

Matt Monjello, AZ

PINE GROSBEAK *Pinicola enucleator*

HABITAT: Open spruce-fir forests. **LOCATION AND STRUCTURE:** Well concealed in dense vegetation, typically close to trunk. Loose and bulky cup is constructed of twigs and roots, with inner cup of smaller vegetation and lining of fine vegetation and feathers. Long twigs are used in construction, and often extend well beyond true diameter of nest structure. Height aboveground is 2–6 m. Outside diameter 15–22 cm; height 7–11 cm; inside diameter 7 cm; depth 3.5 cm. **EGGS:** 3–4; pale blue with drab or lavender splotches at larger end; ovate; 26 x 18 mm; IP 13–14 days (F); NP 13–20 days (F, M). **BEHAVIOR:** Likely monogamous.

Mark Peck, ON

ROSY-FINCHES

GRAY-CROWNED ROSY-FINCH *Leucosticte tephrocotis*

BLACK ROSY-FINCH *Leucosticte atrata*

BROWN-CAPPED ROSY-FINCH *Leucosticte australis*

Three closely related species with limited overlap in otherwise distinct breeding areas; geographic location is easiest way to differentiate among nests. Species overlap in range during winter. Few other passerines nest in this particular habitat.

HABITAT: Alpine or Arctic tundra. **LOCATION AND STRUCTURE:** At or above tree line in tundra, at high elevations in mountains in southern part of breeding range, down to sea level in AK. Often adjacent to snowfields or glaciers. Most nests are built into cliffs or rock outcroppings, less commonly in talus fields. Built in a *rock cavity, within a crack in rock face, or under a large rock,* often with overhanging rock cover. In old buildings in northern part of range. Bulky cup nest. Outer layer and base are constructed of

Gray-crowned Rosy-finch.
Rick Bowers, Bowersphoto.com, AK

Black Rosy-Finch nest. Carl Brown, Wyoming Cooperative Fish and Wildlife Research Unit, Department of Zoology & Physiology, University of Wyoming, WY

Black Rosy-Finch parent feeding nestlings. Ronan Donovan, WY

moss, lichen, grasses, and other plants, sometimes mud. Finer interior lining consists of fine grass, feathers, occasionally hair. Overall shape contours to location. In places where human structures exist, may use rafters of buildings. Outside diameter 7–16 cm; height 7.5 cm; inside diameter 6–7 cm; depth 3–5 cm. **EGGS:** 3–6; white, oval or pyriform; 22–24 mm x 16–18 mm; IP 12–14 days (F); NP 15–22 days (F, M). **BEHAVIOR:** Socially monogamous. Feed extensively on cliff faces, edges and interiors of snowfields and tundra, and in close proximity to cliffs.

HOUSE, PURPLE, and CASSIN'S FINCHES

All 3 socially monogamous species construct a coarse cup of woven fine twigs, stems, rootlets, grasses, and similar, with finer lining including grass, moss, and hair. They favor nesting in conifers when possible in much of their range but will use other substrates as well. Nest site typically has some sort of covering, such as overhanging branches or a nook on a building. Purple and Cassin's nests are often indistinguishable without accessory clues. **DIFFERENTIATING SPECIES:** House Finch nests are lowest on average, often ~2 m off ground, although sometimes much higher. Purple and Cassin's nests and eggs are considered indistinguishable. Where the 2 species overlap, Cassin's is primarily at higher elevation.

HOUSE FINCH *Haemorhous mexicanus*

HABITAT: Originally associated with desert scrublands, grasslands, oak savanna, and open coniferous forests of West, but has expanded into urban and suburban environments across continent. **LOCATION AND STRUCTURE:** Highly variable, conspicuous, adapted to wide array of natural and human environments. Typically 2–5 m up, though may be higher. Often nests in conifers. In desert, often in cholla cactus. May be on human structures, occasionally in abandoned nest of other birds or in tree cavity. In creosote desert almost always incorporates creosote branches with green leaves into outer layer (as mite defense). In urban areas often includes large percentage of green foliage, especially later in season when mites are a larger problem. Human-made items are often in lining. Typically does not remove droppings of nestlings after first several days, and notable buildup occurs. Nest may be bulky or compact, tidy or messy, and measurements vary significantly. Outside diameter 9 cm (7.8–11.5 cm); height 3.5 cm (2.7–4.7 cm); inside diameter 5.2 cm (3.8–7.2 cm); depth 1.9 cm (1.3–2.4 cm). **EGGS:** 4–5; pale blue, sometimes speckled; subelliptical to long subelliptical; 19–20 mm x 14–15 mm; IP 13–14 (12–17) days (F); NP 12–19 days (F, M). **BEHAVIOR:** Socially monogamous. Female builds, often pilfers old nests for material. Sometimes reuses nest for second brood. Before constructing a usable nest, new breeders may build a large "preliminary" platform; made of unanchored material and large, it does not ressemble an actual nest and is abandoned. Egg laying may be delayed up to 10 days after nest is completed. Often semicolonial, multiple-brooded.

Casey McFarland, NM

Casey McFarland, AZ

Young successfully fledged from this House Finch nest. Note the build-up of scat.
Chris Byrd, WA

PURPLE FINCH *Haemorhous purpureus*

HABITAT: Coniferous forests, open woodlands, parkland, shrubby habitat, and developed areas. **LOCATION AND STRUCTURE:** Often high and out on a limb in a conifer tree, especially in northern part of range. In southern part may be found in human structures, deciduous trees, or evergreen shrubs. Height aboveground is 0.75–18 m. Outside diameter 16–17 cm; height 11 cm; inside diameter 10–11 cm; depth 7 cm. **EGGS:** 4–5; pale greenish blue with speckles toward larger end; short subelliptical; 20–22 mm x 14–17 mm; IP 12–13 days (F, M); NP 13–16 days (F, M). **BEHAVIOR:** Monogamous where studied. Solitary during breeding season, unlike House and Cassin's Finches, which are often semicolonial.

Dennis Murphy, NY

CASSIN'S FINCH *Haemorhous cassinii*

HABITAT: Montane coniferous forests. **LOCATION AND STRUCTURE:** Close to top of a conifer tree, often ~1 m from top. Usually out on a branch but occasionally close to trunk. Very rarely documented in deciduous tree or in sagebrush. Nest is typically more than 5 m aboveground (2–27 m). Inside diameter 6–8 cm; depth 2.8–3 cm. **EGGS:** 4–5; blue-green, speckled at larger end, occasionally unmarked; subelliptical to long subelliptical; 18–22 mm x 14–16 mm; IP ~12 days (F). **BEHAVIOR:** Socially monogamous. Semicolonial, with nests ~25 m apart, sometimes as close as 1 m.

Damon Calderwood, OR

REDPOLLS

COMMON REDPOLL *Acanthis flammea*

HOARY REDPOLL *Acanthis hornemanni*

HABITAT: Boreal and Arctic. Open forest, scrub, or tundra with at least some shrub cover in places. **LOCATION AND STRUCTURE:** Close to ground in a small tree or shrub (1–2 m high). Where shrub cover is not readily available, may nest in a rock ledge or in driftwood. A loose base of twigs is laid across branches of a small tree or in crotch of a deciduous shrub. Nest is a coarse cup of small twigs, grass, and plant stems. Lined with finer plant materials, fur, and often ptarmigan feathers. Outside diameter 7–12 cm; height 5–10 cm; inside diameter 4.5–6 cm; depth 3–5 cm. **EGGS:** 4–5; pale blue-green with purplish markings concentrated toward larger end; ovate or short ovate; 17 x 13 mm; IP 10–11 days (F); NP 10–16 days (F, M). **BEHAVIOR:** Both species are apparently monogamous. **DIFFERENTIATING SPECIES:** Where breeding ranges overlap, Common and Hoary may nest in association with one another, and nests are nearly impossible to differentiate. Hoary may nest closer to, even over, water. *Hoary also seems to nest both in more open habitats and in denser forests than Common. Eggs of Hoary have slightly paler background color compared with those of Common.*

Common Redpoll.
Kate Fremlin, Amaroq Wildlife Services, NU

Common Redpoll.
Kate Fremlin, Amaroq Wildlife Services, NU

Common Redpoll. Kent Woodruff, AK

CROSSBILLS

RED CROSSBILL *Loxia curvirostra*

CASSIA CROSSBILL *Loxia sinesciuris*

WHITE-WINGED CROSSBILL *Loxia leucoptera*

HABITAT: Red Crossbill is associated with variety of coniferous forest types, wherever there is abundant cone crop; Cassia Crossbill with mature lodgepole pine; White-winged Crossbill primarily with spruce and larch. *Nests are similar.* **LOCATION AND STRUCTURE:** Red is typically in scattered trees on woodland edge or in open woodlands. Cassia is in mature lodgepole pine stands with dense overhead cover. White-winged is in coniferous forests, often in spruce. Nests are built on side branch of a conifer, typically well up in tree (2–20 m); White-winged typically 2.5–4.5 m up, Red and Cassia often higher. Nests may be close to trunk or farther out on limb. Pairs from a single-species flock often nest in proximity to each other. Groups of Red or Cassia Crossbills are typically spaced 50–100 m apart while White-winged may be more densely clustered. Nest is cup-shaped with exterior of twigs, grass, lichens, and finer inner cup of grass, feathers, moss, etc. Nests may be small and fine or coarse and bulky. Win-

Red Crossbill nest. FLPA/Mike Jones

White-winged Crossbill nest.
Mark Peck, ON

ter nests are bulkier than summer nests, often on southeast side of tree. Outside diameter 10.5–13 cm; height 5 cm; inside diameter 6 cm; depth 3 cm. **EGGS:** 3–4; white, light green, or rose with purplish splotches at larger end; ovate; 21 x 15 mm; IP 13–16 days (F); NP 15–25 days (F, M). **BEHAVIOR:** Monogamous. Irregular nesting season, sometimes starting as early as Jan. and running into June, coinciding with availability of primary food, conifer seeds. White-winged may breed in any month of year. Typically single-brooded. Young hatch without crossed bill; stay with parents for ~1 month after fledging, until crossed bill develops, allowing them to forage successfully on their own.

PINE SISKIN *Spinus pinus*

HABITAT: Primarily coniferous forest, also parks, other suburban landscapes. **LOCATION AND STRUCTURE:** In a tree or shrub, typically a conifer, midway up trunk. Toward end of a branch in a spot well concealed by foliage. Average nest height is 6 m (1–15 m).

Mark Peck, ON

Nest is a thick-walled, well-insulated cup, sometimes quite deep. Often only loosely attached to branch, therefore vulnerable to wind. Built with exterior of twigs, rootlets, leaves, or moss and interior of fur, feathers, grass, bark strips, and other fine plant material. Nest is relatively large for size of bird. Outside diameter 9 cm (6–15 cm); height 4.5 cm (3–6 cm); inside diameter 5 cm; depth 2.5 cm. **EGGS:** 3–4; pale blue with markings toward larger end; short oval; 17 x 12 mm; IP 13 days (F); NP 15 days (F, M). **BEHAVIOR:** Seasonally monogamous. Often in loose colonies, with neighboring nests being a few trees away from each other. During brooding, female stays on eggs and male brings food.

GOLDFINCHES

Social, nomadic, and seasonally monogamous, goldfinches often nest in loose colonies. They create deep, compactly woven cup nests with an exterior of grass and other plant material and an interior of fine plant down and other fibers. Nest is often built around an upright fork in a branch with multiple stems, with material wrapped around branches for support. Like some other finches, goldfinches do not remove droppings from the nest and a ring of them around the nest is common. **DIFFERENTIATING SPECIES:** Nests of Lesser and American Goldfinches are very similar, and eggs are indistinguishable. Nests of American are lower on average than those of Lesser, and Lesser sometimes lacks the thick inner lining characteristic of American. Lawrence's eggs are paler than those of the other 2 species, and it has a limited breeding range. Lawrence's also tends to nest in hotter and drier habitats than Lesser and American, though nests of Lawrence's will still be close to a source of water.

LESSER GOLDFINCH *Spinus psaltria*

HABITAT: Open woodlands and chaparral, suburban areas. **LOCATION AND STRUCTURE:** Variety of deciduous trees and shrubs, less commonly in conifers. Often well away from trunk in a fork or cluster of vertical branches and well concealed by vegetation from sides and above. Nest is compactly woven of plant materials and lined with plant down, fur, feathers, or other fine material. Lining may not be as thick as in nest of American Goldfinch. Nest averages 4–6 m aboveground but may be up to 14 m. Out-

Rick Bowers, Bowersphoto.com, AZ

Casey McFarland, AZ

side diameter 7.5 cm; height 5 cm; inside diameter 4.5 cm; depth 3 cm. **EGGS:** 4–5; pale blue-white or green; ovate; 15 x 12 mm; IP 12–13 days (F); NP 11–15 days (F, M). **BEHAVIOR:** Assumed to be monogamous.

LAWRENCE'S GOLDFINCH *Spinus lawrencei*

HABITAT: Open woodlands, chaparral, areas of brushy cover with adjacent weedy fields and water. **LOCATION AND STRUCTURE:** Variety of deciduous and coniferous trees, often in blue oak. In a fork of multiple, often drooping, branches on a main branch about 1–2 m from end. Outer layer is a loose cup of green leaves, grasses, occasionally flowers. Inner lining consists of fine plant fibers, down, feathers, fur. Lichens are often included in construction or as decoration. Nest is typically 3–7 m high, up to 13 m recorded. Outside diameter 7.7 cm; height 5.3 cm; inside diameter 4.2 cm; depth 2.7 cm. **EGGS:** 5; white (sometimes pale blue); oval to elliptical; 15 x 12 mm; IP 12–13 days (F); NP 13–14 days (F, M). **BEHAVIOR:** Monogamous. Sometimes semicolonial, especially in planted ornamental evergreens.

AMERICAN GOLDFINCH *Spinus tristis*

HABITAT: Fields and agricultural areas, forest edge, often associated with sunflower-family annuals. **LOCATION AND STRUCTURE:** Prefers deciduous shrubs. Built in crotch of a branch that has multiple upright stems under leafy cover, though often visible from below. Nest is 0.5–3 m high, sometimes considerably

Mark Nyhof, BC

Chris Helzer, NE

Maren E. Gimpel, MD

higher, and often high relative to overall height of substrate. Nest is a tight cup woven around supporting twigs with spider silk. Outer layer is constructed of rootlets and other plant material. Thick downy lining of plant fibers is characteristic. *Similar to nest of Yellow Warbler.* Outside diameter 8 cm; height 7.2 cm; inside diameter 5 cm; depth 3.5 cm. **EGGS:** 3–6; pale blue; subelliptical; 17 x 12 mm; IP 12–14 days (F); NP 11–17 days (F, M). **BEHAVIOR:** Socially monogamous, though some females practice polyandry, abandoning first brood to be cared for by male, while beginning a second nest and clutch with a different male. Possibly the latest-breeding N. American songbird, often not commencing until mid-July. Though commonly parasitized by cowbirds, cowbird nestlings do not survive because diet provided by goldfinches is not suitable.

LONGSPURS and SNOW BUNTINGS (Calcariidae)

These small songbirds occupy tundra and grassland habitats. Six species breed in N. America, all but McCown's and Chestnut-collared Longspurs only in or near AK and within the Arctic Circle. All are *ground nesters* (Snow Bunting in crevices), and females are the primary builders. Nests are typically set in depressions or hollows and are tightly woven, mostly of grass leaves and stems, though some species may use other materials such as mosses, dead leaves, bark shreds, rootlets, and similar. Nests are lined with soft materials such as fine grasses, rootlets, and hair, and feathers may be added. Nests are generally of similar dimensions, varying in part because of the size of the depression in which they're set. Outside diameter ~8–9 cm (can be larger); inside diameter ~6 cm; depth ~5 cm.

Calcariids tend toward monogamy or social monogamy, but variations in breeding behavior occur, Smith's Longspur males and females both have 1–3 mates per clutch. Eggs are smooth and glossy and vary in color among and within species. They are typically subelliptical to short subelliptical; pale olive gray, gray, blue-gray, brown, or pinkish cream to cream; and marked with dark speckles, spots, blotches, and scrawls. The female incubates, and both parents care for young. Incubation period 10–15 days, NP 10–14 days, varying by species.

CHESTNUT-COLLARED LONGSPUR *Calcarius ornatus*

HABITAT: Short-grass prairie, but often in grasses or vegetation that is *taller, denser than that preferred by McCown's Longspur,* with which Chestnut-collared shares habitat. **LOCATION AND STRUCTURE:** Placed next to vegetation clump, cow dung, but sometimes in open, commonly with concealment from above, often in tallest, thickest vegetation in vicinity, with more ground debris. Nest rim is often flush with ground surface. See McCown's Longspur. **EGGS:** 3–5; 19 x 14 mm. **BEHAVIOR:** Double-brooded. Found frequently in black-tailed prairie dog towns.

Chestnut-collared Longspur.
John Pulliam, MT

Chestnut-collared Longspur.
Doug Backlund, SD

MCCOWN'S LONGSPUR *Rhynchophanes mccownii*

HABITAT: Restricted to open, sparse, semiarid short-grass prairie or similar (overgrazed pasture). **LOCATION AND STRUCTURE:** Open cup is often placed next to or at base of prickly pear cactus, bunch grass clump, shrub, or other vegetation. Lined primarily with fine grasses, occasionally hair, feathers. *Very similar to nest of Horned Lark (which shares same habitat). Chestnut-collared Longspur and Lark Bunting nests are also similar, but often have sheltering canopy of vegetation, roots, earthen cut, etc.* **EGGS:** 3–4; 20 x 15 mm. **BEHAVIOR:** Typically double-brooded. Build nest in 3–5 days. As in many grassland species, populations and distributions are declining rapidly. This species benefits from grazing that creates short-cropped vegetation for nesting ground.

Sean C. Hauser, CO

Sean C. Hauser, CO

ARTIC-BREEDING
LONGSPURS and SNOW BUNTINGS

LAPLAND LONGSPUR *Calcarius lapponicus*

HABITAT: Common and abundant breeder in Arctic. Moist or wet tundra meadows, also on drier mountain slopes with vegetative cover. **LOCATION AND STRUCTURE:** Usually well concealed, but sometimes open to above on flat, dry ground. Often with water nearby, at base of grass clump, in bank or in hummock or tussock of grass, moss, willow, etc. Nest is like that of other longspurs but usually *thickly lined with feathers or fur.* Does not typically nest in rockier habitats of Snow Bunting, but the 2 species can overlap where rocky areas and vegetated tundra meet. **EGGS:** 5–6; 21 x 15 mm. **BEHAVIOR:** Single-brooded.

Kate Fremlin,
Amaroq Wildlife
Services, NU

SMITH'S LONGSPUR *Calcarius pictus*

Habitat similar to that of Lapland Longspur, but typically more associated with forest-tundra transition zones. Nest and location are also similar, *but nest lining usually has fewer feathers.* **EGGS:** 3–4; 21 x 15 mm. **BEHAVIOR:** Single-brooded.

SNOW BUNTING *Plectrophenax nivalis*

MCKAY'S BUNTING *Plectrophenax hyperboreus*

HABITAT: Snow Bunting: Tundra, in rocky bare areas, rocky patches, boulder fields, scree. McKay's Bunting: Nests only on islands in Bering Sea, similar habitat to Snow. **LOCATION AND STRUCTURE:** Snow *in cavities, crevices, and fissures of rocks; beneath boulders,* or similar settings, typically difficult to see. Readily uses artificial nesting sites and structures (buildings, old cans, barrels, etc.) where less natural sites are less available. McKay's mostly in scree, and fissures, crevices, or holes in rocks. May nest in driftwood, etc. Nests are like that of other Arctic

The crevice in the center of the frame is the entrance to the space where a pair of McKay's Buntings constructed their nest. Rachel M. Richardson, AK

The inner cup of a Snow Bunting nest lined with feathers. Joe Fuhrman, AK

The full cup nest of a McKay's Bunting removed from the crevice in which it was constructed. Rachel M. Richardson, AK

longspurs; often made of mosses or grass, *heavily lined with feathers, fur*. **EGGS:** 4–6, as many as 9; 23 x 16 mm. **BEHAVIOR:** Single-brooded. Males return to their harsh breeding grounds early in season, likely because competition for nesting sites is so high; subzero temperatures and late-season storms may exact a heavy toll.

NEW WORLD SPARROWS (Passerellidae)

Small to medium-sized songbirds, New World Sparrows are primarily insectivorous during the breeding season but feed largely on seeds during nonbreeding months. Forty-three species nest in N. America in a wide range of environments, from deserts to Arctic tundra and from sea level to high-elevation montane ecotypes. Habitats are often dominated by low vegetation and shrubby growth, including coastal and desert scrub, salt- and freshwater marshes, brushy fields, old-growth forests, montane meadows, and woodland edges. Nests are typically open cups (some are loosely domed), often well concealed by vegetation, and can be found on the ground, beneath a shrub or tree, or elevated within thickets or tree branches. Females are the primary builders.

Sparrow nests (excluding those of towhees) are fairly commonly found and can look similar to the nests of members of other families, such as those of some flycatchers. Sparrow nests are typically *built mostly of grasses and grass stems and other long plant materials*, and *contain fewer woody twigs* overall than do nests of many other species. *They also generally lack spider silk.* The inner cup is commonly lined with fine grasses and is more compactly made than the outer cup. Many sparrow nests have a distinctive appearance in that the *upper surface of the outer cup often spreads outward, like a flattish disk or rim that is wider than the cup below it.* The inner cup can be fairly broad, often appearing wider than deep.

Thirty-five sparrow species can be roughly divided into 3 general nesting strategies. About 13 species build almost exclusively on the ground, about 17 build both on the ground or low in shrubs or other vegatation, and about 5 build almost exclusively in trees and shrubs. Some species will build nests on the ground early in the season and move upward as foliage grows and provides better concealment. Species below are ordered taxonomically, not by location or nest placement. Use the Sparrow Chart (pp. 386-87) as a guide when in the field, and also to see where various species (towhees not included) nest and which nests are similar.

Junco nests are primarily on the ground and are similar to those of many ground-nesting sparrows: small, tidy, well concealed, and typically set into a slight hollow that neatly contains the nest.

Towhee nests are *larger than those of sparrows* and can be quite bulky. Some nests are well hidden and difficult to find, whereas others, such as those of Canyon Towhee, are fairly commonly discovered. Where near human habitation, nests *frequently contain bits of garbage and other debris.* Towhees nest on the ground or elevated in shrubs or low in trees.

Most passerellids are monogamous or assumed so, though exceptions occur in several species. Many populations of Savannah Sparrows are polygynous, for instance, and Saltmarsh Sparrows display an unusual form of aggressive promiscuity. Passerellids generally lay 3 or 4 subelliptical eggs. Ground color ranges from white to off-white, pale bluish, or greenish white, and they are either unmarked or marked in shades of browns, olives, grays, and purples. Depending on the species, they may be profusely patterned across the entire egg or primarily at the larger end.

Passerellids may be single- or double-brooded. Females incubate for 10–14 days. The altricial young are cared for by both sexes, and in some species males may provide full parental care of a prior brood when the female renests. Depending on the species, young fledge in 6–14 days, with most species fledging around 8–12 days. Young may fledge prematurely if disturbed.

Left to right: Bachman's Sparrow, Grasshopper Sparrow, Song Sparrow, Lark Sparrow. Sparrow eggs show wide differences in color and markings, which can be useful for identification.

TOWHEES

Left to right: Green-tailed, California, and Canyon Towhees. California Towhee eggs, similar to those of Abert's Towhees, are much less marked than Canyon Towhee eggs.

GREEN-TAILED TOWHEE *Pipilo chlorurus*

HABITAT: Arid, brushy hillsides with scattered trees, up to 3,600 m elevation. Common in areas with understory of sagebrush or other low bushes, such as chokecherry, manzanita, mahogany, snowberry, antelope brush. Also in open pinyon-juniper woodlands, montane forest edges, or burned areas. **LOCATION AND STRUCTURE:** Often hidden from view in *low bush*, usually within 0.6 m of ground. Large, deeply cupped nest has thick outer cup foundation of twigs and stems, and inner cup of grasses, bark, weed stems, pine needles. Lined with fine plant stems, rootlets, hair. Outside diameter ~13–15 cm; height 7.6–8 cm; inside diameter 6–7 cm; depth 4–5 cm. **EGGS:** 3–4; moderately glossy, white, speckled with reddish browns, often at larger end; subelliptical to short oval; 22 x 16 mm; NP 11–14 days (F, M). **BEHAVIOR:** May reuse or build on top of previous year's nest.

Matt Monjello, AZ

SPOTTED TOWHEE *Pipilo maculatus*

HABITAT: Areas with dense brushy growth with thick ground litter, with or without tall trees: chaparral, dry thickets, brushy fields, forests with shrubby understories or edges, coulee or riparian thickets, and canyon bottoms. **LOCATION AND STRUCTURE:** *On ground,* in a depression under a low bush, log, or grass tuft, or under cover along thicket margins. Occasionally 0.6–3.6 m aboveground in woody or herbaceous vegetation. Nest is a firmly built cup of dry leaves, bark strips, grasses, and grass or plant stems. Lined with fine vegetation, occasionally hair. Outside diameter 11 cm; inside diameter 7–8 cm; depth 4–6 cm. **EGGS:** 3–5; smooth and slightly glossy, ground color variable, pale, finely marked with reddish brown, light purple, gray; markings abundant with heavy concentrations at larger end; oval to short oval; 24 x 18 mm.

Mark Nyhof, BC

Chris Byrd, WA

EASTERN TOWHEE *Pipilo erythrophthalmus*

HABITAT: Wide variety with low, brushy understory and thick ground litter, including deciduous or mixed deciduous-coniferous woodlands, forest edges, and interior shrubby clearings, riparian thickets, brushy fields, savanna, montane slopes. **LOCATION AND STRUCTURE:** Typically *on ground* under a low bush or tuft of vegetation. May or may not be well hidden. Sometimes nests aboveground in dense vegetation, on stumps, in root balls of fallen trees, usually within 1.5 m. *Nest is similar to Spotted Towhee's.* Occasionally domed. Lined with fine grasses, rootlets, bark strips, and other plant material, hair. **EGGS:** 3–5; *similar to Spotted Towhee's;* may be wreathed at larger end or marked evenly throughout; 23 x 17 mm.

Kris Spaeth/USFWS, MN

CANYON TOWHEE   *Melozone fusca*

HABITAT: Wide range, including dry riparian scrub and mesquite woodlands, grasslands interspersed with cacti and dense shrubs, brushy canyons, pine-juniper-oak forests, rural areas. *Avoids densely urbanized settlements. Typically in drier habitat than California and Abert's Towhees.* **LOCATION AND STRUCTURE:** Well concealed in vines, small tree, shrub, cactus, 0.6–3.7 m above-ground. Nest is a bulky cup built of grass, weed stems, occasionally trash. Lined with leaves, fine vegetation, bark strips. Outside diameter 14–16 cm; height 14 cm; inside diameter 5–6 cm; depth 7.6 cm. **EGGS:** 3; smooth and slightly glossy, pale blue-white to off-white, usually marked at larger end with dark brown spots and scrawls, fine markings of light purple and pale gray across whole egg; *more heavily marked than California Towhee eggs*; oval to short oval; 23 x 18 mm.

Casey McFarland,
NM

CALIFORNIA TOWHEE   *Melozone crissalis*

HABITAT: Variety with low, brushy growth, including dense coastal and foothill chaparral scrub, riparian areas and canyon bottoms near open, arid hillsides, in pinyon-juniper associations, oak woodlands, densely vegetated parks and gardens. *More common in wetter environs and urbanized areas than Canyon Towhee.* **LOCATION AND STRUCTURE:** Often 1.2–3.7 m aboveground, may be lower or up to 11 m high. Rarely on ground. Nest is often in heavily foliaged area of a shrub or tree, supported by several branches. A loose cup of dried grasses, stems, and flowers, sometimes plastic. Lined with bark strips, fine stems, grasses,

California Towhee. Alan Vernon, CA

California Towhee. Chris McCreedy, CA

hair. Outside diameter 17–22 cm; depth 3.5 cm. **EGGS:** 3–4; pale bluish or creamy white, *with fewer markings than Canyon Towhee eggs*; sparingly marked with spots and blotches of dark brown and purple, sometimes with fine scrawls and light purple markings, typically at larger end; subelliptical to long subelliptical; 25 x 19 mm.

ABERT'S TOWHEE *Melozone aberti*

HABITAT: Common in dense, shrubby, riparian thickets, including understory in cottonwood-willow riparian forests, desert thickets, and mesquite woodlands along perennial rivers and tributaries, irrigation ditches, wetlands, bushy edges of agricultural

Casey McFarland, AZ

Abert's Towhee.
Matt Monjello, AZ

areas, urban environments. *Generally in wetter areas than Canyon Towhee.* **LOCATION AND STRUCTURE:** Usually in low shrubs and small trees (*Baccharis*, mesquite, elderberry, cottonwood, willow), also in mistletoe clumps; 1.5–2.24 m up, sometimes much higher. Nest is a bulky cup of green leaves, bark strips, grasses. Lined with inner bark strips, grasses, sometimes hair. Outside diameter 10–17.5 cm; height 10–12.5 cm; inside diameter 5–10 cm; depth 3–6.9 cm. **EGGS:** 3; *similar to California Towhee's*; smooth and slightly glossy; subelliptical; 24 x 18 mm.

SPARROWS

OLIVE SPARROW *Arremonops rufivirgatus*
HABITAT: Brushy fields and dense thorny thickets, especially mesquite, Texas ebony, huisache. **LOCATION AND STRUCTURE:** *Typically in a low shrub,* within 1.5 m of ground, usually in dense growth. Nest is large for size of bird, *often domed, cylindrical or round in shape.* Made of twigs, stems, grass, bits of bark, leaves. Usually unlined but may contain hair. **EGGS:** 4–5; smooth, glossy, *white to light pink, unmarked*; subelliptical; 22 x 16 mm.

RUFOUS-CROWNED SPARROW *Aimophila ruficeps*
HABITAT: Semiarid, grassy, rock-strewn slopes and canyons with low shrubs or trees; also open, semiarid forests of short pinyon-juniper and oak-juniper, pine-oak; scattered coastal scrub. Avoids dense, pure, unbroken tracts of chaparral and other similar vegetation. To 3,000 m elevation. **LOCATION AND STRUCTURE:** *Nest is usually on ground* in a slight depression at base of grass clump, low shrub, or rock. Often well concealed by overhanging substrate. Rarely up to 45 cm in low shrub. Nest is a loosely built

cup with thick wall. Made of dried grasses and rootlets, occasionally with twigs, plant stems, bark strips. Lined with fine grasses, hair, plant fibers. Outside diameter 7.6–15.2 cm; height 4.8–10.2 cm; inside diameter 2.5–7.6 cm; depth 2.5–6.4 cm. **EGGS:** 3–4; smooth and slightly glossy, *pale bluish white and unmarked*; subelliptical; 19.5 x 15.2 mm.

RUFOUS-WINGED SPARROW *Peucaea carpalis*

HABITAT: Desert scrub and grasslands interspersed with trees, shrubs, and cholla, usually near riparian washes. **LOCATION AND STRUCTURE:** *Conspicuous in fork of a small tree, shrub, or cactus* (palo verde, hackberry, cholla, prickly pear, mesquite), 0.46–2.5 m up. Infrequently in mistletoe clump. Nest is a deep, firm cup of weed stems and coarse grasses, lined with finer grasses, sometimes hair. Outside diameter 8.5–12 cm; height 7 cm; inside diameter 4–5.5 cm; depth 5–6.5 cm. **EGGS:** 3–4; smooth and slightly glossy, *very pale bluish white, unmarked*; subelliptical; 19 x 14 mm; NP 7–10 days (F, M). **BEHAVIOR:** Nesting usually begins in summer with onset of monsoon rains, but may be earlier in wetter years.

BOTTERI'S SPARROW *Peucaea botterii*

HABITAT: In se. AZ and sw. NM: In semiarid habitats, including bottom grasslands interspersed with trees, ocotillo, next to grassy slopes and plateaus. Upland in open mesquite, oak and mixed oak grasslands with scattered shrubs and dense tall grass. Also yucca grassland. In TX: Along saltgrass prairies interspersed with small bushes and cacti, in mixed mesquite-yucca-prickly pear grasslands, in dry pastures with scattered mesquite, along coast in tall bunch grass. See Cassin's Sparrow. **LOCATION AND STRUCTURE:** *On ground, well concealed at base of grass tuft in dense grasses* tucked far back in base or nearer to edge. Always close to perches used for song and lookouts. Nest is a shallow cup of grasses, lined with finer grass. Inside diameter 7 cm; depth 6 cm. **EGGS:** 4; *white and unmarked*; oval to short oval; 20 x 15 mm.

CASSIN'S SPARROW *Peucaea cassinii*

HABITAT: Desert grasslands with mesquite, sagebrush, rabbitbrush, yucca, hackberry, oak, etc. Avoids dense thickets. Also in bunch grass (TX coast), juniper-mesquite grassland (w. TX), along grass-covered cienegas (se. AZ). **LOCATION AND STRUCTURE:** Usually *on ground* or slightly elevated in a grass tuft, *or low (usually below 20 cm) in a small shrub, brush, cholla, or yucca.* Often oriented facing north or northwest. Nest is a cup of dry grasses, forb stems, fibers. Lined with fine grass, rootlets, grass heads, sometimes hair. Outside diameter 8.8–10.2 cm; height 8.2–11 cm; inside diameter 5.3–8.8 cm; depth 4.8–6.5 cm. **EGGS:** 4; slightly glossy, *white, unmarked*; subelliptical; 19 x 14.6 mm; NP 7–9 days (F, M). **BEHAVIOR:** Breeding times and abundance of species vary yearly and throughout range.

Cassin's Sparrow.
Rick Bowers,
Bowersphoto.com,
AZ

BACHMAN'S SPARROW *Peucaea aestivalis*

HABITAT: Prefers moderately open areas with little shrub cover and dense herbaceous groundcover: open pine woodlands, oak savanna, weedy abandoned fields, power line right-of-ways, logged areas, roadcuts, calcareous glades, etc. **LOCATION AND STRUCTURE:** *On ground at base of a grass clump, low shrub, pine seedling, palmetto*; rarely built a few centimeters aboveground in herbage or grass tuft. Nest is a cup of grasses, forb stems, root-lets. Lined with fine grasses, hair. Nests in southern range *often are domed*, with an arch of woven grass covering cup. In northern range, most nests are open cups. Dome length ~15 cm; dome diameter 13 cm (to 19 cm); height (with dome) 10–20 cm; inside

Casey McFarland,
AL

Bachman's Sparrow.
Eric C. Soehren, AL

diameter 6–7 cm. **EGGS:** 3–4; smooth and slightly glossy, *white and unmarked*; oval; 19 x 15 mm. **BEHAVIOR:** Very shy and elusive. A species of concern.

AMERICAN TREE SPARROW *Spizelloides arborea*

HABITAT: Remote areas along transitional zone of northern tree line and tundra: deciduous scrub thickets, open tundra interspersed with low shrubby growth; dwarf spruce forest openings; often near water. **LOCATION AND STRUCTURE:** *Nest is typically on ground* in a grass tuft under a small tree or shrub, sometimes in open tundra on moss hummock; rarely aboveground and up to 1.2 m in dwarf willow or spruce. Nest is a cup of coarse grass, twigs, bark strips, moss, lichen. Lined with finer dried grasses and feathers (often ptarmigan). Outside diameter 11–15 cm; inside diameter 4–5 cm; depth 3–4.5 cm. **EGGS:** 5; pale blue or greenish, marked with speckles, spots, and blotches of brownish reds and purples, often at larger end; 19 x 14 mm.

Mark Peck, ON

CHIPPING SPARROW *Spizella passerina*

HABITAT: Broad range of habitats with shrubby undergrowth that is often adjacent to grassy openings, including open coniferous, deciduous, and mixed forests, woodland edges and clearings, along edges of montane meadows, riparian shorelines, in parks, orchards, farms, suburbs. **LOCATION AND STRUCTURE:** *Often in a conifer*, but will use a large diversity of trees and shrubs. Commonly 1–3 m off ground. May nest on ground or very high in tall trees. Nest is concealed by vegetation, often placed on outer margin of tree or shrub. Nest is a somewhat flimsy cup (light is often visible through walls) of rootlets, weed stems, dry grasses, possibly stringy lichens, similar. Lined with fine grass, hair, sometimes plant fibers. Outside diameter ~6–11 cm; height 4.5–9 cm; inside diameter 4–6 cm; depth 3–5 cm. **EGGS:** 4; light blue, sparingly marked on larger end with streaks, blotches, or spots of black, brown, light purple, occasionally unmarked; subelliptical to short subelliptical; 18 x 13 mm.

Casey McFarland, OR   Mark Nyhof, BC

CLAY-COLORED SPARROW *Spizella pallida*

HABITAT: Grasslands with scattered low shrubs, reseeded grasslands with alfalfa or clover, in shrubby growth along riparian and woodland edges, regenerating second-growth stands and burns, in brushy fields, young conifer plantings. More common in areas unaltered by agriculture and heavy grazing. *Generally in more open habitat than Chipping Sparrow and in denser, more diverse vegetation than Brewer's Sparrow.* **LOCATION AND STRUCTURE:** Generally well concealed. *Nest is on or slightly above ground in a grass clump or built into low dense shrub* (often snowberry and rose); also young pines. To 1.5 m up. Local populations have tendency to specialize in one particular nest site. Nest, *similar to Chipping Sparrow's,* is a cup of woven grasses, thin twigs, rootlets, weed stems. Lined with fine grasses, rootlets, hair.

Clay-colored Sparrow.
Jacqueline Huard, AB

Outer diameter 8–14 cm; height 5.5–9 cm; inside diameter 4–5.4 cm; depth 2.5–4.8 cm. **EGGS:** 4; *similar to Chipping Sparrow's*; 17 x 13 mm.

BREWER'S SPARROW *Spizella breweri*

HABITAT: Ubiquitous in open, arid brushlands (particularly big sagebrush). Also in open in areas of coniferous woodlands, in shrubby, willow, and dwarf birch–dominated areas at or above tree line (northwestern population). **LOCATION AND STRUCTURE:** *Nest is in a low, dense shrub (but usually tallest among others), often sagebrush,* well concealed on outer margins of substrate. Height variable, 8.89–67 m, often increasing as season advances. Nest is a firm cup of dead grasses, weed stems, rootlets, occasionally with sagebrush twigs on outer layer. Lined with fine dried grasses, bark strips, rootlets, hair. *Where range overlaps with Sagebrush and Clay-Colored Sparrows, Brewer's generally places nest in spots with less overhead vegetation than Sagebrush, and occupies less dense and diverse stands of vegetation than Clay-*

Tiffany Linbo, NV

Colored. Outside diameter 7–15 cm; inside diameter 4–7 cm; depth 3–5 cm. **EGGS:** 3–4; smooth and moderately glossy, blue-green, speckled, blotched, and spotted with dark brown, often at larger end; subelliptical; 17 x 13 mm; NP 6–9 days (F, M).

FIELD SPARROW *Spizella pusilla*

HABITAT: Pastures and abandoned fields with scattered shrubs, and edges and in open areas in woodlands; fencerows, roadsides. Avoids similar habitats close to human development. Sometimes in orchards, tree plantations. **LOCATION AND STRUCTURE:** Nest is usually located fairly close to woody vegetation. Early season nests are *on or slightly above ground in a grass clump*; as season advances, often *in a low shrub or sapling*, occasionally as high as 3 m up. Nest is a cup of stems, grasses. Lined with finer grasses, rootlets, and hair. *Nest's grass-stem foundation may have a spokelike radiating pattern when built off ground.* Outside diameter 8.5–21 cm; height 4.7–11 cm; inside diameter 4.3–5.5 cm; depth 2.5–6.2 cm. **EGGS:** 3–5; smooth, slightly glossy, white, creamy, occasionally tinted light blue or greenish, spotted medium brown, reddish brown, purplish brown, with markings usually concentrated at larger end, possibly throughout entire egg; subelliptical; 18 x 13 mm; NP 7–8 days (F, M).

Matt Monjello, ME

Bill Summerour, GA

BLACK-CHINNED SPARROW *Spizella atrogularis*

HABITAT: Dry, brushy, rock-strewn hillsides interspersed with a scattering of shrubs or trees. Up to 2,700 m elevation. **LOCATION AND STRUCTURE:** Often on *south-facing slopes* containing mixed stands of tall and fairly thick brush (1–2 m high), including chamise, chaparral, sagebrush, ceanothus, manzanita, oak scrub, pinyon-juniper, etc. Avoids heavily vegetated areas, instead preferring sites with younger brush and where there are gaps among vegetation. At least somewhat associated with areas that have been burned. Nest is well concealed within a *low, dense shrub* (often sagebrush), up to 1.5 m high. Sometimes in a dead shrub. Nest is a loose cup of long, dead grass stems, occasionally plant fibers, weed stems. Lined with finer grasses, hair, plant fibers, sometimes feathers. Outside diameter 8.3–12.7 cm; height 5.8–11 cm; inside diameter 3.8–5.3 cm; depth 3–5 cm. **EGGS:** 3–4; light blue, *unmarked or sparsely spotted* with dark brown or reddish brown, often at larger end; subelliptical; 18 x 13 mm. **BEHAVIOR:** May nest in small colonies.

VESPER SPARROW   *Pooecetes gramineus*

HABITAT: Wide range of open, dry, sparsely vegetated grasslands interspersed with bare ground, including old-fields, short-grass pastures, hayfields and other cultivated areas, sagebrush steppe, montane short-grass meadows, grown-over surface mines, fencerows bordering fields, weed-covered roadsides, prairies. Also aspen groves, open or arid woodlands and openings. **LOCATION AND STRUCTURE:** *On ground*, often in a slight depression, well concealed at base of a weed or grass tuft or shrub, often in dense vegetation surrounded by sparser vegetation; next to or under a log or branch; sometimes placed in open spot. Nest is a loosely woven, shallow cup of dry grasses, sedges, forbs, rootlets, small twigs, bark strips, moss. Lined with fine grasses, hair, rootlets, sometimes feathers and pine needles. Occasionally domed (see Eastern Meadowlark and Western Meadowlark account). Outside diameter 7–10 cm; height (depends on depth of depression) ~2–5 cm; inside diameter 5–6 cm; depth to ~4 cm. **EGGS:** 3–5; smooth, slightly glossy, creamy white to pale greenish white, spotted, speckled, blotched, and scrawled with shades of brown, purple, gray; subelliptical to oval; 21 x 15 mm; NP 7–14 days (F, M).

R. Adam Martin, WA

John Pulliam, MT

Vesper Sparrow.
Mark Nyhof, BC

LARK SPARROW *Chondestes grammacus*

HABITAT: Upland and lowland prairies, overgrazed pastures, brush-covered pastures, sandhills, shrub-steppe, floodplain edges, open woodlands, mesquite grasslands. Often in open areas with a grass or weedy groundcover and scattered shrubs and trees, including shelterbelts and shrubby areas bordering

Nicholas J. Czaplewski, OK

Casey McFarland, NM

This nest was placed at the base of a cholla cactus. Chris Helzer, NE

old-fields, orchards, etc. **LOCATION AND STRUCTURE:** Usually in a depression *on ground* concealed by a grass or weed clump or shrub. *Also in a shrub or small tree* up to 2.75 m high. Nest is an open cup with a thick wall. Made of grass and forb stems, small twigs, bark strips, lined with fine grasses, rootlets, hair. Elevated nests tend to be bulkier. Outside diameter 9.9–13.7 cm; inside diameter 5.8–7 cm; depth 2.5–4 cm. **EGGS:** 4–5; smooth, glossy, white, creamy, or grayish white, spotted, blotched and scrawled with dark browns, black, purple, often capped; short subelliptical to subelliptical; 20 x 16 mm. **BEHAVIOR:** Will use old nests of other birds (often thrashers, Northern Mockingbird). Male passes twigs to female during copulation.

FIVE-STRIPED SPARROW *Amphispiza quinquestriata*

HABITAT: Limited to and local in se. AZ. Found at 1,060–1,200 m elevation in canyons and on steep rocky hillsides with dense vegetation of grasses and shrubs, such as mesquite, ocotillo, acacia, hackberry; at higher elevations on oak-juniper slopes. *Habitat is always near a perennial water source (tinaja).* **LOCATION AND STRUCTURE:** *In a grass tuft against ocotillo, or in a small shrub* (hackberry, turpentine bush, hopbush, etc.). Nest is a deep, open cup of grasses, stems. Lined with fine grasses, hair. Outside diameter 7.9–12 cm; height 6.4–14 cm; inside diameter 5–7.3 cm; depth 4.7–9 cm. **EGGS:** 3–4; dull *white and unmarked*; subelliptical; 20 x 16 mm. **BEHAVIOR:** Usually begins nesting after onset of monsoon rains.

BLACK-THROATED SPARROW *Amphispiza bilineata*

HABITAT: Variety of arid, open habitats with a scattering of shrubs, trees, and cacti, such as semiarid grasslands, desert washes and canyons, on brushy hillsides with or without grassy understory, along streams in sparsely grassed pastures, desert scrublands, sagebrush steppe, pinyon-juniper woodlands. Up to 2,100 m elevation. **LOCATION AND STRUCTURE:** *Nest is in a low*

Chris McCreedy, CA

shrub, cactus, occasionally a small tree. Common in creosote bush, cholla, creosote and bush-snakeweed tangles, apache plume. Usually placed on a fork. Will also build in acacia, mesquite, and other shrubs. Nest is an open cup of plant fibers, coarse grasses, plant stems, thin twigs. Lined with finer grasses and hair. Typically 10.2–76.2 cm aboveground. Outside diameter 8.2–13 cm; height 6–10.4 cm; inside diameter 5.4 cm; depth 4 cm. **EGGS:** 3–4; smooth and glossy, *white to very light blue, unmarked*; subelliptical; 18 x 14 mm.

SAGEBRUSH SPARROW   *Artemisiospiza nevadensis*

HABITAT: Arid scrublands. Often in habitats dominated by big sagebrush with other brushy growth. Also sagebrush or juniper woodlands bordering open, shrubby flats. Usually below 1,700 m in elevation but may be higher. **LOCATION AND STRUCTURE:** Usually in central portion of shrubs, but sometimes in grass clumps under a low shrub, usually low to ground (24–27 cm). Nest is an open cup of thin twigs, sticks, and stems. Lined with finer grasses, bark shreds, feathers, wool, hair. Outside diameter 9–15 cm; inside diameter 5–8 cm; depth depends on placement, ~2–6 cm. **EGGS:** 3–4; smooth and slightly glossy, light blue or bluish white, speckled, spotted, and blotched with brown, red brown, black, occasionally with lines; markings may be over entire egg or concentrated at larger end; subelliptical; 19 x 15 mm.

BELL'S SPARROW   *Artemisiospiza belli*

HABITAT: Chaparral, coastal and desert scrub-dominated areas. Also in big sagebrush (southern mountains). Typically less abundant in taller chaparral than in shorter regenerating stands. **LOCATION AND STRUCTURE:** *Nest and location are similar to that of Sagebrush Sparrow, but inner cup may be smaller.* Size variable; outside diameter 9–13 cm; inside diameter 5–7 cm; depth 3–5 cm. **EGGS:** *Similar to Sagebrush Sparrow's but smaller.*

Garrit Vyn, WY

LARK BUNTING *Calamospiza melanocorys*

HABITAT: Short-grass prairies, grassy and weedy sagebrush steppe, cultivated hayfields. **LOCATION AND STRUCTURE:** *Ground nest,* usually in a depression with rim level with ground or slightly elevated. Typically beneath overhanging grass or weed clump, low shrub, or cactus. Often oriented to side of substrate that receives least amount of sun or wind exposure. Nest is a simple, loose cup of dried grasses, blades, forb stems, rootlets. Lined with finer grasses, hair. Outside diameter 9 cm; inside diameter 7 cm; depth 4 cm. **EGGS:** 4–5; smooth, slightly glossy, *similar to those of Dickcissel, pale blue to light greenish blue and unmarked;* short oval to oval; 22 x 17 mm.

Rick Bowers,
Bowersphoto.com,
CO

SAVANNAH SPARROW *Passerculus sandwichensis*

HABITAT: Open habitats: grass-covered meadows, pastures, prairies and coastal dunes, alfalfa and other agricultural fields, saltmarsh, estuaries, sedge bogs, tundra, roadside edges. Northernmost populations: Often associated with open dwarf willow or birch habitat. Arid populations: In grasses along pond edges and other irrigated sites. *Generally avoids heavily wooded areas.* **LOCATION AND STRUCTURE:** *On ground* in a shallow depression beneath tufts of grass or at base of low shrubs, well concealed beneath grass or forbs. Outer cup of coarse grasses with inner cup of finer grasses, rarely feathers. Outside diameter 7.5 cm; inside diameter 4.9–8.3 cm; depth 2.5–6.6. **EGGS:** 4; smooth, slightly glossy, pale greenish, bluish, or dirty white, speckled, blotched, spotted with shades of brown, often concentrated at larger end; subelliptical; 19 x 15 mm.

Matt Monjello, ME

GRASSHOPPER SPARROW *Ammodramus savannarum*

HABITAT: Commonly in relatively dry, open grasslands and prairies of intermediate height with patches of bare ground and sparsely scattered shrubs. Also cultivated fields, and in FL in palmetto prairie. Inhabits less vegetated and drier sites in Midwest and East, and more verdant and shrubbier areas in drier grasslands of Southwest and West. *Generally avoids grasslands that are densely vegetated with shrubs. Usually in drier and more open sites than Savannah Sparrow.* **LOCATION AND STRUCTURE:** *On ground* in a slight depression with nest rim level or slightly elevated from ground, at base of a grass or weed tuft or shrub, well concealed by overhanging vegetation. Nest is an open cup of grasses or stems, *commonly domed at back and sides.* Lined with fine grasses, rootlets, and hair. Outside diameter 11–14 cm; height 5–7 cm; inside diameter 6–8.5 cm; depth 3–4 cm. **EGGS:** 4–5; smooth, slightly glossy, creamy white, spotted, blotched, and speckled with reddish brown, may be scattered over entire egg or concentrated at larger end; short elliptical to subelliptical; 19 x 14 mm. **BEHAVIOR:** May nest in small colonies.

Megan Milligan, MT

Chris Smith, MT

BAIRD'S SPARROW *Centronyx bairdii*

HABITAT: Ungrazed to fairly well grazed native prairies mixed with scattered low woody growth and mats of dead herbaceous vegetation, sometimes in hayfields. Also dry shallow ponds and channels during dry seasons. **LOCATION AND STRUCTURE:** *Prefers sites well covered with dead or taller vegetation, a good amount of litter.* Nest is *on ground* in a natural depression or scrape, concealed by overhanging grass clump, a grass clump and shrub, or in more open spot with *no overhanging vegetation.* Nest is a shallow cup of coarse herbaceous stems, leaves. Lined with fine grasses, rootlets, sometimes hair, moss, twine. Inside diameter

Baird's Sparrow.
John Pulliam, MT

~6 cm; depth 4–4.6 cm. **EGGS:** 4–5; slightly glossy, grayish white, heavily spotted, blotched, and speckled with reddish brown, usually covering entire egg; oval to subelliptical; 19.4 x 14.6 mm. **BEHAVIOR:** May breed in loose colonies.

HENSLOW'S SPARROW   *Centronyx henslowii*

HABITAT: Often in tall, thick, grassy fields and meadows with dense dead or upright vegetation, a thick litter layer, with or without scattered shrubs. Also weedy pastures and hayfields, reclaimed surface mines; sometimes salt marsh. **LOCATION AND STRUCTURE:** *More common in wetter and denser sites than Grasshopper and Savannah Sparrows.* Nest is *on ground*, often in dense litter, against or close to base of a grass tuft. Occasionally elevated off ground, in grass tuft, 2–8 cm up and attached to surrounding herbage; rarely builds nest into a depression, or in low vegetation without overhead cover. Nest is a deep, loosely woven cup of coarse grasses and weeds. Lined with finer grasses. When built in low vegetation, *may be saucer-shaped.* Outside diameter 9–13 cm; height 4–7.5 cm; inside diameter 4.5–6 cm; depth 2.8–5 cm. **EGGS:** 3–5; creamy white to pale greenish white, speckled, spotted, sometimes blotched with reddish browns, with purple-gray or gray undermarkings, more concentrated at larger end; subelliptical; 18 x 14 mm. **BEHAVIOR:** May breed in loose colonies.

LECONTE'S SPARROW   *Ammospiza leconteii*

HABITAT: Wet grasslands and meadows, along dry and densely vegetated margins of marshes and bogs, in hayfields, fallow fields, large pastures. *Often in drier sites than Nelson's Sparrow.* **LOCATION AND STRUCTURE:** Nest is *on ground or slightly elevated* in a dense grass tuft, 0–20 cm up. A well-built, deep cup of grasses, rush stems. Lined with finer grasses, sometimes hair. Outside diameter 9–12 cm; height 6.5 cm; inside diameter 5.5 cm; depth 4.5 cm. **EGGS:** 4–5; smooth, slightly glossy, pale greenish or grayish white, spotted, speckled, and blotched with brown, reddish browns, over entire egg; subelliptical; 18 x 14 mm.

SEASIDE SPARROW *Ammospiza maritima*

HABITAT: Tidal marsh. "Cable Sable" (*A. m. mirabilis*) subspecies occupies freshwater prairie sites, salt marsh in FL. **LOCATION AND STRUCTURE:** Along openings and edges in zones above high tide. Generally avoids low marshes with tall cordgrass. *Nest is built very near to ground (but to 4 m up)* in grasses or emergent vegetation, also in herbaceous growth, woody substrate, or under tidal wrack. Nest is a cup of grass and stems *concealed by canopy of overhanging vegetation*, lined with finer grass blades. Occasionally domed (Everglades, FL). Outside diameter 8–13 cm; height 4–11 cm; inside diameter 4–7 cm; depth 5 cm. **EGGS:** 3–4; smooth, slightly glossy, bluish white to pale gray, speckled and blotched with browns, with pale purples and grays undermarkings, often more concentrated at larger end; short subelliptical to subelliptical; 21 x 16 mm.

Bill Summerour, AL

NELSON'S SPARROW *Ammospiza nelsoni*

HABITAT: Freshwater prairie wetlands; interior bays above high-tide mark in sedge bogs mixed with willow and birch, and along coasts in high saltmarsh grasslands and unused diked fields, or in freshwater marshes and floodplain grassland and meadow. *Where range overlaps with LeConte's Sparrow, Nelson's is often in wetter sites.* **LOCATION AND STRUCTURE:** Usually *raised slightly above ground or water in grass layer, but up to 25 cm.* Nest is a well-rounded cup that may be domed when built in previous

Nelson's Sparrow. David L. Slager, NS

Nelson's Sparrow.
David L. Slager, NS

year's vegetation. Outer layer of coarse and fine grasses. Lined with finer grasses. **EGGS:** 4–5; smooth, slightly glossy, greenish white to greenish blue, usually well speckled throughout with reddish browns, occasionally wreathed; short subelliptical to subelliptical; 19 X 14 mm **BEHAVIOR:** Both sexes are promiscuous. May build multiple nests before choosing one.

SALTMARSH SPARROW *Ammospiza caudacuta*

HABITAT: Restricted to tidal marshes above high-tide mark. Common in areas dominated by cordgrass. **LOCATION AND STRUCTURE:** Often well concealed by overhanging vegetation, with entrance to nest commonly oriented east, south, or southeast; sometimes in open. Usually *within 1 m of ground, supported in upright grasses*, but may be *on ground* at base of vegetation or woody substrate. Nest is a bulky, open cup (rarely partially domed) of coarse dead grass stems, blades, and seaweed. Lined

Garth McElroy, ME

with finer grasses. Outside diameter 7.6–10.8 cm; height 5–8.9 cm; inside diameter 5–6.4 cm; depth 5–6.4 cm. **EGGS:** 3–6; *similar to those of Nelson's Sparrow;* 19 x 14 mm. **BEHAVIOR:** Nests in loose colonies. Both sexes are aggressive, promiscuous; a female appears to choose a male based on his ability to suppress her resistance to his "ritualized mounting."

FOX SPARROW *Passerella iliaca*

HABITAT: Common in areas with dense, shrubby cover, including woodland edges, riparian and deciduous thickets, brushy montane meadows, scrubby woodlands, clearings, regenerating burns, stunted coniferous forests, in alder and hellebore habitat above timberline. **LOCATION AND STRUCTURE:** Variable. *On ground,* well concealed in a dense thicket, under a shrub or tree. *Also in a bush or low tree or in upturned roots,* usually no higher than 2.7 m, occasionally as high as 6 m. Nest is a deep, well-built cup of small twigs, bark shreds, rotted wood, coarse grasses, moss, lichens. Lined with finer dried grasses, rootlets, moss, lichens, hair, occasionally feathers. Elevated nests are larger than ground nests. Nest size varies greatly. Outside diameter 6.9–35.5 cm; height 6.7–13.9 cm; inside diameter 6.3–7.6 cm; depth 2.5–5 cm. **EGGS:** 3; smooth, slightly glossy, pale bluish green, heavily spotted and blotched with reddish browns; 23 x 16 mm.

Mark Peck, ON

SONG SPARROW *Melospiza melodia*

HABITAT: Wide range of open, brushy areas, often near water (especially in arid regions). Includes salt- and freshwater marshes; grasslands; stream, river, and lake shorelines; woodland edges and clearings; brushy fields; desert and coastal scrub; beach dunes; coniferous, deciduous, and mixed forests; grassland shelterbelts; suburban areas; and gardens. **LOCATION AND STRUCTURE:** Variable. Nests are often placed on leeward side of substrate. *Often on ground,* well concealed in a grass or weed tuft or shrub. *Also builds elevated nests, commonly near or over*

Song Sparrow. David Moskowitz, WA

Song Sparrow. Matt Monjello, ME

Song Sparrow.
Casey McFarland,
WA

water in a shrub, small tree, or marshy vegetation (sedge, cattail), often within 1.2 m of ground, infrequently to 9 m. Rarely in tree cavities, rotted logs, human-made structures, nest boxes. Nest is a well-built cup of variable material: generally made of dry grasses, weed stems, bark strips or fibers, leaves, occasionally twigs. Lined with fine grasses, rootlets, sometimes hair. Outside diameter 11.4–19.7 cm; height 4.9–9.2 cm; inside diameter 5–8.4 cm; depth 3.2–5.5 cm. **EGGS:** 3–5; smooth, slightly glossy, pale blue, greenish blue, or gray-green, marked with speckles, spots, and blotches of brown to reddish browns, often concentrated at larger end; subelliptical to short subelliptical; 22 x 17 mm.

LINCOLN'S SPARROW   *Melospiza lincolnii*

HABITAT: Montane habitats. Often in dense, shrubby areas, including brushy margins of bogs, shrubby cut-over areas, riparian habitats, streamside thickets, mixed deciduous groves. Avoids riparian zones in dense woodlands and open areas lacking shrubs. **LOCATION AND STRUCTURE:** *Usually on ground*, slightly

elevated on a mound and set well back under a willow bush or birch (or similar) with ample vegetation for concealment. Rarely elevated in a shrub. *Nest rim is often bulkier at entrance. Nest often has entrance tunnel that is oriented east-northeast.* Outer cup of coarse grass stems, sedge leaves, sometimes fine bark. Lined with finer blades, occasionally hair. Size varies with placement, smaller when built in upright stems. CO nests: outside diameter ~9–11 cm; height 4–6 cm; inside diameter 5.4–6.7 cm; depth 3–4 cm. **EGGS:** 3–5; *similar to Song Sparrow's;* can be sparsely marked; 19 x 14 mm.

Lincoln Sparrow. Mark Peck, ON.

SWAMP SPARROW *Melospiza georgiana*

HABITAT: Freshwater and tidal marshes, bogs and swamps with emergent vegetation and often interspersed with low shrubby growth, marshy vegetation bordering lakes and ponds, brackish marshes, cedar forests. **LOCATION AND STRUCTURE:** *Often found nesting in inner grassy portions of marshes, unlike Song Sparrow, which typically inhabits brushy edges.* Usually well concealed *in a tussock of grass or built to cattail stalks, sometimes in shrubs.* Commonly elevated above ground or water; occasionally on ground. Nest is a bulky cup of dead, coarse sedges or grasses, stalks, cattail, leaves, twigs, rootlets, ferns. Lined with finer grasses and sedges, sometimes with hair, plant down, fibers, and rootlets. Outside diameter 8–15 cm; height 4–11 cm; inside diameter 4.7–7.3 cm; depth 2.4–6.6 cm. **EGGS:** 4–5; smooth, slightly glossy, pale blue-green to greenish white, spotted, blotched, and scrawled with reddish brown, with pale gray undermarkings; similar to Song Sparrow's but more heavily blotched; short subelliptical; 19 x 15 mm. **BEHAVIOR:** Male may be polygynous.

Chris Young, IL

WHITE-THROATED SPARROW *Zonotrichia albicollis*

HABITAT: Coniferous, deciduous, and mixed woodlands. Commonly found in openings with dense shrubby growth, including clearings and burned-over areas, woodland edges, margins of ponds, wetlands, rivers and lakes, roadcuts, power-line cuts, overgrown fields. Mostly absent in thick spruce forest. **LOCATION AND STRUCTURE:** *Often set on flat ground*, infrequently on sloped terrain. Nest is well concealed by a low shrub, grass clump, dead ferns, or tree branch. *Sometimes aboveground* in shrubs, ferns, uprooted stumps, brush piles, low tree branches; rarely up to 3 m high in conifer. Nest is a bulky cup of coarse grasses, twigs, bits of wood, pine needles, rootlets. Lined with finer grasses, rootlets, hair. Outside diameter 7–14 cm; height 4.6–11.4 cm; inside diameter 4.4–10 cm; depth 2.5–6.4 cm. **EGGS:** 4–6; smooth and slightly glossy, pale blue to greenish blue, marked with spots, speckles, and blotches of reddish brown and light purple, concentrated on larger end or profuse; subelliptical to long subelliptical; 21 x 16 mm.

Matt Monjello, NH

Mark Peck, ON

HARRIS'S SPARROW *Zonotrichia querula*

HABITAT: Northern stunted spruce-larch forests mixed with open tundra. Common along woodland edges and burns, in clearings. **LOCATION AND STRUCTURE:** *Nest is a scraped depression* on or next to a hummock or under a shrub or tree, especially birch, alder, spruce, Labrador tea. Sometimes builds underneath an overhang such as a rock or patch of ground. Nest is a bulky cup of moss, twigs, lichens. Lined with dead sedges, grasses, sometimes hair. *Generally prefers sites with less vegetative density and height than closely related White-crowned Sparrow.* **EGGS:** 4; smooth, slightly glossy, pale green, spotted and blotched with brown, usually covering entire egg; subelliptical; 22 x 17 mm.

Peter Blancher, NT

WHITE-CROWNED SPARROW *Zonotrichia leucophrys*

HABITAT: Common in habitats that contain a mixture of grasses, bare ground, shrubs, and small trees. Northern forests bordering tundra, high-montane meadows, scrubby woodland edges, in forest clearings, roadcuts, along coastal thickets; urban gardens. **LOCATION AND STRUCTURE:** Coastal species generally place nest *in a small shrub* such as sage, saltbush, or berry thicket, often within 1.2 m of ground. More northern and higher-elevation breeders generally place nests *on ground* under a shrub or low tree, in or adjacent to overhanging grass clump and other herbage; infrequently build elevated nest. Nest is an open cup of grass and weed stems, fine twigs, pine needles, fine ferns, bark strips, moss. Lined with fine stems, some hair possible. Outside diameter 11–12 cm; inside diameter 5.5 cm (postfledging nests can be considerably larger, to 8.5 cm); depth 4–5 cm. **EGGS:** 2–5; smooth, slightly glossy, greenish, greenish blue, pale blue, spotted, blotched with reddish brown, usually profuse; subelliptical; 21 x 16 mm.

R. Adam Martin, WA

A White-crowned Sparrow nest, lined with feathers, from the northern end of this species' range in NT. Peter Blancher, NT

GOLDEN-CROWNED SPARROW *Zonotrichia atricapilla*

HABITAT: Common in willow and alder thickets and stunted conifers near timberline in mountains and along coast in similar tundra scrub. **LOCATION AND STRUCTURE:** *Usually on ground* in a slight depression at base of a low deciduous shrub, grass tuft, or other herbage. *Sometimes in a small conifer or shrub.* Nest is a thick, well-built cup of small twigs, pieces of bark, ferns, moss, leaves, grasses. Lined with fine grasses, hair, feathers (ptarmigan). Outside diameter 12–18 cm; height 5.5–9 cm; inside diameter 6.5–7.5 cm; depth 3.5–5 cm. **EGGS:** 3–5; smooth and slightly glossy, pale blue to greenish blue, profusely marked with fine speckles, spots, and mottles of reddish brown; subelliptical to long subelliptical; 23 x 16 mm.

Use this chart to see at a glance the variety of nesting habitats and nest placements for each species, and as an aid in the field if you think you've found a sparrow nest. Nest placement is shown at left (gray), habitat at right (white

	Ground	Ground, Shrub or Tree, or Elevated in Upright Grasses, Forbs	Shrub, Tree, or Similar	Arctic or Far North Only	Arid or Semi-arid Regions (desert scrub, woodlands)
Olive Sparrow			X		X
Brewer's Sparrow			X		X
Black-throated Sparrow			X		X
Black-chinned Sparrow			X		X
Chipping Sparrow			X		
Rufous-winged Sparrow		X			X
White-crowned Sparrow		X			
Clay-colored Sparrow		X			
Field Sparrow		X			
Cassin's Sparrow		X			X
Lark Sparrow		X			X
Bell's Sparrow		X			X
Five-striped Sparrow		X			X
Fox Sparrow		X			
Song Sparrow		X			X
White-throated Sparrow		X			
American Tree Sparrow		X		X	
Golden-crowned Sparrow		X		X	
Seaside Sparrow		X			
Saltmarsh Sparrow		X			
Sagebrush Sparrow		X			X
Swamp Sparrow		X			
Baird's Sparrow	X				
Grasshopper Sparrow	X				
Savannah Sparrow	X				
Vesper Sparrow	X				
Lincoln's Sparrow	X				
Henslow's Sparrow	X				
Botteri's Sparrow	X				X
Harris's Sparrow	X			X	
Nelson's Sparrow	X				
Lark Bunting	X				X
LeConte's Sparrow	X				
Rufous-crowned Sparrow	X				X
Bachman's Sparrow	X				
Dark-eyed Junco	X				
Yellow-eyed Junco	X				X

and blue); note that column content overlaps to cover species that place nests in a variety of locations and habitats. At far right (green) are species whose eggs are typically unmarked, a useful identification characteristic.

Wetlands, Marshes, Salt Marsh	Arid or Semiarid Grassland, Scrub Grassland, Wet Grassland	Broad Range of Habitats (old-fields, brush and scrubland, coniferous and deciduous forests, savanna, cropland, tundra, montane, bog, riparian areas, etc.)	Eggs White and Unmarked (may have tint of blue, green, or pink)
			X
		X	
	X	X	X
			X (or sparsely marked)
		X	
	X		X
		X	
	X	X	
		X	
	X		X
	X		
			X
		X	
X		X	
X		X	
		X	
		X	
X			
X			
X			
	X		
	X	X	
X		X	
	X	X	
		X	
	X		
	X		X
		X	
X			
	X		X
X			
			X
		X	X
		X	
		X	

JUNCOS

DARK-EYED JUNCO *Junco hyemalis*

HABITAT: Coniferous, deciduous, and mixed woodlands (both wet and dry) with early to late-stage growth, along clearings, woodland edges, on rocky slopes, densely wooded coastal islands, tree plantations, roadcuts, urban wooded parks, gardens. **LOCATION AND STRUCTURE:** Variable. *Usually on ground in a slight depression and well concealed*: against or among grass tufts, ferns, and other herbaceous vegetation, at base of bush or tree, under an overhanging rock, log, or fallen trunk, upturned roots, on bank or rock slope under overhanging cover. Occasionally builds elevated nests in shrubs, trees, vines (atop horizontal branches), on building ledge, supports on ledges beneath elevated human structures. Nest is typically an open cup with outer layer of rootlets, dead leaves, fine twigs, mosses, bark fibers,

A Dark-eyed Junco nest revealed by momentarily holding the overhead grasses aside. Casey McFarland, NM

Casey McFarland, NM

Jacqueline Huard, BC

with inner cup of grasses, rootlets, rhizomes. Lined with fine grass or moss stems, hair. When built in rock cavity, nest may simply consist of a lining. *When built in tree, nest may be more substantial in size.* Outside diameter 11 cm; outside cup height (often set into depression in ground) 5.8 cm; inside diameter 6 cm; depth 4–7 cm. **EGGS:** 3–5; pale bluish white, white, grayish, light greenish white, marked with brown to reddish brown spots, speckles and blotches, often concentrated at larger end; short subelliptical to subelliptical; 19 x 14 mm.

YELLOW-EYED JUNCO *Junco phaeonotus*

HABITAT: Open coniferous and pine-oak forests, at 1,200–3,500 m elevation. **LOCATION AND STRUCTURE:** Commonly on sloped terrain. Built on ground and hidden by a grass clump, other vegetation, log, or rock, with opening to nest oriented downhill. Infrequently built aboveground in a conifer, tree cavity, herbage, or artificial structure. *Similar to Dark-eyed Junco nest but can be larger.* **EGGS:** 3–4; *similar to Dark-eyed Junco's but sparsely marked*; 20 x 15 mm; NP 10–13 days (F, M).

YELLOW-BREASTED CHAT (Icteriidae)

YELLOW-BREASTED CHAT *Icteria virens*

HABITAT: Broad range of lowland and upland habitats, but generally low, shrubby thickets, or scrubby woodland with open to semiopen canopy: second growth and scrub in old pastures, clear-cuts, power lines; stream and pond edges, swamps, forest edges, regenerating burns, floodplains, similar. In West often in dense shrubs along riparian areas; mesquite, salt cedar, cottonwood, and willow. **LOCATION AND STRUCTURE:** Nest is often low (0.6–2 m up), but up to 3 m or higher, on a horizontal branch or in a fork or crotch, usually supported from below. Nests vary regionally but often are bulky, made of coarse grasses and stems, bark strips, thin vines, weed stems, dead or skeletonized

Casey McFarland, AL

Casey McFarland, NM

leaves. Lined with finer grasses and fine stems. Outside diameter 12.7–15 cm; height 7.6 cm; inside diameter 6.4–8.9 cm; depth 5–6.5 cm. **EGGS:** 3–5; typically white or creamy white, with numerous bold reddish markings and blotches over entire egg, concentrated mostly at larger end; 22 x 17 mm. **BEHAVIOR:** Double-brooded. May nest in loose colonies. If a nesting attempt fails, pair bond generally ends.

MEADOWLARKS, ORIOLES, BLACKBIRDS, and ALLIES (Icteridae)

Twenty-two species of icterids commonly breed in N. America. They are generally well known, and well adapted to landscapes and habitats altered by humans. Icterids use a wide variety of nest designs and locations: some nests are hidden on the ground among grasses and herbaceous plants, some are built above water in swamps and marshes, and others are tucked high in tree canopies. They vary from woven bulky cups to dome-shaped structures to pendulous socklike pouches.

GROUND-NESTING ICTERIDS

BOBOLINK and MEADOWLARKS

BOBOLINK *Dolichonyx oryzivorus*

HABITAT: Originally in tall-grass and mixed-grass prairies, now also in cultivated lands of clover, hay, alfalfa, thick weeds, irrigated fields/meadows, commonly with mix of forbes. **LOCATION AND STRUCTURE:** On ground in a natural (or scraped out) bare-soil depression (female clears plants), often at base of a large nonwoody plants. *Female builds outer nest wall of coarse grasses and weed stems, leaving a bare floor that is then lined with fine*

Bobolink nest with a Brown-headed Cowbird egg.
Chris Helzer, NE

grasses. Typically open above. Inside diameter ~9 cm (6–11 cm); depth 3.5 cm (1.8–4.4 cm). **EGGS:** 5–6; smooth and moderately glossy, with markings and color highly variable; pale blue-gray to gray, or pale reddish brown, with various sizes of spots or blotches of browns, purples, often more at larger end; oval to short oval; 22 x 16 mm; IP 11–13 days (F, M); NP 10–14 days (F, M). **BEHAVIOR:** Single-brooded. If first attempt fails, may renest, with only female incubating. Nestlings are fed occasionally by additional adults. Nests are very challenging to find, and females rarely flush.

EASTERN MEADOWLARK *Sturnella magna*

WESTERN MEADOWLARK *Sturnella neglecta*

HABITAT: Eastern Meadowlark: In native grassland habitats (including desert grassland), meadows, pastures, savannas, hayfields, open woodlands, old farm fields, weedy borders of farmlands, also in sparse woodlands or orchards. Western Meadowlark: Habitat similar but includes mountainous woodlands, open sagebrush; less abundant in areas where vegetation is dense, tall. *Eastern and Western Meadowlarks are nearly identical in appearance, and where they overlap are most easily identified by song.* **LOCATION AND STRUCTURE:** Builds nest in a natural depression or hollow, or female scrapes it out. Coarse grasses, plant stems, and strips of fine bark make up outer layer, which is lined with finer grasses, possibly hair. *A domed or hoodlike roof is added by loosely weaving and pulling over surrounding vegetation, leaving a side entrance (often oriented away from prevailing winds).* Vesper Sparrow, p. 372. Trails and tunnels may be evident near nest from birds coming and going. Outside diameter 14–21 cm; inside diameter 8–15 cm; depth 5–8 cm. If tunnel is present, usually ~5–12 cm in diameter. **EGGS:** 3–5 (Eastern), 5 (Western); base color white, but marked with speckles, spots, and blotches with shades of browns, purples; 28 mm x 20–21 mm; IP 13–15 days (F); NP 10–12 days (F, M). **BEHAVIOR:** Female

Western Meadowlark. Bill Summerour, SK

Eastern Meadowlark. John Pulliam, FL

builds alone in 3–8 days. Young are tended primarily by female in Eastern, by both parents in Western. Male meadowlarks typically have 2 mates at a time. Where Eastern and Western overlap, they rarely hybridize. Nests are extremely difficult to find (Westerns are often in dense vegetation), and birds land and walk to nest, hidden from view.

ORIOLES

New World Orioles rank among the world's most accomplished nest architects, building remarkable, readily recognizable nests, generally woven from lengths of fine plant materials (grasses, palm fibers, twine, similar) into an intricate, suspended basket. Seven species commonly breed in N. America, though only 5 nest over broad ranges. Orioles tend toward social monogamy, though extra-pair copulation or occasional polygyny have been reported.

Nests are pensile (hanging and attached only at the rim) or semipensile (with additional attachment at the sidewalls) and can be loosely separated into 2 categories: those that are classically *pendulous and socklike,* typically longer than wide with a narrower top, and those that are like a *shallow, open gourd,* the opening of which is often fairly wide and more broadly secured, in part due to the span of the fork or branches to which the rim is fastened. Nests of Bullock's, Baltimore, and Altimira Orioles fall into the first category: Altimira nests are the longest of any species in N. America. Spot-breasted Oriole *(Icterus pectoralis)* and Streak-backed Oriole *(Icterus pustulatus),* not included below, also have distinctly long nests. Nests of Scott's, Audubon's, Orchard, and Hooded Orioles fall into the second category. Note, however, that overlap occurs between categories, and nests may be difficult to differentiate. Nests are typically built by the female in 3 days to 3 weeks (often about 6-10 days); more elaborate nests generally take longer. In many regions multiple species overlap; note which species occur in the area and potential differences in design and placement.

Eggs are fairly long and pointy, ranging from long subelliptical to long elliptical or long oval. Markings and coloration vary, but eggs are generally smooth and slightly glossy, pale blue or grayish blue-white, typically with combinations of fine scrawls (like hair-thin lines of paint drizzle), streaks, spots, speckles, and blotches ranging in color from black, dark purple, to browns. Descriptions for individual species are not included; use nests, region, and adults as identifiers. Markings are often concentrated at the

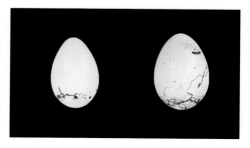

Side view, Bullock's Oriole (left) and Orchard Oriole (right). Oriole eggs are often pointy, sometimes biconical, and can also have spots, speckles, blotches, and streaks.

larger end or in an irregular wreath. Female incubates, and both parents feed. Incubation period ~12–15 days, NP ~11–14 days, varying by species. Many species are single-brooded.

ORCHARD ORIOLE *Icterus spurius*

HABITAT: Greenbelts, shade trees around farms and suburbs, orchards, open woodlands, park-like areas (grassy camp-grounds, similar) particularly along waterways, riparian zones. Mostly in trees (favors deciduous), occasionally in shrubs. **LOCA-TION AND STRUCTURE:** Nest is 1–21 m aboveground, mostly 3–6 m. *Usually not as pendulous as Baltimore or Bullock's Oriole nests*; nest is generally wider than deep. Nest is typically suspended from a forked twig or branch at tips of branches. Thin-walled, well woven, *sometimes made exclusively of long green grasses (when fresh), which may be diagnostic.* Lined with fine grasses, fur, plant down, and feathers. Bottom may be woven loosely enough to see through. Outside diameter 8.9–10.2 cm; height 6.4–10.2 cm; inside diameter 6.4–7.6 cm; depth 5.7–7.6 cm. **EGGS:** 4–5; 20 x 15 mm. **BEHAVIOR:** Often nests in groups, with

Casey McFarland, AL

Casey McFarland, AL

multiple nests in same tree; solitary nesting may indicate poorer breeding habitat. Also nests among several other species, including other orioles and kingbirds; the latter may be beneficial because of their aggression toward predators and cowbirds.

HOODED ORIOLE *Icterus cucullatus*

HABITAT: Various woodlands, mesquite, especially streamside or along other watercourses. In shade trees, palms, shrubs; parks, suburbs. **LOCATION AND STRUCTURE:** Typically semipensile. *Cup is frequently fastened directly to underside of a palm or other large tree leaves*, but can also be suspended in twigs or a small fork. Commonly 1.5–13.7 m aboveground. Nest has a small top or side entrance. Built of grasses, shredded palm, or other long fibers into a sturdy, thick-bottomed cup with strong sides. Sometimes lined with plant down, feathers, similar. Height 10 cm; inside diameter ~7 × 5 cm; depth 6.4 cm. **EGGS:** 4; 22 x 15 mm. **BEHAVIOR:** Double- and occasionally treble-brooded. Female sews or rivets nest to the surface of large leaves, jabbing material through with her bill (see Chapter 2, p. 46).

Note how the nest is sewn into place.
Thomas P. Brown, AZ

BALTIMORE ORIOLE *Icterus galbula*

BULLOCK'S ORIOLE *Icterus bullockii*

HABITAT: Baltimore: Farms, wooded residential areas, shade trees, orchards, open deciduous woodlands. Bullock's: Similar, also mesquite, sycamore, cottonwood, riparian belts. **LOCATION AND STRUCTURE:** Baltimore: Often in isolated trees. Commonly attached by rim to a thin, drooping branch or suspended by a small fork. Thin twigs and leaves that hang down may also be woven to nest sides. Height aboveground is 1.8–27 m, typically 7.5–9 m. Nest is a deep woven pensile pouch of string, hair, yarn, or similar material (bailing twine common), grapevine, bark strips, Spanish moss (in South). Lined with soft materials: plant cotton, fine grasses, hair. *Nests of Baltimore may more often be attached to drooping or vertical branches than those of Bullock's, which often builds in twigs that grow upward. Bullock's nests may be less deep and have wider opening.* Outside diameter 9–21 cm (average 12 cm); length of nest similar; oval-shaped entrance 5–8 cm across, inside diameter 6.4 cm; depth ~11 cm. **EGGS:** Baltimore: 4–5; 23 x 16 mm. Bullock's: 4–5; occasionally biconical; 24 x 16 mm. **BEHAVIOR:** Baltimore is single-brooded (Bullock's assumed so). In Great Plains, ranges overlap and the 2 species frequently hybridize. Bullock's may nest in association with Yellow-billed Magpies and other aggressive birds, presumably benefiting from antipredator defense.

Bullock's Oriole. David Moskowitz, WA

An old Bullock's Oriole nest, made largely of fishing line. David Moskowitz, WA

ALTIMIRA ORIOLE *Icterus gularis*

Lower Rio Grande Valley. *Nest is extremely recognizable: long, pendulous, to nearly 1 m in length.* Eggs are largest of all orioles, 29 x 19 mm.

Bettina Arrigoni, TX

AUDUBON'S ORIOLE *Icterus graduacauda*

In mesquite, thickets, open woodlands. Among the most cryptic of the orioles; its nest can be difficult to find. *They are more cup-shaped than those of most other orioles* and are built in centralized portion of the tree or shrub, rather than at branch tips. **EGGS:** 3–5; 26 x 19 mm.

SCOTT'S ORIOLE *Icterus parisorum*

HABITAT: Arid to semiarid mountains, deserts. Pinyon-juniper, yucca tree, Joshua tree, oak, and similar habitats; scrubby grasslands, foothills. *Often in more arid deserts unsuitable for many other oriole species.* **LOCATION AND STRUCTURE:** Typically a semipensile cup commonly *tucked beneath crown of yucca trees*, with which this species is closely associated, woven at top and sides to hanging, dead yucca leaves (often mostly hidden, well shaded). Also in palms or other trees (pinyon, juniper, sycamore,

Scott's Oriole
Casey McFarland, TX

cottonwood, etc.). Generally 1.5–3 m high, sometimes higher. Outside diameter 12 cm; height 10 cm; inside diameter 10 cm; depth 9 cm. **EGGS:** 3; 24 x 17 mm. **BEHAVIOR:** Double-brooded.

"WETLAND" BLACKBIRDS

RED-WINGED, YELLOW-HEADED, and TRICOLORED BLACKBIRDS

These well-known, attractive birds build similar-looking nests in similar locations. Nests are easily recognized by their placement in marshes and wetlands, their design, and often their sheer numbers. Each species demonstrates varying degrees of polygyny, but Red-winged Blackbird is considered the most polygynous species in N. America; a single male may hold a territory that contains nests of 15 females. Tricolored Blackbird males defend smaller territories, mating with only 1–4 females.

Blackbirds nest colonially, and Tricolored Blackbirds form the largest breeding colonies of N. American passerines, with nests traditionally numbering in the hundreds of thousands. Yellow-headed Blackbirds also form large colonies, whereas Red-wingeds are more semicolonial. Nesting is generally highly synchronized: the first eggs in thousands of Tricolored Blackbird nests, for instance, are all laid within a single week.

Nests are generally constructed in 3–6 days by weaving long, pliable materials (grasses, cattails, rushes, forb stems, etc.) into upright stems of emergent vegetation to create a platform and cup of coarser materials. Mud or rotted materials often make a firm inner layer, which is lined with fine grasses and other materials. Nests are built exclusively by the female, who is also the sole incubator. Nest dimensions among the 3 species are similar: outside diameter ~11–14 cm; height 8–19 cm (varying greatly depending on placement); inside diameter ~6.5–7.5 cm; depth ~6.5–7.5 cm.

Red-winged young are cared for by both parents; female Yellow-headed and Tricolored Blackbirds may tend young alone. Incubation period 10–13 days, NP 10–13 days, though young are incapable of flying for a week or more.

RED-WINGED BLACKBIRD *Agelaius phoeniceus*

HABITAT: Extremely varied, more so than in Yellow-headed or Tricolored Blackbirds. Found in wetlands, saltwater marshes, wet meadows, roadside thickets, agricultural areas, dry fields and pastures, some urban and suburban settings;. **LOCATION AND STRUCTURE:** Occasionally nests on or near ground. Nest is neatly woven, deep-cupped, often with inner layer lightly plastered with wet vegetation and mud, though not always present. Lined with fine grasses. **EGGS:** 3–4; glossy, smooth, pale blue or green, may be washed in pinks or purples; sparsely blotched, spotted, and scrawled in browns, purples, and blacks, concentrated primarily at larger end; elliptical to subelliptical; 25 x 18 mm. **BEHAVIOR:** Male may help feed nestlings, but to widely varying degrees in different populations. Frequent host of Brown-headed Cowbird; studies have shown that three-quarters of broods may be parasitized.

Eric C. Soehren, AL

Matt Monjello, GA

This nest is built mostly of grasses.
Matt Monjello, NC

TRICOLORED BLACKBIRD *Agelaius tricolor*

HABITAT: Freshwater marshes, wetlands, swamps, sloughs, but sometimes in agricultural areas. Also in habitat dominated by willow, nettles, or thistles. Usually in larger wetlands and marshes with thicker vegetation than Red-winged Blackbird. **LOCATION AND STRUCTURE:** Nests *are densely placed*, as close as every square meter, and may spread beyond more typical habitats into thickets, crops, shrubs, and trees. Otherwise *similar to nests of Red-winged*. **EGGS:** 4; *similar to Red-winged's*; 28 x 20 mm. **BEHAVIOR:** Nesting locations tend to change yearly, and Tricolored Blackbirds are itinerant breeders, nesting more than once during the breeding season, mostly in different areas. This may optimize food sources and quality nesting sites, particularly important for species that nest together in enormous colonies. Rarely parasitized.

YELLOW-HEADED BLACKBIRD
Xanthocephalus xanthocephalus

HABITAT: Mostly in prairie marsh and wetland settings, often in deep water; frequently shares habitat with Red-winged Blackbird. Also in mountain wetlands and meadows, and arid landscapes. **LOCATION AND STRUCTURE:** In aquatic vegetation, *consistently over water*, often 15 cm–1 m high, up to 1.8 m. *Similar to others above*. Long, water-soaked leaves are woven tightly around uprights and may shrink as nest dries, pulling the uprights inward. Inner layer may contain rotted leaves or green plants, rootlets, and fine grasses, *but lacks mud*. **EGGS:** 4; grayish white, whitish green, heavily speckled with browns, grays, as if sprayed with a paint can from afar; markings across entire egg, may be heaviest on larger end; long subelliptical to long oval; 26 x 18 mm. **BEHAVIOR:** Rarely parasitized. Will displace slightly smaller Red-wingeds, as well as Marsh Wrens.

Casey McFarland, WA

Casey McFarland, WA

COWBIRDS

BROWN-HEADED COWBIRD *Molothrus ater*

BRONZED COWBIRD *Molothrus aeneus*

SHINY COWBIRD *Molothrus bonariensis*

As brood parasites, cowbirds use a wide range of nests and nest sites. They do not build their own nests, but instead parasitize those of other birds. Watchful females often locate host nests during their construction. This may be done from a perch or by patrolling on foot, hunting for movements of nesting hosts. Nest finding may also be a conspicuous affair, flapping about from shrub to shrub to flush up nesting females of another species. Once such females are discovered, female cowbirds covertly visit and inspect a nest while the maker is away, and when the time is right, lay an egg—usually only 1 per nest—and often toss out or peck one of the host's eggs. Several female cowbirds may parasitize the same nest, resulting in a clutch of multiple cowbird eggs.

Despite a low percentage of eggs actually resulting in adult cowbirds, cowbird populations have exploded and extended in range. In addition to the large-scale landscape alteration that created suitable cowbird habitat, cowbirds' success is due in part to the unique physiology of the females. Long reproductive cycles and the ability to produce multiple clutches allow a female cowbird to exploit dozens of nests over a period of months—a Brown-headed Cowbird can lay 40 eggs a season for 2 years. Breeding strategy varies too. Some cowbirds are monogamous, or polygynous, or promiscuous. They maintain home ranges but don't defend them.

Cowbirds pose considerable threats to numerous hosts. They target more than 220 species, mostly preferring species whose eggs are smaller than their own. Research indicates that individual females generally specialize on a particular host species.

A Dickcissel nest containing 4 Brown-headed Cowbird eggs.
Chris Helzer, NE

Larger birds may be unaffected by cowbird parasitism, but populations of smaller, more vulnerable species have undergone alarming declines. Eggs can vary in size, color, and shape but are easily identified because they are often distinct from those of the host. Brown-headed and Shiny Cowbirds: 21 x 16 mm. Bronzed Cowbird: 24 x 17 mm. Cowbird eggs often hatch before the host's eggs, and nestlings grow rapidly, generally outperforming their smaller nest mates. Incubation period 10–13 days, NP 10–12 days.

RUSTY BLACKBIRD *Euphagus carolinus*

HABITAT: The northernmost breeding blackbird. To tree line in wet coniferous and mixed forests; bogs, sphagnum swamps, muskegs, beaver ponds, river oxbows, creek sides, similar, to edges of grasslands and deciduous wooodlands. **LOCATION AND STRUCTURE:** In a coniferous or deciduous tree, 0.5–6 m up, usually near water. Nest is a bulky, well-built cup, often built against trunk, with outer layer of twigs, grasses, string, lichens, or moss. *A hardened, thick inner cup of rotted leaves or muck is often present.* Lined with plant fibers, fine grass. Outside diameter 13–20 cm; height 8–10 cm; inside diameter 8.5–9.5 cm; depth 5–6 cm. **EGGS:** 28 x 21 mm; IP 10–13 days (F); NP 9–12 days (F, M). **BEHAVIOR:** Nests built in trees are often reused by Solitary Sandpipers. Populations have declined by as much as 90 percent, due largely to alteration and loss of habitat in wintering and breeding ranges.

Mark Peck, ON

BREWER'S BLACKBIRD *Euphagus cyanocephalus*

HABITAT: Highly variable, often associated with wide range of human habitations: parks, golf courses, agricultural areas, roadsides, irrigation ditches, etc. Also in sagebrush, marshes, bogs, cleared areas, and others. Often near water. **LOCATION AND STRUCTURE:** Highly flexible in choice of nest location. Commonly

Brewer's Blackbird. Matt Monjello, CA

Brewer's Blackbird. Ground nest on a shallow ditch bank at the base of vegetation. Casey McFarland, OR

Brewer's Blackbird nest. The three smaller eggs are those of Brown-headed Cowbirds. Casey McFarland, WA

in conifers (1–12 m, occasionally to 45 m), but also deciduous trees, placed in forks, along branches, or at branch tips. When built on ground, nest is often placed where topographical features change abruptly: at base or top of ditch banks, ravines, or a small rise. Also where distinctive changes or edges in vegetation occur. Nest is a solidly built cup of fine and coarse grasses, stems, pine needles, thistle, similar, with a possible layer of mud or dung. Lined with fine grasses, horsehair, or cottony materials, soft plants. Outside diameter 15 cm; height 10–12 cm; inside diameter 9.5–10 cm; depth 6–9 cm. **EGGS:** 5–6; highly variable, pale gray to greenish white, with variety of markings: blotching, speckles, streaks or scrawls, spots, of browns, purples, yellows, grays; may cover egg entirely or concentrate at larger end; 25 x 19 mm; IP 12–13 days (F); NP 13 days (F, M). **BEHAVIOR:** Nests singly or in colonies. Nesting timing is adjusted to food availability, predation, and nest parasitism (as many of 50 percent of nests may be parasitized by cowbirds in some regions).

GRACKLES

These 3 conspicuous species build in a variety of locations, often near water. Grackles are polygynous, and males are considerably larger than females: female Boat-tailed Grackles appear about half the size of males and are differently colored. Boat-tailed Grackles defend harems and territories, but Common Grackles are typically socially monogamous. Grackle nests, typically built by the female in 4–8 days, are bulky cups of plant materials and mud or dung. Wet or rotted vegetative material forms a firm inner cup that is lined with fine grasses, forb stems, or rootlets. *The construction process and appearance of grackle nests built in emergent vegetation in marshes and wetlands are like those of Red-winged Blackbird, but grackle nests are larger.* Nests are also built in shrubs, trees, cactus, and human structures. Great-tailed and Boat-tailed Grackles *frequently nest in colonies.* Nests of all 3 species are similar in dimensions and vary in size depending on placement. Outside diameter 17–22 cm or larger; height 10–33 cm; inside diameter 10–13 cm; depth 7–10 cm. Eggs are striking in appearance: pale blue to pale greenish blue-gray *(Great-tailed eggs can be bright blue)*, boldly scrawled (looking similar to paint drizzle), and marked with blotches and large spots of blacks, browns, purples, and grays. Short to long subelliptical to long oval. Female alone incubates. Sometimes double-brooded, Boat-tailed occasionally treble-brooded.

COMMON GRACKLE *Quiscalus quiscula*

HABITAT: Coniferous groves, open woodlands, thickets in swamps and marshes, cultivated land, islands, open woodlands; is well adapted to nesting in cities and suburbs. **LOCATION AND STRUCTURE:** *Prefers conifers* but will use deciduous trees. To 18 m high but often low or at ground level in marshes or wet areas. May also use ledges, cavities in trees, nest boxes, and human structures. Nest is bulky, constructed of forbs, grasses, pine needles, seaweed (on coasts), with inner cup of rotted, mucky vegetation, mud, or soft dung. Lined with fine grasses, feathers, hair, or other fine materials. **EGGS:** 4–5; glossy; 28 x 21 mm; IP 11–14 (F);

Common Grackle marsh nests.
Matt Monjello, CT

Matt Monjello, CT

NP 10–17 days (F, M). **BEHAVIOR:** An adaptive, successful species that has increased rapidly in population and range. Does considerable damage to eggs and nestlings of other species and is known to kill and eat adult birds. Nests singly or in small, loose colonies.

BOAT-TAILED GRACKLE *Quiscalus major*

HABITAT: Islands and open coastal zones, fresh- and saltwater marshes. Also cities, suburbs, agricultural lands. **LOCATION AND STRUCTURE:** *Nest is usually built near or above water in vegetation* (cattail, sawgrass, bulrush, willow, etc.) and usually low, often 30 cm–1 m, but occasionally to 15 m in trees, including oak, cedar, mangrove. Nest is a bulky, somewhat loosely constructed cup of coarse grasses, vines, rushes, bark strips, woven to upright twigs or stems of supportive vegetation, with firm inner cup of

Matt Monjello, NC

Matt Monjello, NC

mud and rotting plants. Lined with fine grasses, rootlets, forb stems. Outer dimensions of nest are highly variable. **EGGS:** 3–4; 32 x 22 mm; IP 13–14 days (F); NP 20–30 days (F). **BEHAVIOR:** Remains separated by sex throughout year, rejoining during nesting season. Females nest in large colonies, and mating opportunities among males are driven by rank, though females often breed with nondominant or non-colony males. Rejects cowbird eggs.

GREAT-TAILED GRACKLE *Quiscalus mexicanus*

HABITAT: Human habitations (towns, parks, golf courses, etc.), agricultural areas, pasturelands, mangrove, orchards; generally in open areas, avoids dense forest. Also in mesquite, chaparral, yuccas, cactus. **LOCATION AND STRUCTURE:** *Nests are commonly built above or near water in trees and bushes* (usually 3–4 m up) or in marsh vegetation where trees or thickets aren't available. Also in human structures. *Nests often have little to no mud.* In trees, nests are made with an assortment of long grasses or other plant materials, twigs, string, paper, or other soft garbage such as plastic bags and cloth, bark strips, feathers, and Spanish moss. **EGGS:** 3–4; 33 x 22 mm; IP 13–14 days (F); NP 20–23 days (F, M). **BEHAVIOR:** Occasionally double-brooded. Females steal materials from other nests, and males may guard a nest to prevent pilfering. Frequently nest close together in large colonies, and may nest in heronry, sometimes using heron nests. Rejects cowbird eggs.

WOOD-WARBLERS (Parulidae)

Warblers are very small to small songbirds, primarily insectivorous. Roughly 50 species breed in N. America. Most species occupy eastern forests, though a variety are found in West, mostly in mountains and in coastal habitats. Breeding habitats are generally wooded, brushy, and usually where moisture occurs. Almost all wood-warblers are migratory, and some species travel remarkably long distances from boreal forests to Neotropical regions (some over vast stretches of open ocean) and rely on healthy habitat on both their wintering and breeding grounds. Numerous species, such as Golden-winged Warbler, exhibit strong site fidelity and return each spring to the exact location where they nested in previous years. Many warblers are habitat specialists; deforestation, habitat degradation, and landscape alteration have caused sharp population declines among many species.

Some species are generally monogamous (at least within a breeding season), though polygyny and extra-pair copulations occur. Other species are socially monogamous. Males are often wonderful songsters, and the songs of species among *Setophaga*, *Vermivora*, and *Parula* are considered to serve different purposes: dawn territorial advertisements and territorial flight songs may typically be directed at other males, and simpler songs throughout the day may serve as mate attraction. Females of some species also sing, perhaps to announce boundaries to other females or to communicate with a new mate.

Nests are mostly open cups, but Prothonotary and Lucy's Warblers use cavities, parulas make a pendulous-like nest, and Ovenbird builds a domed or roofed structure in forest litter. Twenty or more species nest mostly on the ground (or very near it), while others build at various heights in shrubs or tree canopy, from roughly knee level to 30 m or higher. Nests generally contain a small inner cup, but the outer structural layers range in design from thick, loose, and bulky to thin and feltlike. They are constructed in a great variety of ways out of leaves, fibrous material, fine twigs and stems, grasses, mosses, etc. To simplify the overwhelming diversity of their nest placements and designs, we have divided warbler nests into 10 categories: (1) Ground nests built with an outer shell of leaves (often shaped when damp); (2) Ground nests composed primarily of grasses, typically in moist northern forests and bogs or similar habitats; (3) Ground nests in the mountainous, arid Southwest; (4) Other ground or near-ground nests; (5) Nests typically placed low in a shrub, bush, or small tree (often deciduous); (6) Hanging moss nests; (7) Nests typically placed low in conifers, pines (often lower than 3 m); (8) Nests typically placed high in conifers, pines (higher than 6 m); (9) Nests in coniferous or deciduous trees; (10) Nests on or close to ground, situated in a recessed or elevated cavity, or with a domed roof. Note that many species have large ranges or nest differences and may fit into more than the single category in which they've been placed. Nests of species that build both on the ground and higher in shrubs can also look remarkably different depending on placement. Read descriptions carefully, and know which warblers frequent your region. Wood-warbler nests are small, with inner cup diameters typically 4–6 cm (those of Ovenbird and waterthrushes are larger).

The female chooses the site and usually builds the nest in ~4–7 days. The male typically does not help but accompanies her as she works. The female is also the primary incubator, often fed by the male. Incubation period 9–15 days, NP 8–12 days, varying by species. Young are tended by both parents, who often split the brood and care for their portion individually. Eggs are short subelliptical to subellipitcal and range in size and color. The typical appearance (referred to below as "typically white/creamy white with markings," or a variation thereof) is generally smooth, slightly glossy, pale whitish or cream-colored to pale pink, with speckles, spots, and blotches of browns, dark reds, or purples, often concentrated at the larger end, occasionally wreathed. Exceptions are noted below.

GROUND NESTS BUILT WITH AN OUTER SHELL OF LEAVES (OFTEN SHAPED WHEN DAMP)

WORM-EATING WARBLER *Helmitheros vermivorum*

HABITAT: Generally large unfragmented tracts of deciduous or mixed deciduous-evergreen forest, on brushy (rhododendron, similar) slopes and hillsides. Also coastal areas (MD, NC) along riverbanks and in lowland pine forests. LOCATION AND STRUCTURE: On ground, *typically on a hillside or ravine bank (slope is apparently important for nesting)*, often near water. Usually concealed from above and in accumulation of leaves at base of thick

Worm-eating Warbler.
Bob MacDonnell, CT

shrubs, saplings, tree roots, but also against large downed logs over abundant leaf litter and in crevices of outcroppings with lush vegetation. Nest is typically bulky, large for warblers. Foundational cup is usually built of moist skeletonized or dead leaves. *A lining of long spore stems of mosses may be present,* red in color when fresh; fine grasses, deer hair or horsehair, pine needles. **EGGS:** 4–5; typically white or creamy white with markings, sometimes pinkish, occasionally unmarked; 17 x 14 mm. **BEHAVIOR:** Single-brooded. Female may wet her breast feathers to facilitate pliability while shaping inner and outer cups. Low percentage of males are bigamous. Females may join in territorial scuffles.

GOLDEN-WINGED WARBLER   *Vermivora chrysoptera*

HABITAT: Wide variety over a broad range. Similar features and plentiful overlap with Blue-winged Warbler, but may choose wetter or swampier areas. Primarily deciduous forest edge habitat, shrubby hillsides, pasture, recovering surface coal mining sites, etc. In s. Appalachian Mts., requires mostly herbaceous openings in early successional forests with scattered shrubs and widely spaced overstory trees (usually 850–1,200 m elevation). **LOCATION AND STRUCTURE:** *Nest is very similar to Blue-winged's.* Foundation

Mark Peck, ON

may be built *around large upright stem* at ground level. Outside diameter 9–15 cm; height 7.6–12.7 cm; inside diameter 4.3–6.4 cm; depth 3.3–6.4 cm. **EGGS:** 4–5; *similar to Blue-winged's,* white with markings but more heavily blotched, may have dark spots or fine streaks; 16 x 12 mm. **BEHAVIOR:** Likely single-brooded. Populations in decline across much of range, having vanished altogether from some regions.

BLUE-WINGED WARBLER *Vermivora cyanoptera*

HABITAT: Forest edge habitat with dense shrubs or thickets: clear-cuts, overgrown fields and pasture, brushy swamps, wetland edges, often in shade from trees, usually well away from forest edge. Also along small cuts through underbrush, such as paths or deer trails. **LOCATION AND STRUCTURE:** Nest is on or close to ground, sometimes to 30 cm; usually at base of herbage or a small bush or built low in upright stems, sometimes in a grass clump. Nest is a deep, narrow cup, built within a solid foundation of dead leaves (stems oriented to project outward), of coarse grasses, grapevine bark and similar, dead leaves. Lined with finer bark fibers, fine grass stems, occasionally horsehair. Usually hidden by leaves, some of which may cap nest. *Nest is similar to Golden-winged's but may be less bulky, more often elevated; tree canopy may be denser than that of Golden-winged, but varies geographically.* Outside diameter 8.7–14.5 cm; height 7–12.7 cm; inside diameter 4–6.4 cm; depth 3.3–6.2 cm **EGGS:** 5; *similar to Golden-winged's* but not as heavily marked; white, finely and sparsely dotted, speckled with browns, purples, grays; 16 x 12 mm. **BEHAVIOR:** Single-brooded. Hybridizes with Golden-winged.

Mark Peck, NJ

BLACK-AND-WHITE WARBLER *Mniotilta varia*

HABITAT: Primarily deciduous woodlands, but also mixed forests. In north-central portion of breeding range also in predominantly coniferous habitat. Favors mature forest. **LOCATION AND STRUCTURE:** *Location similar to that of Worm-eating Warbler:* on ground, often on hillsides or banks, but more variable; also in swamps,

Black-and-white Warbler.
George Peck, ON

flat terrain. Usually well concealed from above, in hollow at base of a tree, shrub, stump, rock, below or against log or broken limb, or elevated in hollow in low stump, upturned root, crevice, to 1.8 m up. *Nest is similar to that of Blue-winged and other species in this category*: skeletonized leaves, grasses, similar. Lined with grasses, rootlets, shredded grapevine bark, occasionally hair. **EGGS:** 4–5; typically white or creamy white with markings; 17 x 13 mm. **BEHAVIOR:** Single-brooded. Frequently parasitized by Brown-headed Cowbird. A long-distance migrant and one of the earliest wood-warblers to return north in spring. Quite aggressive on breeding territory, scuffling with several species: Red-breasted Nuthatch, chickadees, warblers, and others.

KENTUCKY WARBLER *Geothlypis formosa*

HABITAT: Moist, mature bottomland deciduous forests (often near streams), requiring dense understory and good ground cover. **LOCATION AND STRUCTURE:** Mostly on ground, but may also be slightly elevated, built a few centimeters above ground in

Matt Monjello, AL

twigs or upright stems, to 1 m up. When on ground, nest place-ment and concealment are similar to those of other ground-nesting warblers that use leaves as foundation or outer wall, *though may more often build nest wall into surrounding upright stems of vegetation. May also be concealed at base of upright fern fronds.* Nest is similar to others in this category, with possible heavy use of wetted leaves in outer structure, and large, deepish cup of grasses or stems, leaf stems, similar, lined with fine blackish rootlets. Outside diameter 7–9 cm; height 6–9.4 cm; inside diameter 4–6 cm; depth 3–5 cm. **EGGS:** 4–5; typically white or creamy white, marked, but *larger than those of many warbler species,* often with fine speckling concentrated on larger end; 20 x 15 mm. **BEHAVIOR:** Rarely double-brooded. Long bouts of song, to 15 minutes in length, throughout morning. Once mated, male sings less; steady song may be an indicator that there is no active nest in immediate vicinity. Males often reclaim same territory each year.

GROUND NESTS COMPOSED PRIMARILY OF GRASSES, TYPICALLY IN MOIST NORTHERN FORESTS AND BOGS OR SIMILAR HABITATS

TENNESSEE WARBLER *Oreothlypis peregrine*

HABITAT: A common and abundant northern breeder, at fairly low elevations in open deciduous, coniferous, or mixed forests, com-monly in small clearings with grasses, abundant shrubs, small stands of small deciduous trees. **LOCATION AND STRUCTURE:** Similar to that of other ground-nesting warblers (concealed by overhanging herbage, at base of small tree, stump, etc.), often in a moss hummock or tussock of grass or sedges. Nest is usually set in a hollow (often in sphagnum moss), so that rim is level with surrounding substrate. Nest is made almost entirely of grasses, sedges, plant stems, lined with finer grass, fine rootlets, hair, mosses. Occasionally uses leaves in walls. *Often grassier than similarly located nest of Nashville Warbler.* **EGGS:** 4–6; typically white or creamy white, marked; 16 x 12 mm. **BEHAVIOR:** Single-brooded. Sometimes nest in small "colonies." When spruce bud-worm is abundant, breeding population densities can increase dramatically.

NASHVILLE WARBLER *Oreothlypis ruficapilla*

HABITAT: Wide variety of mostly northern second-growth forest in East and West (clear-cuts, burns, etc.) and open forests (deciduous and mixed), usually with plentiful shrubby under-story, open canopy, usually at edges. Also along bogs, streams, wetlands, edges, and in old-fields and pastures in herbaceous ground cover, thick stands of saplings. **LOCATION AND STRUC-TURE:** Concealed on ground, usually in open areas, set into a depression beneath shrubs or grass clumps or overhang of club-moss, ferns, etc. *Not as frequently against trees, logs, or on raised moss hummocks as nests of other warbler species.* Nest is made of grasses, mosses, rootlets, shredded bark, leaves, pine nee-dles, grasses. Rim is often bordered with mosses. Lined with finer, similar materials and hair. *Nest is tidy, and not made with as*

many grasses as Tennessee Warbler nest. Outside diameter 7–10.5 cm; height 2–4.7 cm; inside diameter 4.2–6 cm; depth 2.3–3.7 cm. **EGGS:** 4–5; typically white or creamy white, marked; may have scrawls and large blotching; 15 x 12 mm. **BEHAVIOR:** Single-brooded.

CONNECTICUT WARBLER *Oporornis agilis*

HABITAT: Wide variety, at low elevations, in mix of moist forest (spruce and tamarack, coniferous or deciduous), muskeg, bogs, sedge meadows, but also in drier pine or oak-pine or pine-aspen-poplar; usually at margins or edges, in forest openings, or among widely spaced trees with good understory. **LOCATION AND STRUCTURE:** Sunken into sphagnum moss in a bog or similar, or on dry ground at base of shrubs, saplings, logs, beneath dry grasses, under ferns. Nest is a deep, rounded cup of fine dry grasses, rootlets, weed and sedge stems, lined with fine grass,

rootlets, hair. *Mourning Warbler usually nests in earlier successional second growth, and nests are larger; Connecticut's nest is often more thin-walled.* Outside diameter 12–14 cm; inside diameter 5 cm; depth 4–5.5 cm. **EGGS:** 4–5; typically white or creamy white, *boldly marked and blotched*; 19 x 14 mm. **BEHAVIOR:** Single-brooded.

MOURNING WARBLER *Geothlypis philadelphia*

HABITAT: Northern boreal forests and similar habitats in mountainous South, in early successional clear-cuts, slashings, road cuts, pipelines, mining, or other natural or human-made clearings; old brushy fields, bogs, swamp borders. **LOCATION AND STRUCTURE:** On or *near ground*, to 30 cm up, hidden in or below thin-stemmed briar tangle, bush or dense herbaceous plants, sedge clumps or grass tussock, ferns. *Nest walls are often built into surrounding uprights.* Nest is large, bulky, similar to other species in this category, but is often built of coarser plant stalks, sedges, grasses, bark shreds, and lined with rootlets, fine grasses or sedges, hair. See MacGillivray's Warbler (its e. counterpart). Outside diameter 9–22 cm; height 7–10 cm; inside diameter 4.4–6.4 cm; depth 3.8–5.7 cm. **EGGS:** 4; typically whitish, with markings, *occasionally with black scrawls*; 18 x 14 mm. **BEHAVIOR:** Likely single-brooded.

Mark Peck, ON

PALM WARBLER *Setophaga palmarum*

HABITAT: Northern bogs, muskeg, in open to sparse conifer forest or similar habitat with sphagnum moss, scatterings of trees (coniferous or deciduous), dense undergrowth or heathlike ground cover. **LOCATION AND STRUCTURE:** Set into sphagnum moss at bog edges, concealed at base of small tree (often coniferous, but also deciduous) or beneath low-growing shrubs. Sometimes up to 0.6 m high in a conifer sapling. Nest is similar to others in this category: often made almost entirely of grasses, weed stems, but also can include bark shreds, rootlets, fern

Jim Zipp, ME

fronds. Lined with fine grasses, feathers. Hair possible. **EGGS:**
4–5; typically white or creamy white, marked, sometimes
scrawled, wreathed; 17 x 13 mm. **BEHAVIOR:** Single-brooded. May
nest in loose "colonies" where habitat allows.

WILSON'S WARBLER *Cardellina pusilla*

HABITAT: Abundant and widespread, in broad variety of habitat,
including riparian areas in shrubby thickets, alder or willow
thickets. In various forest types from far north to NM with clearings or openings (ponds, lakes, creeks, bogs, wet meadows,
clear-cuts); moist bog habitat with scattered or young or stunted
trees; coastal scrub and woodland. **LOCATION AND STRUCTURE:**
Similar to that of other ground- or bog-nesting warblers: usually
on ground or in depression *(except in coastal CA and OR, where
often low in shrubs)*, concealed, tucked against shrubs, small
trees, grass or sedge clumps, in sphagnum moss hummock.
Nest is bulky and large for bird's small size; outer layer *often has
more varieties of fibrous material* than other nests in this category.
Varies geographically: moss, weed and grass stems, grasses,

Mark Nyhof, BC

bark fibers, some leaf litter. Lined with fine grasses, possibly fine rootlets and hair. Some nests reported to be built mostly of blackberry leaves, nettle, oak, twigs, rotted leaves (CA). Outside diameter 7–17 cm (varies by region, substrate); height 5–6 cm; inside diameter 4.5–6 cm; depth 3–4 cm. **EGGS:** 5; typically white or creamy white, marked, often wreathed; 16 x 12 mm. **BEHAVIOR:** Single-brooded. Favorable nest depressions may be used for years, though frequently by different females.

CANADA WARBLER *Cardellina canadensis*

HABITAT: Wide variety. In low to mountainous moist mixed forests with heavy undergrowth, often near water and forest openings (streams, ravines, clearings). Also in wet, low habitats typical of other bog-nesting warblers. Rhododendron in southern range. **LOCATION AND STRUCTURE:** On or near ground in broad range of sites typical of others in this category, *but also in dirt banks, upturned tree root cavities, and often on slopes, rise on landscape, and in rocky substrates.* Nest is on foundation of leaves, bulky, *different than others in this category,* made of grasses, leaves, bark shreds, plant down, moss, twigs, stems, possibly pine needles, green leaves. Lined with fine grasses, rootlets, hair, leaves. Outside diameter 9–14 cm; height 5.5–10 cm; inside diameter 5–7.6 cm; depth 2.5–5 cm. **EGGS:** 4; typically white or creamy white with markings, often wreathed; 16 x 12 mm. **BEHAVIOR:** Single-brooded. In breeding range for only a short time, arriving late and leaving early.

Mark Peck, ON

GROUND NESTS IN THE MOUNTAINOUS, ARID SOUTHWEST

(See also Orange-crowned Warbler, p. 417)

COLIMA WARBLER *Oreothlypis crissalis*

HABITAT: Extremely localized, common only in Chisos Mts. of TX, in pinyon-juniper-oak woodland, usually with abundant grass or shrub understory, at 1,500–3,200 m elevation. **LOCATION AND STRUCTURE:** Similar to that of other ground-nesting warblers, concealed at base of shrubs, rocks, roots, etc., tucked into hollow or slight cavity, often on slopes. Nest is usually a foundation of dry oak leaves, cup of grass, leaves, shredded bark (juniper, grape, similar), possibly moss or lichen. Lined with fur or hair, fine grasses. Inside diameter 5–5.7 cm. **EGGS:** 4; typically white or creamy white, marked; 18 x 14 mm. **BEHAVIOR:** Possible second brood.

VIRGINIA'S WARBLER *Oreothlypis virginiae*

HABITAT: Foothills, mountains, mostly in pinyon-juniper or oak woodlands, but also in pine-oak, or mostly conifer, but all in association with deciduous shrubs. Also in chaparral thicket (mountain mahogany and similar scrub). **LOCATION AND STRUCTURE:** On or near ground, often on a slope, often in hollow at base of a shrub, grass clumps, or set in drift of dead leaves at base of stones, log, roots, slight overhang. Nest is loose but well woven, made of shredded bark, grass stems, roots, mosses; lined with fine grass or similar materials as in wall, and fur or hair. Outside diameter 9 cm; height 5 cm; inside diameter 5.5 cm; depth 3.5 cm. **EGGS:** 4; typically white or creamy white, marked, finely speckled, spotted; 16 x 12 mm. **BEHAVIOR:** Single-brooded. Males are extremely territorial. Brown-headed Cowbird may be regular nest parasite in parts of range.

Matt Monjello, AZ

PAINTED REDSTART *Myioborus pictus*

HABITAT: Oak, pine-oak, or juniper-oak woodlands of riparian canyons or mountains, with dense canopy and understory, usually with nearby water. **LOCATION AND STRUCTURE:** Usually on ground, on steep slope (but also on level ground), or on steep or vertical streambank or rock wall, rocky hillsides. Beneath grass clumps, vegetation, small shrubs or roots, rock, or layer of pine needles. May be in slight ground cavity or hollow in open. *Nest is similar to Red-faced Warbler's, possibly with more shallow cup and consisting of more grass.* Dry leaves may form base; outer wall made of coarse grass, pine needles, small leaves, fibrous bark or plant materials. Lined with fine grasses and hair. **EGGS:** 4;

A just-built Painted Redstart nest, ready for eggs. Matt Monjello, AZ

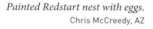

Painted Redstart nest with eggs.
Chris McCreedy, AZ

typically white or creamy white with little to no gloss, markings fine and mostly on larger end; 16 x 13 mm. **BEHAVIOR:** Double-brooded. Male and female seem to both contribute to nest site selection. Both sexes are competent songsters and sing back and forth during pair formation. Males are sometimes polygynous.

RED-FACED WARBLER *Cardellina rubrifrons*

HABITAT: At moderate to high elevation in mountains (to 2,800 m), in fir, spruce-fir, and mixed coniferous-deciduous woods (including pine-oak, spruce-aspen). Also in steep riparian canyons. **LOCATION AND STRUCTURE:** Often toward bottom of slopes or can-

Matt Monjello, AZ

yons where habitat is more moist, but also on higher ground. On ground, often in a small cavity or hollow, or in dry leaf litter. Nest is often placed at base of a grass clump, small woody plants, sapling or conifer, forbs, rock, or log, but may also be set with more exposure from above. Nest is made of soft, fibrous materials: a foundation and outer wall of bark (often aspen or juniper), also grass, leaves, and pine needles. Inner cup is made of grasses, lined with finer grasses, needles, hair possible. Outside diameter 9.6–11.8 cm; height 5–6 cm; inside diameter 3.5–5.5 cm; depth 3–4 cm. **EGGS:** 5; white, finely speckled, with possible blotching at larger end, often wreathed; 16 x 13 mm. **BEHAVIOR:** Single-brooded. Generally monogamous, but extra-pair copulations are frequent. Secretive, difficult to flush from nest.

OTHER GROUND OR NEAR-GROUND NESTS

ORANGE-CROWNED WARBLER *Oreothlypis celata*

HABITAT: Wide variety. In deciduous or mixed forests, often in patches or at edges of second growth, old clearings, overgrown pastures, burns, each with plentiful shrubby growth. Streamside thickets, chaparral. **LOCATION AND STRUCTURE:** Typically in shaded, moist areas (or similar microclimates in drier landscapes). Most often on ground, but also to 1 m. Nest and placement are similar to those of other ground-nesting warblers; frequently on shady wooded slopes, hillsides, canyons, or steep banks. Generally builds foundation and then adds grasses, mosses, fine twigs, bark shreds, stems, and hair to build or line cup. Dimensions are similar to those of other warblers, and elevated nests are often larger. **EGGS:** 5; typically white or creamy white, markings vary, commonly across much of egg, concentrated at larger end, may be blotched; 16 x 13 mm. **BEHAVIOR:** Single-brooded.

Mark Nyhof, BC

A nest set in deep, dead leaves, as is common among many ground-nesting wood-warblers. Matt Monjello, AZ

COMMON YELLOWTHROAT *Geothlypis trichas*

HABITAT: Wide variety, from wetlands to dry areas, but always with thick, low undergrowth. Sites include swamps and saltwater marsh edges, overgrown fields, drainage ditches, wet bottomlands, dry pine forests, riparian areas, orchards, clear-cuts, young second-growth deciduous stands, similar. **LOCATION AND STRUCTURE:** On or very near ground, occasionally over water, secured in surrounding upright vegetation (reeds, sedges, cattail, low bush, herbaceous plants). *Nest can look like that of a miniature Red-winged Blackbird nest*: bulky, woven of coarse grasses, sedges (sometimes of matted deciduous leaves or plant stems), reed shreds, and lined with similar, finer materials, sometimes bark fibers, hair. General dimensions (considerable size variation occurs): outside diameter 8.6 cm; height 7.8 cm; inside diameter 4.5–5.7 cm; depth 4–4.6 cm. **EGGS:** 4; typically white or creamy white, but with sparse dark specks, spots, *scrawls,* mostly at larger end, may make thin band or wreath; 17 x 13 mm. **BEHAVIOR:** Double-brooded.

Matt Monjello, ME

Casey McFarland, AL

KIRTLAND'S WARBLER *Setophaga kirtlandii*

HABITAT: Very rare, breeding only in nearly pure stands of jack pine in n. MI and small areas of WI. **LOCATION AND STRUCTURE:** On ground, concealed beneath or behind grass clump, other vegetation, usually within 1 m from edge of dense jack pine stand or trunk (periodic fire is important to maintain suitable habitat structure to promote jack pine regeneration). Female creates a depression in loose sand, in which a foundational cup is made mostly of grass or sedge leaves and pine needles, but also rotted wood, small twigs, oak leaf fragments, and may include spider silk. Lined with fern down, hair, moss, grass fibers, rootlets. Outside diameter ~10 cm; inside diameter 5.8 cm; depth 3.8 cm. **EGGS:** 4–5; typically white or creamy white, marked, but may be pinkish with blotching; 18 x 14 mm. **BEHAVIOR:** Single-brooded, occasionally double-brooded. Kirtland's Warbler was almost extinct by 1970s, but careful and intensive management of its nesting habitat, as well as rigorous Brown-headed Cowbird control, has enabled population numbers to rise: Kirtland's Warbler was officially delisted in 2019; 2018 estimates placed mature adult numbers at 4,500–5,000.

NESTS TYPICALLY PLACED LOW IN SHRUB, BUSH, OR SMALL TREE (OFTEN DECIDUOUS)

SWAINSON'S WARBLER *Limnothlypis swainsonii*

HABITAT: Large, moist forest habitats with dense understory with leaf litter floor but generally lacking herbaceous plants. From bottomlands (swampy areas, canebrakes, and privet thickets) to Appalachian Mts. (wooded ravines, rhododendron thickets, similar). Also in dense pine plantations. **LOCATION AND STRUCTURE:** Commonly in cane where available, but in variety of thickets, vine tangle, shrub, small tree, especially where numerous stems come together. Nest is cryptic, appearing like a mess of leaves, made of leaves, tendrils, vine. Cup is built compactly of skeletonized or dead leaves, plant and weed stems, pine needles, mosses, leaf stems (*stems often pointing outward*). Smoothly lined, various items possible: fern stems, hair, rootlets, Spanish moss, needles, grasses. The *largest nest* of all shrub- or tree-nesting warblers, similar in size to nests of larger songbirds but with small cup and less woody material. Outside diameter 13.7 cm; height 8.5 cm; inside diameter 6 cm; depth 4.3 cm. **EGGS:** 3; *white, unmarked*; 19 x 15 mm. **BEHAVIOR:** Single-brooded. Nest is difficult to locate and may be placed at edge of or outside male's territory.

Mia R. Revels, OK

MACGILLIVRAY'S WARBLER *Geothlypis tolmiei*

HABITAT: Wide variety, but generally riparian thickets in various habitats, especially in mountainous mixed conifers, but also deciduous or mixed woodlands, including cottonwood bottomlands; brushy moist hillsides or deep ravines, recovering clearcut or burn thickets. **LOCATION AND STRUCTURE:** Mostly low in brush, usually 0.5–1 m, may be higher. Sometimes on ground on grass clump or similar. Nest is well hidden, typically built into several upright stems. Fairly loose, made of coarse grasses, willow or other bark strips, dead weed stems, leaves (often tattered by builder). Lined with fine grasses, rootlets, hair. In shrubs, nests are neater, more compactly built. See Mourning Warbler. (its w. counterpart). Outside diameter 12–14 cm; height 5.7–7.6 cm; inside diameter 5–6 cm; depth 3–4 cm. **EGGS:** 4; typically white or creamy white with markings, but with unique dark blotches, spots that are larger than those of many other warbler species; 18 x 14 mm. **BEHAVIOR:** Double-brooded.

David Moskowitz,
WA

HOODED WARBLER *Setophaga citrina*

HABITAT: Primarily mature, moist deciduous forests with gap habitat (small opening or edges) and shrubby undergrowth or thicket, but also in unbroken shaded slope forest, and wooded swamp or bottomland (sweetgum and cypress swamp common). **LOCATION AND STRUCTURE:** Usually low, 0.3–1.4 m, in bush or in dense thicket within forest or at edge, in thin upright stems of brambles, small bushes, cane, herbaceous plants, saplings, palmetto. Nest is compact, neat, fibrous, made of bark fibers, fine grasses, fine twigs, plant down, forbs, some silk; nest is often fastened in place with spider silk. Outer and lower portion of nest may have *shell of dead or skeletonized leaves.* Lined with black rootlets, fine grasses, occasionally hair. Outside diameter and height ~7–8 cm; inside diameter 4–5 cm; depth 3–5 cm. **EGGS:** 3–4; typically white or creamy white, with bold, scattered reddish or brown markings, mostly at larger end; 18 x 14 mm. **BEHAVIOR:** Double-brooded. Extra-pair copulations are common.

Scott Somershoe, TN

Liz Tymkiw, WA

AMERICAN REDSTART *Setophaga ruticilla*

HABITAT: Broad range, generally in moist deciduous forest, may be mature but more often in second growth, where there are openings and dense understory. Also in mixed woodlands, alder or willow thickets, orchards, roadside thickets and trees, gardens, similar. **LOCATION AND STRUCTURE:** Typically 1–9 m aboveground in a tree (usually deciduous, sometimes white cedar) or leafy shrub, in upright, pronged fork or crotch or against trunk on limb with upright twigs or branchlets. Occasionally out on limb with similar upright supports. Nest is strong, neat, fibrous, often smooth-walled, compactly built, made of fine materials, including plant or bark fibers, plant down or seed hairs, fine grasses and stems, lichens, mosses, wasp nest paper, etc. *Nest is usually visibly bound with spider silk and has a camouflaging outer layer of bark (birch, etc.), occasionally lichens or bud scales, or similar.* Outside diameter 7 cm; height 6–7. 6 cm; inside diameter 4.5 cm; depth 3.5–4 cm. *American Goldfinch nests are generally more puck-shaped, and Yellow Warbler nests generally have thicker walls, not as tidily built.* **EGGS:** 4; typically white or creamy white, may be tinted pale green, gray, boldly marked, blotched, wreathed

Casey McFarland, CT

Casey McFarland, CT

or capped; 16 x 12 mm. **BEHAVIOR:** Single-brooded. Female primarily finds and selects nest site, but male may show her several potential sites during initial courtship. Susceptible to high rates of nest parasitism by Brown-headed Cowbird.

YELLOW WARBLER *Setophaga petechia*

HABITAT: Widely distributed; common in willow or similar thickets in riparian areas, swamp or marsh edges, deciduous woodlands; also in thickets at shorelines, overgrown pastures, orchards, gardens, roadsides, hedges; mangrove. **LOCATION AND STRUCTURE:** Nest is in upright fork or crotch of a bush, sapling, tree, or briar, usually low, 1–2.4 m, sometimes to 3 m. A unique nest, often with 2 or 3 distinct layers: a foundation, an upper structural cup (if nest is tall), and lining. Durable, compact,

Matt Monjello, ME

David Moskowitz, WA

Kate Fremlin,
Amaroq Wildlife
Services, NU

deeply cupped, outer wall is made of plant fibers, grasses, and plant down; often thickly lined with plant down, possibly hair and feathers. *Nest is more thick-walled, messy (walls less smooth) than that of American Redstart, often lower than nest of Yellow-rumped and Black-throated Gray Warblers. Smaller than Willow Flycatcher's. See also American Goldfinch.* Spider silk is sometimes present. Outside diameter 5–7.6 cm; height 5–13 cm (including multiple nests stacked); inside diameter 4–6 cm; depth 3–5 cm. **EGGS:** 4–5; grayish white or tinted pale green, and pecks and blotches can be bold, large; may be long subelliptical; 17 x 13 mm. **BEHAVIOR:** Single-brooded. Frequent host of Brown-headed Cowbird. On discovering intruder's egg, Yellow Warbler simply builds a new nest atop the old (burying her own clutch), sometimes several times. This creates an especially tall, distinctly layered structure. May nest in small colonies, with small territories.

CHESTNUT-SIDED WARBLER *Setophaga pensylvanica*

HABITAT: Shrubby, early successional deciduous woodlands: old clear-cuts, fields, burns; open woods with thick underbrush, and thickets along paths, road cuts, field edges, similar. **LOCATION AND STRUCTURE:** Low, 0.3–1 m, well hidden in shrub, brambles, or very small tree (mostly deciduous), woven into crotch or small upright stems (can be bulkier, flattish, when not in crotch). Foundation is made of coarse fibers of bark, plants, grasses, and grass stems, woven and bound in place with bits of spider or caterpillar silk. Upper walls are built of same materials, and sometimes plant down is lightly added. Lined with fine grass, possibly hair, rootlets. Outside diameter ~7 cm; height 6.7 cm; inside diameter 5 cm; depth 3.7 cm. **EGGS:** 4; like other warbler eggs but may have pale greenish tint, with fine or bold markings, mostly in ring on larger end; can be long subelliptical; 17 x 12 mm. **BEHAVIOR:** Single-brooded, occasionally double-brooded.

Matt Monjello, ME

BLACK-THROATED BLUE WARBLER *Setophaga caerulescens*

HABITAT: Largish tracts of deciduous or mixed boreal forest, with dense undergrowth or secondary growth, usually of deciduous shrub, dense saplings. **LOCATION AND STRUCTURE:** Low to ground, usually 0.3–1.5 m, often in rhododendron, laurel, saplings or young trees (spruce, hemlock, deciduous), or other shrubs and thickets. Usually in an upright fork or similar supportive crotch of upright stems. Nest is bulky, well hidden, often fibrous, made of *strips of papery bark, bits of rotted wood,* vine bark, twigs, fastened and bound with spider silk and possibly saliva. Lined with bark fibers, horsehair fungus, moss, needles, hair. Outside diameter 8–9.6 cm; height 5–7.8 cm; inside diameter 4.5–5 cm; depth 2.5–4 cm. **EGGS:** 4; typically white or creamy white, markings variable, sometimes wreathed; 17 x 13 mm. **BEHAVIOR:** Double-brooded, occasionally treble-brooded.

Matt Monjello, NH

Bill Summerour, NC

PRAIRIE WARBLER *Setophaga discolor*

HABITAT: Low brushy undergrowth in open forests (with open canopies), or secondary growth of clearings, roadsides, old burns, edge habitat, mangrove (where nest is usually over water), and scrubby dunes. Also in open pine barrens and young pine stands, Christmas tree farms. *Usually in drier habitats than other warblers in area.* **LOCATION AND STRUCTURE:** Usually in a dense shrub or small tree, in an upright fork or crotch or twigs, small horizontal branch, briar or vine tangle. Usually low, 0.3–3 m, occasionally higher (to 8 m). Nest is fibrous, foundation and outer wall made of long plant fibers or bark strips, grasses, plant down bound and woven to supports with spider silk. Inner layer is made of plant down, feathers, and fur, and nest is lined with fine grasses, feathers, moss; regularly incorporates British soldier lichen, a good indicator for identifying vacated nests. Outside diameter 6–7.6 cm; height 6–8 cm; inside diameter 4–5 cm; depth 5 cm. **EGGS:** 4; typically white or creamy white, marked, blotched, sometimes with scrawls, often wreathed; may be long subelliptical; 16 x 12 mm. **BEHAVIOR:** Occasionally double-brooded.

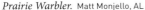

Prairie Warbler. Matt Monjello, AL

Prairie Warbler. Matt Monjello, AL

GOLDEN-CHEEKED WARBLER *Setophaga chrysoparia*

HABITAT: Rare, endangered, and endemic to TX (found in Hill Country of cen. TX), in mature ashe juniper woodland (often with mix of other tree species), in dense to moderately dense stands in canyons, on slopes, upland plateaus. **LOCATION AND STRUCTURE:** Often well hidden in juniper, but also in other deciduous trees (oak, elm, etc.), usually on a horizontal branch near trunk, also in smaller branches and forks, commonly in upper third of tree, 1.5–7 m, or higher, depending on tree. Occasionally on ground. Nest is a compact, thickly walled, soft or fibrous cup built primarily of *juniper bark strips*, bound loosely with spider silk or cocoons, and may include other items such as twigs, lichen, oak leaves, forbs. Lined with fine grass, hair, feathers possible. **EGGS:** 4; typically white or creamy white with fine markings; 18 x 13 mm. **BEHAVIOR:** Single-brooded.

James Hearn, TX

James Hearn, TX

NORTHERN PARULA *Setophaga americana*

HABITAT: Across northern and southern ranges in mostly mature coniferous or deciduous forests with stringy, hanging mosses or lichens. Often near swampy or boggy areas, rivers, or other water. **LOCATION AND STRUCTURE:** Usually in a hanging mass of moss or lichen (*Usnea*, Spanish moss), wherein the bird opens a cavity through a side entrance and builds a cup. Typically at tip of branch, 2–30 m aboveground. Cup is lined with mosses, tendrils, plant down, hair, rootlets, fine grasses. Where such epiphytic growths are not common, nests are reported in clumps of flood debris, or constructed of other materials into a pendulous nest. Inside diameter ~4–5 cm; depth ~2.5–5 cm. **EGGS:** 4–5; typically white or creamy white with markings usually concentrated on larger end; 16 x 12 mm. **BEHAVIOR:** Double-brooded. One of the smallest wood-warblers.

Matt Monjello, NC

Bill Summerour, AL

The entrance to this old Northern Parula nest has been opened slightly to reveal the soft inner cup.
Casey McFarland, AL

TROPICAL PARULA *Setophaga pitiayumi*

HABITAT: Deciduous riparian areas, usually with ball mosses for nesting. **LOCATION AND STRUCTURE:** *Like that of Northern Parula,* in hanging Spanish moss. **EGGS:** 3–4; *like those of Northern Parula, but often wreathed and often with markings over entirety;* 16 x 12 mm.

NESTS TYPICALLY PLACED LOW IN CONIFERS, PINES (OFTEN LOWER THAN 3 M)

(See also Prairie Warbler, p. 424)

MAGNOLIA WARBLER *Setophaga magnolia*

HABITAT: Northern coniferous forests of spruce, fir, hemlock (also with deciduous mix), in young, dense stands: second growth or recovery in old pastures, burns, timbering, etc. **LOCATION AND STRUCTURE:** Nest is generally low (often 1–2 m, sometimes to 10 m) and usually near trunk of a *small bushy conifer,* but may be toward tips, supported by horizontal branches or forks. Nest is a loose foundation and cup mostly of fine twigs, also of grass and weed stems, bark fibers, possibly conifer needles, *Usnea.* Lined with moss stems and horsehair fungus, needles. Outside diameter 7.5–12.7 cm; height 4–9 cm; inside diameter 4–6 cm; depth 2.5–4 cm. **EGGS:** 4; typically white or creamy white, marked, heavy blotching possible, and may have greenish tint; 16 x 12 mm. **BEHAVIOR:** Single-brooded.

Matt Monjello, NH   Kent McFarland, KPMcFarland.com, VT

BAY-BREASTED WARBLER *Setophaga castanea*

HABITAT: Northern coniferous or coniferous-deciduous forests, commonly with water nearby, favoring mature stands. **LOCATION AND STRUCTURE:** Typically in *lower portion* of a bushy, dense conifer, often on horizontal limb 1.5–3 m from trunk with a sheltering limb close above. Commonly 4.6–7.6 m, but as high as 20 m. Sometimes placed in a pine or shrub. *Nest is similar to those of Magnolia and Blackburnian Warblers but may include spider silk and plant down*, and may be lined with horsehair fungus, pine needles, fine grasses, hair, rootlets. **EGGS:** 4–5; may be tinted

Bay-breasted Warbler.
Mark Peck, ON

pale blue or greenish, marked and blotched across entirety, concentrated at larger end; long subelliptical; 18 x 13 mm. **BEHAVIOR:** Single-brooded. Very territorial, and mostly does not overlap with other warbler species.

BLACKPOLL WARBLER *Setophaga striata*

HABITAT: In low, young conifers (spruce, fir) but also in mixed forest (fir, paper birch, maple, similar). **LOCATION AND STRUCTURE:** Nests in a conifer, usually 0.5–2 m aboveground, but can be higher. Commonly against a trunk supported by one or more horizontal branches, sometimes toward tips. Nest is broad, bulky, thick-walled, with wide cup, with outer wall of fine or coarse twigs, but sometimes fine grass stems, and inner layer often largely *Usnea* lichen, fine twigs, rootlets. Lined with grasses, rootlets, and occasionally numerous feathers. **EGGS:** 4–5; like other warblers, but sometimes tinted green, blotched heavily at larger end; 18 x 13 mm. **BEHAVIOR:** Occasionally double-brooded in some regions. An incredibly long-distance migrant: some of population flies from AK to Brazil.

Mark Peck, ON

BLACK-THROATED GREEN WARBLER *Setophaga virens*

HABITAT: Wide variety. Generally a species of coniferous forests, favoring large unfragmented tracts. Also mixed forests (particularly where coniferous and deciduous forests interface), and in some regions (se. Canada) may be found in deciduous forest. **LOCATION AND STRUCTURE:** Mostly low, 1–3 m, but to 25 m. Usually in conifers supported where 2 or more thin branches create a fork at trunk, sometimes farther out where branches similarly fork. Occasionally in deciduous trees. Nest is often made of *long fine twiglets*, stems, bark strips *(often birch)*, lichens, bound lightly with spider silk. Lined with plant or bark fibers, moss, rootlets, hair, occasionally feathers. Outside diameter 7.6–10 cm; height 5 cm; inside diameter 5 cm; depth 4 cm. **EGGS:** 4–5; grayish white with typical markings, but also blotched and possibly scrawled; 17 x 13 mm. **BEHAVIOR:** Single-brooded. Both male and female build. One of earliest wood-warblers to return north in spring, especially in southern part of its range.

Paul Rossi, MI

NESTS TYPICALLY PLACED HIGH IN CONIFERS, PINES (HIGHER THAN 6 M)

CAPE MAY WARBLER *Setophaga tigrina*

HABITAT: Mostly in northern mature or second-growth coniferous forest; may be densely wooded or sparse. Also in coniferous bogs. **LOCATION AND STRUCTURE:** Usually in a spruce, sometimes fir, *in top of tree* (unlike most other species in this category), concealed completely from below and supported in spread of branchlets or needles near or against trunk, often in open forest or at edges. Height varies by tree size, 10–18 m. *Nests of Black-burnian Warbler are usually well out on large limbs.* Nest is bulky, made of spruce twigs, grass stems, pine needles, cedar bark or bark fibers, intermixed with plant down, outside *sometimes covered in moss.* Lined with grasses, possibly hair or fur, rootlets, feathers. Outside diameter ~9–11.5 cm; height 5–7 cm; inside diameter 4.5–5.5 cm; depth 2.5–5 cm. **EGGS:** 5–6; typically creamy white, with bold markings, blotches, occasionally dark scrawls; 17 x 13 mm. **BEHAVIOR:** Single-brooded.

BLACKBURNIAN WARBLER *Setophaga fusca*

HABITAT: Primarily in mature coniferous and mixed coniferous-deciduous forests. **LOCATION AND STRUCTURE:** Almost always in larger conifers far out from trunk near tips of branches, supported by forking twigs. From 1.5 m to higher than 25 m, but *typically high*, often higher than nests of other warbler species. May also nest in deciduous trees, particularly in southern, Appalachian range. *See Bay-breasted and Magnolia Warblers.* Cup is made of fine twigs, bark, rootlets, *Usnea* lichen where present; nest is often *fastened with spider silk.* Lined with pine needles, fine grasses, moss, rootlets, horsehair fungus. **EGGS:** 4; may be tinted pale greenish white, and with possible scrawls; 17 x 13 mm. **BEHAVIOR:** Single-brooded.

Mark Peck, ON

YELLOW-RUMPED WARBLER *Setophaga coronata*

HABITAT: Mature mixed coniferous-deciduous or coniferous forest. **LOCATION AND STRUCTURE:** Usually on horizontal branch(es) of conifer, 1–15 m up, usually ~6 m. Typically near trunk supported where 2 or more branches sprout from bole, or out on limb. Sometimes in a deciduous tree. Nest is well made and bulky, compact, thick-walled, made of twigs, needles, bark strips, grass, plant down, rootlets, and lined with hair, stringy

Matt Monjello, AZ

Mark Peck, ON

lichens, and numerous feathers, of which the *tips commonly are woven into cup wall so that the feathers protrude inward and provide a ceiling of sorts for the eggs; distinctive.* Outside diameter 7.5–9 cm; height 5.7 cm; inside diameter 5 cm; depth 3.8 cm. **EGGS:** 4–5; pale cinnamon, light gray, to almost white, with typical markings, sometimes scrawled, paler markings often prominent; usually wreathed or capped; 17 x 13 mm. **BEHAVIOR:** Occasionally double-brooded.

PINE WARBLER *Setophaga pinus*

HABITAT: Common. Primarily in pine woodlands, but also in mixed forest of pine-hardwoods. **LOCATION AND STRUCTURE:** Usually high in a large or mature pine tree (height ranges widely among tree species, but often 10–15 m up), saddled (woodpewee–like) to horizontal limb in a fork or in a twig or needle cluster, usually well away from trunk or toward branch tip; well hidden and very difficult to find. Nest is compactly built of plant or bark fibers, rootlets, weed stems, fine twigs and pine needles, often bound lightly with silk. Lined with plant down, feathers, hair, needles, fibers, bark strips. **EGGS:** 4; like those of other warblers, but sometimes pale gray greenish or bluish white, sometimes scrawled, markings often forming a band or wreath; 18 x 13 mm. **BEHAVIOR:** Double-brooded. Winters and breeds almost exclusively within U.S.

GRACE'S WARBLER *Setophaga graciae*

HABITAT: Pine forests (primarily ponderosa) and mixed pine-oak, in mountains, canyon bottoms. **LOCATION AND STRUCTURE:** High up near crown or upper half of pine (6–18 m), far out on a horizontal limb, hidden in thick clusters of needles or twigs. Nest is soft and fibrous, *flattish*, made of plant fibers, fine grass, catkins, hair, bound with spider silk. Lined with fine grasses, hair, rootlets. Nest is small, reflecting species' small size. Outside diameter 6–7.6 cm; height 3.8–5.8 cm; inside diameter 4.4–5 cm;

depth 3.2 cm. **EGGS:** 3–4; white to light gray or pale greenish white, with typical markings, mostly in wreath at larger end but possible across egg; sometimes scrawled; 17 x 13 mm. **BEHAVIOR:** Single-brooded.

TOWNSEND'S WARBLER   *Setophaga townsendi*

HABITAT: Tall, mature coniferous and mixed-coniferous forests of Northwest, from temperate rainforest to drier inland mountains. **LOCATION AND STRUCTURE:** Typically in a fir or spruce, often high (to 34 m) but as low as 1.8 m; varies with tree age and height. On a horizontal limb or spreading branchlets, close to or far from trunk. Bulky, thick-walled, possibly shallow-cupped. Nest is usually soft and fibrous, made of cedar bark, small twigs, inner bark or plant fibers, dried grasses, lichens, spider silk or cocoons possible. Lined with moss stems, fine grasses, deer or elk hair possible. Dimensions vary geographically; apparently larger in colder or wetter climes. Outside diameter 7.6–10 cm; height 5.7–6.8 cm; inside diameter 5–6 cm; depth 2–3.8 cm. **EGGS:** 3–5; typically white with markings, boldly specked and spotted at larger end; 17 x 13 mm. **BEHAVIOR:** Single-brooded. Hybridizes with Hermit Warbler.

HERMIT WARBLER   *Setophaga occidentalis*

HABITAT: Generally in young or mature, densely canopied coniferous forests in Pacific Northwest and Sierra Nevada (predominantly in Douglas-fir in north, pines in Sierra Nevada). **LOCATION AND STRUCTURE:** Usually high in tall conifers, often 6–12 m up, but also much higher. *Placement similar to that of Townsend's Warbler.* Nest is thick-walled, made of weed stems, twigs, rootlets, pine needles (southern range), cedar bark, bound lightly with spider silk. Lined with cedar bark, plant fibers, dried grasses, hair, feathers, plant down. *Dimensions are similar to those of Townsend's.* **EGGS:** 3–5; typically creamy white, heavily marked with dark spots, speckles, blotches at larger end in broad wreath, but markings throughout; 17 x 13 mm. **BEHAVIOR:** Single-brooded. Hybridizes with Townsend's Warbler.

NESTS IN CONIFEROUS OR DECIDUOUS TREES (SEE THE FOLLOWING RANGES OF SPECIES)

CERULEAN WARBLER   *Setophaga cerulea*

HABITAT: Wet bottomlands and riparian zones, as well as high on dry slopes, mountains, generally in mature deciduous forests with large trees, often in trees near canopy gaps created by blowdowns. **LOCATION AND STRUCTURE:** Usually high in tallest trees in area, in middle or upper canopy, 5–35 m up, often hidden from above by low leaves. May be saddled or placed at a fork near ends of horizontal limbs. Nest is *unique among warblers*, resembling that of *Blue-gray Gnatcatcher, Eastern Wood-Peewee, Least Flycatcher.* Nest is compact, *shallow*, dark gray, woven tightly with silk and grapevine bark, bark or plant fibers, fine grasses, mosses, catkins, sometimes lichens. *Outside is decorated with* *gray or white materials.* Lined with fibers, mosses, hair. Often very small. Outside diameter 7 cm; height 3.3–5 cm; inside diameter 4. 4–5 cm; depth 2–3 cm. **EGGS:** 4; creamy white to pale grayish or tinted greenish white, typical markings but bold, blotched; marking often across entire egg, but concentrated at larger end; 17 x 13 mm. **BEHAVIOR:** Single-brooded. Male may contribute to initial building, primarily collecting silk, of which much is used. Has seen alarming declines in large portions of breeding range.

YELLOW-THROATED WARBLER *Setophaga dominica*

HABITAT: Variety, ranging from pine forest to riparian deciduous (commonly sycamore) or mixed forest; cypress swamps; live oak woodlands with abundant Spanish moss. Favors mature trees. **LOCATION AND STRUCTURE:** Usually well hidden high in canopy of large trees, but varies, 3–30 m, often 10–15 m. *Nest is in a clump of Spanish moss* where occurring (see Northern and Tropical Parulas) far out on a horizontal branch. Elsewhere in a *cluster of pine needles or leaves* near end of branches, saddled to branch in open, or neatly wedged or fastened in upright forks. Nest is made of weed stems and bark strips, leaf fragments, moss, *lined heavily with plant down*, or lined with grass, weeds, moss, feathers possible. Outside diameter 6.4–7 cm; height 4.5–5 cm; inside diameter 4–4.7 cm; depth 3.3–4.4 cm. **EGGS:** 4; white or creamy white with bold markings, blotching of reds, browns, purples, which can be very dark; 17 x 13 mm. **BEHAVIOR:** Double-brooded in southern range. Among earliest returning wood-warblers in spring in Southeast, rivaled only by Louisiana Waterthrush.

Mikael Behrens, TX

BLACK-THROATED GRAY WARBLER *Setophaga nigrescens*

HABITAT: Wide variety of primarily mountainous forest types over broad range; generally in open coniferous or mixed woodland with plentiful brush (including Douglas-fir and oak, pinyon-juniper-oak, oak-madrone, etc.), also in chaparral, scrub oak habitats. **LOCATION AND STRUCTURE:** In variety of conifers and deciduous trees (fir, oak, pinyon-juniper, etc.), *often on a horizontal branch*, but also in a crotch in limb or near main stem in shrub. Height varies with tree species, 1–11 m. Nest is deeply cupped, made of fibrous materials, varies regionally: grasses, weed stems, bark or plant fibers, mosses, fastened in place and bound with spider silk. Lined with hair, fine grasses, plant down, and *sometimes many feathers*. Species is small, and nest dimensions are similar to those of *Grace's Warbler*, but cup is deeper. **EGGS:** 4; typically white or creamy white, with markings, finely marked with specks, small blotches, may be wreathed; 6 x 12 mm. **BEHAVIOR:** Single-brooded.

Bettina Arrigoni, AZ

NESTS ON OR CLOSE TO GROUND, SITUATED IN RECESSED OR ELEVATED CAVITY, OR WITH DOMED ROOF

OVENBIRD *Seiurus aurocapilla*

HABITAT: Generally in large tracts of unfragmented, mature deciduous or mixed forests with mostly open, deeply leaf-covered forest floors. **LOCATION AND STRUCTURE:** Built in a hollow on ground. Nest is a readily identifiable *domed structure*, immaculately blended into forest floor, roofed with leaves and surrounding materials or vegetation, with a small side entrance usually invisible unless viewed from the right angle. Nest and dome above are built as a single structure of dead leaves, grasses, stems, twigs, mosses, bark, and other materials from surrounding area. Lined with fine rootlets, fibers, hair. Outside diameter 16–23 cm; height from ground to dome top 11–13 cm; inside diameter 7.5–8 cm; depth, 4.8 cm. Side entrance usually wider than high, and smaller than inner cavity: 5.7–6 cm x 3.8–4.5 cm. **EGGS:** 4–5; creamy white or yellowish white with typical markings; *similar to those of Lousiana and Northern Waterthrushes*; 20 x 15 mm. **BEHAVIOR:** Single-brooded. Frequently parasitized by Brown-headed Cowbird. The authors have discovered nests by watching the woods around singing males and waiting for females to leave or return to nests. Females will scurry, rodent-like, away from nest site if intruder comes much too close; they are otherwise impressively difficult to flush.

Matt Monjello, CT

Matt Monjello, CT

LOUISIANA WATERTHRUSH *Parkesia motacilla*

HABITAT: Along flowing streams in deciduous or mixed coniferous forests, commonly in mountainous or hilly topography, but also in flatter bottomlands and swampy areas. **LOCATION AND STRUCTURE:** Near a stream in hollows or crevices in steep bank, upturned tree roots, beneath log, usually well hidden by varying overhanging substrates with entrance to one side. Nest is bulky, well insulated, usually with a foundational structure of dead, mucky leaves (mud helps bind nest) and dry leaves, twigs, needles. Cup consists of fine stems, small grasses, rootlets, mosses, hair. A pathway is sometimes built of dead leaves. Dimension

Eric C. Soehren, AL

Eric C. Soehren, AL

depends on size of hollow or cavity: outside diameter ~13–18 cm; height 7–20 cm; inside diameter 6.5–7.5 cm; depth 4–5 cm. **EGGS:** 4–6; typically white or creamy white, marked, though highly variable, commonly wreathed, *similar to those of Ovenbird and Northern Waterthrush*; 20 x 15 mm. **BEHAVIOR:** Single-brooded. Both sexes search for nest sites, and both build. Among the earliest wood-warblers to arrive on breeding grounds in spring, and soonest to leave in fall. Its streamside territories are linear, usually ~250–400 m in length, and males provoke and chase each other frequently.

NORTHERN WATERTHRUSH *Parkesia noveboracensis*

HABITAT: Boreal forests and southern woodlands, deep in thick, cool, shady wooded swamps, bogs, and wetlands (coniferous and deciduous habitats); riparian thickets at borders of lakes, rivers, mountainous streams. In dense, low-growing understory near water. **LOCATION AND STRUCTURE:** Location similar to that of Louisiana Waterthrush: in cavities in upturned tree roots, root of living trees, stumps, beneath overhanging banks, fern clumps *(where it overlaps with Louisiana, Northern's nest is often located where fern density is higher, with more moss and coniferous trees)*. Nest is *similar to Louisiana's, but may be composed mostly of mosses, liverworts, grasses or stems*, skeletonized or dead leaves possible. Inner bowl is made of fine grass stems or similar, mosses, possibly pine needles; lined with moss, animal hair. **EGGS:** 4–5; white with typical markings, *similar to those of Louisiana Waterthrush and Ovenbird*; 19 x 13 mm. **BEHAVIOR:** Single-brooded. Interestingly, while this species is aggressive toward other birds, it tolerates Louisiana Waterthrush. These closely related songbirds use different niches, and so apparently are unconcerned with each other.

PROTHONOTARY WARBLER *Protonotaria citrea*

HABITAT: Wooded bottomlands, swamps, and other wet, low-elevation shaded forests in flat terrain with water, often slow moving (flooded areas, swamp, ponds, lakes, etc.), but also near creeks, streams, rivers. Favors areas where undergrowth is

Mark Peck, NJ

minimal. **LOCATION AND STRUC-TURE:** The only truly cavity-nesting warbler in e. U.S. Nest is near (usually within 5 m) or over water in variety of cavities: woodpecker holes (often on small-diameter trees, in holes of *Downy Woodpecker*, but also other species), natural tree cavities (cypress knees, snags, branch malformations, etc.), small bird boxes, and other cavities. Sometimes in old bird nest. Variable: nest is a bulky foundation or outer layer (typically built to within 10 cm of hole) of *moss,* with cup of rootlets, bark shreds, pine needles, plants stems, skeletonized leaves. Lined with fine grasses or sedges, leaf stems, tendrils of poison ivy, possibly feathers, and fishing line can be a common addition. **EGGS:** 4–6; typically white or creamy white, but boldly marked over entirety with spots and blotches,

Prothonotory Warbler.

Mark Musselman, USFWS, SC

and can be pinkish in base color; 18 x 15 mm. **BEHAVIOR:** Double-brooded. Male searches out nest cavities prior to female's arrival, and places moss inside potential sites; female chooses. Specific requirements for breeding grounds in U.S. and wintering habitat in Cen. and S. America make this species especially vulnerable as essential habitat is lost.

LUCY'S WARBLER *Oreothlypis luciae*

HABITAT: Most abundant in riparian areas, particularly mesquite thickets; also in riparian cottonwood and willow, and tamarisk thickets. Also in drier mesquite (or similar) scrub habitats of dry

Scott Olmstead, AZ

grassland, along desert washes. Less commonly in higher elevations in dry deciduous mixed woodland. **LOCATION AND STRUCTURE:** *A cavity nester. Nest is often placed behind loose bark (like Brown Creeper) or beneath dead leaves of yucca trees, in woodpecker hole in tree or saguaro, in Verdin nest, natural tree cavities, exposed roots along banks, rarely in a hole in bank.* Nest is often low, but to 17 m. Nest is small with a small cup, somewhat frail, well woven of fine twigs, weed stems, grasses, leaves, mesquite leaf stems, flower clusters. Lined with plant or bark fibers, hair, feathers possible. Rough dimensions: outside diameter 10 x 7.6 cm; height 7.6 cm; inside diameter 5 x 3.4 cm; depth 4.4 cm. **EGGS:** 4–5; typically white or creamy white, marked; 15 x 11 mm. **BEHAVIOR:** Double-brooded. One of smallest wood-warblers in N. America. Many breeding populations have been lost as limited desert riparian habitats (and other types) have been destroyed or altered.

CARDINALS and ALLIES (Cardinalidae)

TANAGERS

Four species of these small, lovely songbirds breed in N. America, and they share similar nest characteristics and breeding behaviors. Where species overlap, they tend toward different habitats, which aids in nest identification. All species are presumably socially monogamous. Except in Western Tanager, nests are typically *flimsy, thin, shallowly built or saucerlike cups,* generally placed well out on a horizontal limb in a fork and ~2.4–18 m up. They are similar in size and typically built by the female in 4–5 days, though the male may accompany her. Eggs are similar: subelliptical to short subelliptical, smooth and semiglossy, light blue to blue-green, and spotted, speckled, blotched with brownish tones, and ranging in size only slightly, from 23 x 16 mm to 24 x 18 mm. Tanagers are usually single-brooded and lay 3–5 eggs in a clutch, varying by species. **DIFFERENTIATING SPECIES:** Where Summer and Scarlet Tanagers overlap, Summer tends to breed in shorter and more open woodlands. In the West, Western and Hepatic Tanagers use coniferous forests at higher elevations, while Summer Tanager breeds in lowlands.

HEPATIC TANAGER *Piranga flava*

HABITAT: Coniferous or deciduous woodlands of Southwest; in montane canyons, mostly in open pine, pine-oak, at 1,524–2,137 m elevation. **LOCATION AND STRUCTURE:** In a fork toward end of horizontal branches, 5.5–15 m up (usually high). Nest is a flat, saucer-shaped, loosely built cup of plant fibers, grasses, stems, thin twigs; lined with finer grasses and other soft plant materials, sometimes pine needles or rootlets, hair. **EGGS:** 4; IP, NP not well known; likely similar to those of other tanagers.

SUMMER TANAGER *Piranga rubra*

HABITAT: Dry, deciduous habitats and mixed (pine-oak) forests, typically preferring open gaps and forest edges. Also orchards, arid riparian groves. **LOCATION AND STRUCTURE:** On horizontal limb, in a fork or leaf cluster, well out from trunk and often over an opening below, 3–10.6 m up, sometimes lower, or higher than 20 m. In East, nest is a flat, flimsy cup, sometimes ragged (eggs may be visible through bottom), in West, nests are more solidly built. Made of forb stems and other dry herbaceous material, bark, Spanish moss in South, some spider silk. Lined with fine grasses. Three nest layers may be visible: a base, bowl, and inner bowl. Outside diameter ~8.5 cm; height 5.5–11 cm; inside diameter 6–7.6 cm; depth 3.5 cm. **EGGS:** 4; IP 11–12 days (F); NP 10 days (F, M).

Male Summer Tanager with chick.
Damon Calderwood, CA

David Pierce, MO

SCARLET TANAGER *Piranga olivacea*

HABITAT: Deciduous, sometimes mixed woodland, preferring large tracts of mature forest. Often in closed canopy with taller trees *(Summer Tanager is in more open, shorter forests, though both species use similar habitat in some regions).* **LOCATION AND STRUCTURE:** Placement is similar to that of other tanagers, in a leaf cluster, often with an unobstructed view to ground below. Nest is often thin enough to see eggs from below. Cup may be slightly oblong. Dimensions are similar to those of other tanagers. **EGGS:** 4; *markings may not appear as bold as on eggs of Western and Summer Tanagers, may be finer than on Summer*; IP 12–14 days (F); NP ~15 days (F, M). **BEHAVIOR:** The smallest *Piranga* species in N. America and, because of habitat preference, sensitive to forest fragmentation.

WESTERN TANAGER *Piranga ludoviciana*

HABITAT: The northernmost-breeding tanager. Wide variety: open coniferous and mixed woodlands, mostly breeding in mountains, from high-elevation forest to pinyon pine. Often in conifers. Also in riparian woodlands, oak woodlands, aspen forests. **LOCATION AND STRUCTURE:** Placement is similar to that of other tanagers, 2.7–19.8 m up, commonly 6–9 m. Nest is a compact (occasionally flimsy), shallow, flattish bowl or saucerlike cup of conifer twigs, rootlets, coarse grasses, sometimes stems, mosses, bark, needles, etc. Lined with finer rootlets, hair when available, sometimes other soft materials. Cup may be slightly oblong. Three layers may be distinct (see Summer Tanager). Dimensions are similar to those of Summer's nest but *may be slightly larger*. **EGGS:** 3–5; *markings across entire egg but concentrated mostly at larger end*; IP 13 days (F); NP 10–11 days (F, M). **BEHAVIOR:** In far north, may spend only 2 months on breeding grounds before returning south.

Mark Nyhoff, BC

CARDINALS

NORTHERN CARDINAL *Cardinalis cardinalis*

HABITAT: Wide variety, typically dense shrubbery, tangled vines, briars, also small deciduous and coniferous trees. Overgrown clearings, woodland edges, suburban yards and parks, or in mesquite or other thickets in riparian areas and small waterways in arid habitats. Avoids nesting in deep forest. **LOCATION AND STRUCTURE:** Typically in a fork or in small branches, 0.3–6 m up, but usually below 3 m; often about head height. Nest is a loosely built cup of twigs, vines, leaves, strips of grapevine bark, rootlets, weed stems; lined with fine grasses, sometimes hair. Often contains bits of garbage in outer wall, sometimes snakeskin. Nest may have 4 layers: outer structure of twigs (kinked by female with her beak), inner mat of leaves, inner cup of grapevine bark, and lining of fine grasses. Outside diameter 10–11 cm; height 6–7.5 cm; inside diameter 6.8–7.7 cm; depth 3–4 cm. **EGGS:** 3–4; subellipitical to short subelliptical; 25 x 18 mm; IP 11–13 days (F); NP 10 days (F, M). **BEHAVIOR:** Begins nesting early, 3 or 4 broods possible. Male manipulates nesting materials in display while the pair searches together for nest sites. Female is primary builder. Mostly socially monogamous. Both sexes sing; female sings while sitting on nest, possibly to communicate to male that it's time to be fed.

Matt Monjello, NC   Matt Monjello, NC

PYRRHULOXIA *Cardinalis sinuatus*

HABITAT: Open desert scrub, thicket, commonly in mesquite. Also in other desert shrub (catclaw, condalia, graythorn, etc.), mistletoe clumps, palo verde, occasionally in elm and salt cedar. *Where it overlaps with Northern Cardinal, Pyrrhuloxia nests in more open portions of shrubs.* **LOCATION AND STRUCTURE:** Commonly 1.5–2.4 m aboveground, loosely placed in a fork of twigs and small branches, often toward shrub's outer edge, generally away from main branches and trunk, but may be placed against trunk.

Nest is a neat, fairly compact cup of twigs (often thorny), strips of inner bark, coarse grasses, weed stems, rootlets, and spider silk. Lined *primarily with rootlets*, mostly on bottom more than on sides, occasionally with fine grasses, hair. *Very similar to Northern Cardinal nest but may be slightly smaller, more compact, and contain spider silk.* **EGGS:** 2–3; *similar to those of Northern Cardinal*; 24 x 18 mm; IP 14 days (F); NP ~10 days (F, M). **BEHAVIOR:** Likely single-brooded. Apparently monogamous.

Pyrrhuloxia. Scott Olmstead, AZ

Pyrrhuloxia. Scott Olmstead, AZ

GROSBEAKS

ROSE-BREASTED GROSBEAK *Pheucticus ludovicianus*

HABITAT: Wide variety: shrubby growth, thickets; moist or dry deciduous, mixed, and coniferous forests; gardens, orchards, overgrown fields. **LOCATION AND STRUCTURE:** *Similar to nest of Black-headed Grosbeak*: very flimsy, may be slightly smaller.

Matt Monjello, ME

Placed at roughly the same height, though it can be much higher in large trees. **EGGS:** 3–5; *similar to Black-headed's, possibly less oval*; 25 x 18 mm; IP and NP also similar to Black-headed. **BEHAVIOR:** Apparently monogamous. Will hybridize with Black-headed Grosbeak where ranges overlap, and the 2 species share similar nesting and breeding behavior. Double-brooded.

BLACK-HEADED GROSBEAK *Pheucticus melanocephalus*

HABITAT: Wide variety: thickets and densely foliaged trees, particularly along streams in riparian areas, moist canyons, floodplains. Also cottonwoods and willows, aspen, mature pine forest, pinyon-juniper woodlands, orchards, and urban and suburban areas. Often associated with robust understory and forest edges. **LOCATION AND STRUCTURE:** Commonly wedged in fork of a shrub or tree 1.2–3.6 m up, may be higher, usually well concealed but not always. Nest is a loose, thinly built (eggs are often visible through nest bottom), bulky cup of interlaced small twigs, forbs, bark strips, stems, and rootlets. Sparsely lined with finer rootlets, fine grasses, and other materials. Outside diameter 14 cm; height 8 cm; inside diameter 7–8 cm; depth 4–5 cm. **EGGS:** 3–4; short subelliptical to subelliptical or oval; 28 x 18 mm; IP 12–13 days (F, M); NP 12 days (F, M). **BEHAVIOR:** Female builds in 3–4 days (one of the authors observed a male building in NM). Socially monogamous. Male takes part in sitting on eggs but lacks a brood patch. Both sexes sing frequently, and their songs can be used to find a nest's general location during incubation; male will sing while sitting atop eggs. Mostly single-brooded.

David Moskowitz, WA

Casey McFarland, NM

BUNTINGS and BLUE GROSBEAK

Five species of these small, colorful songbirds nest in N. America. Males and females look strikingly different, and a female flushing quickly from a nest into the brush can give little indication of the species. They are mostly monogamous or presumed so, though polygyny has been reported in Painted and Lazuli Buntings. Nests share many characteristics: small, tidy, and somewhat densely built cups, largely of grasses, small stems, and plant fibers. Grasses may be wound to stems to securely attach the nest in place, and the nest may contain spider or caterpillar silk. Except in Indigo Bunting, nests are all built low (usually below 2 m) but can be 3–4.5 m and up. Dimensions are similar: outside diameter ~7.6–10 cm; height 7.6 cm; inside diameter ~5 cm; depth 3.8–5 cm. Blue Grosbeak nest is considerably larger (inside diameter to 6.5 cm). Eggs are similar: pale bluish white or white and unmarked except in Painted and Varied Bunting, two species whose eggs are finely speckled with reddish browns, mostly at the larger end. Bunting eggs: 18–19 mm x 14 mm. Blue grosbeak eggs: 22 x 17 mm. There are usually 3 or 4, sometimes 5, eggs in a clutch, with slight variation across species. Males may feed fledged young while females renest.

BLUE GROSBEAK *Passerina caerulea*

HABITAT: Though widespread, this species is relatively scarce. Old overgrown fields, along streams, hedgerows, brushy woodland edges, open slash (post-logging); scrubby thickets of mesquite, salt cedar, etc; pine forests in s. U.S. **LOCATION AND STRUCTURE:** In small trees or bushes, tangled vines and briars, weeds. Usually low, near ground (15 cm), occasionally as high as 7 m. Often fastened to 2 or 3 upright stems. Nest is compact, deeply cupped. *Similar to Indigo Bunting's nest but larger.* May contain snakeskin, paper, and similar materials. **EGGS:** 4; IP 11–13 days (F); NP 9–13 days (F, M). **BEHAVIOR:** Female builds, male occasionally. Often double-brooded. Likely a frequent host of Brown-headed Cowbird.

Bill Summerour, GA

LAZULI BUNTING *Passerina amoena*

HABITAT: Brushy habitat in wide variety of regions; riparian thickets, brushy hillsides, chaparral, sagebrush, but also in deciduous tree thickets (aspen, cottonwood, etc.). **LOCATION AND STRUCTURE:** Often in outer edges of chosen shrub, typically well shaded, 0.3–3 m up, but often 0.6–1.2 m. Nest is similar to those of other bunting species, somewhat thick-walled. **EGGS:** 4; sometimes white; IP 12 days (F); NP 10–15 days (F, M, male feeding varies). **BEHAVIOR:** Double-brooded.

Chris McCreedy, CA

Chris McCreedy, CA

INDIGO BUNTING *Paserrina cyanea*

HABITAT: Brushy, weedy habitats along wooded edges; fields, roadside thickets, brushy swamps. In West in brushy canyons, thickets in floodplains, riparian zones. **LOCATION AND STRUCTURE:** Built low to ground, 0.6–4.6 m up, often 0.9 m. In understory shrubs, plants, vine tangles, fastened into upright stems or twigs in forks or crotches. Nest is similar to those of other bunting species, often begun with green plant material and a foundation of skeletonized leaves. Occasionally snakeskin. **EGGS:** 3–4; usually white; IP 12–13 days (mostly F); NP 9–13 days (F). **BEHAVIOR:** Female builds, may take 8–10 days. Frequently parasitized by Brown-headed Cowbird.

Casey McFarland, AL

Casey McFarland, AL

VARIED BUNTING *Passerina versicolor*

HABITAT: Thorny brush in arid canyons and washes; thick vegetation in riparian areas. **LOCATION AND STRUCTURE:** In shrubs at outer edges, or in a small tree, tangled vines, often with one side exposed, supported at sides (rarely from bottom) by 2 or 3 small twigs or branches. Shaded from above, 0.3–3.6 m up, often low. Made mostly of dry grass, small stems, plant down. Rim and nest attachments may be bound with spider silk. Lined with small stems, rootlets, sometimes horsehair, hair. Occasionally snakeskin. **EGGS:** 3–4; speckled (AZ); IP 12–13 days (F); NP 12 days (F, M). **BEHAVIOR:** Nest is built by female, perhaps male as well. Single-brooded. Egg laying is timed with first good rains (AZ).

PAINTED BUNTING *Passerina ciris*

HABITAT: Partly open landscapes of scattered bushes or trees, woodland edges, roadside and streamside thickets, hedgerows. **LOCATION AND STRUCTURE:** In shrubs and bushes, low trees, vine tangles, rank herbage. Commonly 0.9–1.8 m up, but sometimes 6 m or higher. Nest is firmly attached by weaving plant materials and silk to supporting twigs. A neat, deep cup is well built of grasses, thin forb stems, leaves or leaf skeletons, bark strips, perhaps tissue and similar, bound in spider silk. May be thin-walled. Lined with fine grasses, rootlets, hair. **EGGS:** 3–4; speckled; IP 11–12 days (F); NP 8–9 days (F, M). **BEHAVIOR:** Built by female in as few as 2 days.

DICKCISSEL *Spiza americana*

HABITAT: Prairie grasslands, agricultural areas (hayfields, etc.), meadows. **LOCATION AND STRUCTURE:** Built on or near ground, in dense grasses, rank forbs, clover, alfalfa, tall weeds. Some nests are higher up, to 3.6 m. Nest is a bulky, often thick-walled cup of

Mark Peck, ON

forbs, coarse weed stems, grasses or stems, leaves. Lined with fine grasses, rootlets, occasionally hair. Loosely placed; not fastened to supporting vegetation. **EGGS:** 4; *like eggs of Lark Bunting*; pale blue, unmarked; oval to long oval; 21 x 16 mm; IP 11–13 days (F); NP 7–10 days (F). **BEHAVIOR:** Female builds in 2–4 days. Usually single-brooded. Demonstrates resource-defense polygyny, with quality territories seemingly chosen by number of nest sites rather than food supply.

MORELET'S SEEDEATER (Thraupidae)

MORELET'S SEEDEATER *Sporophila morelleti*
Rio Grande Valley of s. TX. Commonly near water, in small trees or shrubs in habitat with thick brush/ground cover in savanna, pasture, old fields, marshes possible. Often in small colonies. Nest is a small, deep cup of rootlets, fine grass stems, plant fibers, fastened in place with spider web. May appear frail, in that light can be visible through nest. 2–4 eggs; glossy, light blue or greenish blue with dark speckles; 16 x 13 mm.

ACKNOWLEDGMENTS

Since time immemorial, North America has been home to many nations of indigenous peoples. Our fieldwork for this book, and all of the research we reference here, occurred on their traditional territories. We are grateful for the countless generations of study and stewardship, continued to this day, that these communities have dedicated to the lands and wildlife of this continent.

This book is possible only because of lifetimes of tireless field research by countless individuals. We are indebted to the authors of many books on the subject of birds' nests that not only informed the text, but also guided us in the field and shaped our perspective on bird life and avian architecture. In particular they are Hal H. Harrison, Paul J. Baicich, and Colin H. Harrison, Richard Headstrom, Mike Hansell, Nicholas E. and Elsie C. Colias, and James E. and Carol G. Gould. To study nests is to study birds, and *Birds of North America* (BNA), the world's most comprehensive resource for life histories of North American birds, was invaluable to our research and the foremost resource for the species accounts in this book. Originally a collaborative effort between the Cornell Lab of Ornithology, the Academy of Natural Sciences, and the American Ornithological Society, BNA started first as an exhaustive printed series that compiled academic articles of thousands of authors into extensive species accounts, and was later converted into an ever-evolving online resource. (See Selected Bibliography for more information.) In 2020, BNA expanded to *Birds of the World* and now provides the same wealth of information for bird species globally. The efforts of countless ornithologists that dedicated much of their lives to the study and conservation of birds cannot be commended enough, and we are so grateful for their work and what they have shared with all of us.

The idea for this book was first conceived in 2013, and the project has absorbed a large part of four years of our lives. It has been a long road, and many people made this work possible—from those we've known for years to strangers who generously reached out to lend support, knowledge, and advice. We would like to recognize the ongoing support of our partners, family, and friends. Darcy Ottey, Micaela Monjello, and Rachael Nickerson backed us throughout endless writing sessions, heavy travel loads, and "retreats," where the three of us took over our respective home spaces with laptops, books, notes, old dilapidated nests, constant bantering, and long hours of work. Thanks for your feedback, prompting, love, and encouragement. Our family members and friends cheered us on through the multiyear process and provided many moments of inspiration with their continued excitement to see the finished book. They were welcome reminders that the project was valuable, exciting, and worth the grind when we sometimes lost sight of that. Special thanks to Johanna Goldfarb, Ralph Moskowitz, Marianne Moskowitz, and Rosa Levin, Mary Kiesau, the birding community of the Methow Valley, Emily Plott, Lauren Rodin, Beverly McFarland and the New Mexico "bird gang," Gary McFarland, Zora O'Neill, Patrick O'Neill, Karen Monson, Julie Nickerson and Benjamin Hall, Don Monjello, Jeff Monjello, and Katherine Sotzing.

Emily Gibson, mentioned at the front of the book, has her fingerprints all across this project. From her uncanny ability to find nests in the field (which was a huge part of the inspiration of this project to begin with) to her monumental contribution of combing through the bulk of our species accounts, she not only tightened the writing overall but found discrepancies, areas to improve, and added valuable information from her own experience in the field with many species. Along with helping us (guiding us might be a better description) on numerous field days, she joined us for some long days photographing eggs at the Western Foundation of Vertebrate Zoology, and for comparing our descriptions with nest specimens in hand. Her contributions helped make this project far more manageable, and certainly made the book better.

Our fieldwork was supported by numerous people and organizations in large and small ways. Thanks to Steve Engel, Leonard Reitsma, Audubon Project Puffin, Susan Schubel, Eric Soehren, Bill Summerour, John Trent, Stratton Island Audubon, Zeke Smith, Audubon North Carolina, Walker Golder, Lindsay Addison, Anna Parot, Dan Gusset, Jason Fidora, Dan Gardoqui, Haley Andreozzi, Robert A. Behrstock and Karen LeMay, Lori and Scott Kehoe, Travis Hamilton, Terry and Celeste Crabtree, Tim and Anne Crabtree, Tony Gallace, Linda Shearer Whiting, Dave and Felicia Feldman, Ray Robertson, and the Tolson Family. Thanks also to the

Ahousaht First Nation for granting a permit to visit and photograph nests in their traditional territory.

Many others helped us with the research and literature review and organization of the material in this book. Kristi Dranginis, of Bird Mentor.com, assisted with various aspects of research, including literature review, outreach to subject area experts, review of our draft material, and tracking down images. Matthew Young and the Macaulay Library at Cornell University assisted us with photo research. The Western Foundation of Vertebrate Zoology, René Corado, and Linnea S. Hall were invaluable in our research, providing warm welcome to an extensive collection of nests and eggs to review, as well as expertise, inspiration, and encouragement for the project. Tim Bowman, author of *Field Guide to Bird Nests and Eggs of Alaska's Coastal Tundra,* provided invaluable insight into how to treat numerous species. He also provided images of various nests and duck feathers of his own and assistance with tracking down other nest images. Additional feathers for us to photograph for the duck feathers plate were provided by Arnold and Debbie Schouten of Dry Creek Waterfowl, and the Pinola Conservancy. Thanks also to Dr. Mark Colwell, Peter Hodum, Scott Pearson, and Adam Bailey.

Thanks to the many subject area experts who reviewed parts of this manuscript and provided critical and invaluable feedback: Eric Soehren, Bill Summerour, Nicholas Czaplewski, Katherine Gura, Peter Lowther, Chris Smith, Kevin Ringleman, Carl Brown, Brett Lovelace, Kim Nelson, Kent Woodruff, Laura McDuffie, Adam Martin, George Divoky, James Tucker, Daniel Baldassarre, Heather Mathweson, Taza Schaming, David Ligon, Melody Talcott, and Stephen Hiro.

Many people made financial contributions to support our research for this book. These include Julie Nickerson, Benjamin Hall, Wyatt Harris, Elizabeth Kligge, John Brossard, Bethany Haley, Anna Kusler, Bob Ollerton, Michelle Peziol, Simon Schowanek, Bard Edrington, Chris Smith, Dorothea Sotiros, Eric Downes, Stephen Leckman, Melissa Howard, Beverly McFarland, Karen Herzenberg, Warren and M'Liss Moon, Richie, Natalie, and Ruby Rivera-Booth, Andy and Shari Franjevic, Fiona Clark, Liz Snair, Ben Hagedorn, Caitlyn Trautmann, Jonathan Goff, Larita June Rohla, Leah Houghton, Brooke Nelson, Brandon Allum, Ralph Moskowitz, Sarah Schieron, Aaldrik Pot, the McBride Family, Jakub Galczynski, Chris Carter, Kim Cabrera, Karen Monson, Jeremy Williams, Brooks Thomas, Michele Cappel, Nicholas Sharp, Nancy Price, Nicholas Czaplewski, Diane Carney, and Garth Olson.

Numerous others made in-kind contributions of various sorts. Mallory Clarke and Ralph Moskowitz each let us take over their homes for a week for writing retreats. Linda Walsh and Terrell Dixon also provided their home and support. Nate Bacon and Christina Stout and Sam Bowman lent us watercraft.

Thanks to Mark Kang-O'Higgins for his enthusiasm for the project and for generously donating the illustrations used in the text, and to Rowan Kang-O'Higgins for creating the final images of egg plates.

More than 150 photographers helped illustrate this guidebook. The following are those who generously donated the use of their images: Aaron Van Geem, Adam Martin, Alan J. Vernon, Alix d'Entremont, Amelia DuVall, Bettina Arrigoni, Bill Bumgarner, Bill Summerour, Bob Armstrong, Brendan Higgins, Bri Benvenuti, Brian McConnell, Carroll Henderson, Charles Robertson, Chris Byrd, Chris McCreedy, Chris Smith, Chris Young, Cyndi Shepherd, Dan Fontaine, Dana Visalli, Daniel Baldassarre, Dave Slager, David Pierce, David Sherer, Delta Waterfowl, Dennis Murphy, Don Gorney, Douglas Mason, Ed Hess, Earl Deickman, Elisabeth Ammon, Elizabeth Tymkiw, Eric Soehren, Gae Henry, Garth Olson, George Divoky, Gus Lane, Hannah Specht, Heidi Blankenship, Hillary Allen, Ian Tait, J. Brett Lovelace, Jacqueline Huard, Jake Scott, James Hearn, Janet Bauer, Jason Fidorra, Jason Kleinert, Jean Knowles, Jeffrey Stratford, Jens Kirkeby, Joe Kosack, Joe Smith, Joel Jorgensen, John Alstrup, John Jacobs, John Pulliam, Johnathan Mays, Jonah Evans, Justin Sweitzer, K. P. McFarland, Kate Fremlin, Katherine Hocker, Ken Collis, Kent Woodruff, Kim Cabrera, Kim Shelton, Kris Spaeth, Laura McDuffie, Laurie Pocher, Les Dewar, Libby Megna, Linda H. Godwin, Lindsey Sanders, Lois Manowitz, Marcus Reynerson, Maren Gimpel, Mark Musselman, Mary Alice Tartler, Mary Kiesau, Matthew Jung, Megan Milligan, Mia Revels, Michael A. Schroeder, Mikael Behrens, Nate Bacon, Neal Wight, Nick Czaplewski, Nick Hatch, Pat Leigh, Paul Suchanek, Peter Blancher, Peter Hodum, Phil Blair, Quinn Bailey, Richard Besser, Rick Wright, Robert Gundy, Robert Kaler, Robert Powers, Robin Diaz, Rohan Kensey, Roy Peacock, Sandra Young, Sarah Borealis, Sarah Hegg, Scott Olmstead, Scott Somershoe, Sean C. Hauser, Sigurdur H. Stefnisson, Stephanie Beh, Steve Smith, Steven Poole, Sue Hirshman, Susan Felage, Sylvia M. Robertson, Thomas Erdman, Tiffany Linbo, Tim Bowman, and Timothy Lawes.

Thanks to Anne Hawkins, for helping to make this possible, to all those at Houghton Mifflin Harcourt for providing this opportunity, and to Lisa White for the careful editing and coordination required to turn a massive manuscript with many moving parts into a book. Liz Pierson, our copy editor, honed the text with her fantastic eye for detail and incredibly helpful knowledge of bird life and natural sciences.

There are many others who were part of the broad network of support for this book that we have failed to name. Thank you for every way you helped see this project through.

Last, and so important, we owe everything to the birds themselves. We cannot imagine a world without them—the support, inspiration, and comfort their presence provides is immeasurable. It was such a gift to dedicate so much time to learning more about their lives.

As with any human endeavor, despite the many hands that supported the three of us, errors no doubt remain. These are ours alone. We are grateful that life provided us the opportunity to jump into the project, and what we have learned up to this point has been immense. Similarly immense are the insights and knowledge that we as individuals, and all of us as a species, have waiting for us in our continued study of the complex and beautiful world of bird nests.

The following short list includes books that we found essential to learning about bird nests and their contents, and that we recommend to those who wish to learn more about animal architecture, nest design, building and breeding behavior, and related topics.

Baicich, Paul J., and Colin J. O. Harrison. *Nests, Eggs, and Nestlings of North American Birds,* 2nd ed. Princeton, NJ: Princeton Univ. Press, 2005.

Collias, Nicholas E., and Elsie C. Collias. *Nest Building and Bird Behavior.* Princeton, NJ: Princeton Univ. Press, 1984.

Ehrlich, Paul, David S. Dobkin, and Darryl Wheye. *Birder's Handbook.* New York: Simon and Schuster, 1988.

Goodfellow, Peter. *Avian Architecture: How Birds Design, Engineer, and Build.* Princeton, NJ: Princeton Univ. Press, 2011.

Gould, James L., and Carol Grant Gould. *Animal Architects: Building and the Evolution of Intelligence.* New York: Basic Books, 2012.

Hansell, Mike. *Bird Nests and Construction Behaviour.* Cambridge, UK: Cambridge Univ. Press, 2000.

Harrison, Hal H. *Eastern Birds' Nests.* Boston: Houghton Mifflin, 1998.

Harrison, Hal H. *A Field Guide to Western Birds' Nests.* Boston: Houghton Mifflin, 2001.

Lovette, Irby J., and John W. Fitzpatrick, eds. *Handbook of Bird Biology,* 3rd ed. Chichester, West Sussex: John Wiley & Sons, 2016.

SELECTED
BIBLIOGRAPHY

This selected bibliography includes all of the material we referenced for the introductory text. We've also included references that were essential to many of the species accounts in this book (some of which are also listed in the preceding Recommended Reading section). Regrettably, space does not allow us to include every one of the many hundreds of authors and journal articles that informed the species accounts. For thorough bibliographical information on bird nest research, we encourage you to access *Birds of the World* (see Acknowledgments for more information). It is an extraordinary resource on the life histories of birds, and also provides distributional maps and models, sounds, images, and video. There too you can learn about the authors who compile the individual BNA species accounts, and see lists of the authors and research cited for each, and the important journals that featured them, including *The Auk, The Condor, The Wilson Journal of Ornithology* (formally *The Wilson Bulletin*), *Journal of Field Ornithology, The American Naturalist, The Canadian-Field Naturalist, Ibis, Journal of Avian Biology, Journal of Raptor Research, Journal of Wildlife Management, North American Birds,* and many others. Visit birdsoftheworld.org.

For a quick look at local bird life, allaboutbirds.org (Cornell Lab of Ornithology) and Audubon.org are also excellent resources that not only inform about birds but offer ways to get involved in counts and conservation efforts. Visit nestwatch.org (Cornell Lab of Ornithology) to learn about nest monitoring and how to contribute beneficial data. The American Ornithological Association (americanornithology.org) and the American Birding Association (aba.org) also provide publications and other resources.

Ackerman, Jennifer. *The Genius of Birds.* New York: Penguin, 2016.

Arcese, Peter, and James N. M. Smith. Effects of population density and supplemental food on reproduction in song sparrows. *Journal of Animal Ecology* 57, no. 1 (1988): 119–136.

Bailey, Ida E., Felicity Muth, Kate Morgan, et al. Birds build camouflaged nests. *Auk* 132, no. 1 (2015): 11–15.

Ballentine, Barbara, Jeremy Hyman, and Stephen Nowicki. Vocal performance influences female response to male bird song: an experimental test. *Behavioral Ecology* 15, no.1 (2004): 163–168.

Beadle, David, and Jim Rising. *Sparrows of the United States and Canada.* London: Academic Press, 2001.

Behrensmeyer, Anna K., John D. Damuth, William A. DiMichele, et al. *Terrestrial Ecosystems through Time: Evolutionary Paleoecology of Terrestrial Plants and Animals.* Chicago: University of Chicago Press, 1992.

Bell, C. P. The relationship between geographic variation in clutch size and migration pattern in the Yellow Wagtail. *Bird Study* 43, no. 3 (1996): 333–341.

Berg, Mathew L., Nienke H. Beintema, Justin A. Welbergen, et al. The functional significance of multiple nest-building in the Australian Reed Warbler *Acrocephalus australis*. *Ibis* 148, no. 3 (2006): 395–404.

Bergstrom, Peter W. Incubation temperatures of Wilson's plovers and killdeers. *Condor* 91, no. 3 (1989): 634–641.

Blackburn, Daniel G., and Howard E. Evans. Why are there no viviparous birds? *American Naturalist* 128, no. 2 (1986): 165–190.

Böhning-Gaese, Katrin, Bettina Halbe, Nicole Lemoine, et al. Factors influencing the clutch size, number of broods and annual fecundity of North American and European land birds. *Evolutionary Ecology Research* 2, no. 7 (2000): 823–839.

Böhning-Gaese, Katrin, Mark L. Taper, and James H. Brown. Are declines in North American insectivorous songbirds due to causes on the breeding range? *Conservation Biology* 7, no. 1 (1993): 76–86.

Bowman, Timothy Dale. *Field Guide to Bird Nests and Eggs of Alaska's Coastal Tundra.* Fairbanks: Univ. Alaska Fairbanks: Alaska Sea Grant College Program, 2004.

Bradbury, Jack W., and Robert M. Gibson. Leks and mate choice. Pp. 109–138 in *Mate Choice*, Patrick Bateson, ed. Cambridge, UK: Cambridge Univ. Press, 1983.

Brown, Charles R., and Mary Bomberger Brown. Ectoparasitism as a cost of coloniality in cliff swallows (*Hirundo pyrrhonota*). *Ecology* 67, no. 5 (1986): 1206–1218.

Brown, Jerram L. The evolution of diversity in avian territorial systems. *Wilson Bulletin* 76, no. 3 (1964): 160–169.

Brown, Jerram L. Territorial behavior and population regulation in birds: a review and re-evaluation. *Wilson Bulletin* 81, no. 3 (1969): 293–329.

Bruinzeel, Leo W., and Martijn Van de Pol. Site attachment of floaters predicts success in territory acquisition. *Behavioral Ecology* 15, no. 2 (2004): 290–296.

Budden, Amber E., and Janis L. Dickinson. Signals of quality and age: the information content of multiple plumage ornaments in male western bluebirds *Sialia mexicana*. *Journal of Avian Biology* 40, no. 1 (2009): 18–27.

Burley, Nancy. Sex ratio manipulation and selection for attractiveness. *Science* 211, no. 4483 (1981): 721–722.

Cardoso, Gonçalo C., Paulo Gama Mota, and Violaine Depraz. Female and male serins (*Serinus serinus*) respond differently to derived song traits. *Behavioral Ecology and Sociobiology* 61, no. 9 (2007): 1425–1436.

Carpenter, F. Lynn. Food abundance and territoriality: to defend or not to defend? *American Zoologist* 27, no. 2 (1987): 387–399.

Cassey, Phillip, Gavin H. Thomas, Steven J. Portugal, et al. Why are birds' eggs colourful? Eggshell pigments co-vary with life-history and nesting ecology among British breeding non-passerine birds. *Biological Journal of the Linnean Society* 106, no. 3 (2012): 657–672.

Charnov, Eric L., and John R. Krebs. On clutch-size and fitness. *Ibis 116*, no. 2 (1974): 217–219.

Chi, Kelly Rae. Mapping the "Big Bang" of Bird Evolution. *USA Today Magazine* 144, no. 2842 (2015): 58–60.

Clark, Larry, and J. Russell Mason. Effect of biologically active plants used as nest material and the derived benefit to starling nestlings. *Oecologia* 77, no. 2 (1988): 174–180.

Collias, Nicholas E. On the origin and evolution of nest building by passerine birds. *Condor* 99, no. 2 (1997): 253–270.

Corman, Troy E., and Cathryn Wise-Gervais, eds. *The Arizona Breeding Bird Atlas*. Albuquerque: Univ. of New Mexico Press, 2005.

Croston, R., and M. E. Hauber. The ecology of avian brood parasitism. *Nature Education Knowledge* 1, no. 3 (2010): 56

Davis, Stephen K. Nest-site selection patterns and the influence of vegetation on nest survival of mixed-grass prairie passerines. *Condor* 107, no. 3 (2005): 605–616.

Dawson, Russell D., and Gary R. Bortolotti. Carotenoid-dependent coloration of male American kestrels predicts ability to reduce parasitic infections. *Naturwissenschaften* 93, no. 12 (2006): 597–602.

Dixon, Charles. *Birds' Nests: An Introduction to the Science of Caliology*. London, UK: G. Richards, 1902.

Emlen, Stephen T., and Natalie J. Demong. Adaptive significance of synchronized breeding in a colonial bird: a new hypothesis. *Science* 188, no. 4192 (1975): 1029–1031.

Erickson, Laura, and Marie Read. *Into the Nest: Intimate Views of The Courting, Parenting, and Family Lives of Familiar Birds*. North Adams, MA: Storey Publishing, 2015.

Feduccia, Alan. Explosive evolution in tertiary birds and mammals. *Science* 267, no. 5198 (1995): 637–638.

Fogarty, Dillon T., R. Dwayne Elmore, Samuel D. Fuhlendorf, et al. Influence of olfactory and visual cover on nest site selection and nest success for grassland-nesting birds. *Ecology and Evolution* 7, no. 16 (2017): 6247–6258.

Fretwell, Stephen Dewitt. On territorial behavior and other factors influencing habitat distribution in birds. *Acta Biotheoretica* 19, no. 1 (1969): 45–52.

Gauthier, Marc, and Donald W. Thomas. Nest site selection and cost of nest building by Cliff Swallows (*Hirundo pyrrhonota*). *Canadian Journal of Zoology* 71, no. 6 (1993): 1120–1123.

Grant, Gilbert S. Avian incubation: egg temperature, nest humidity, and behavioral thermoregulation in a hot environment. *Ornithological Monographs* 30 (1982): 1–75.

Green, David J., and Elizabeth A. Krebs. Courtship feeding in Ospreys *Pandion haliaetus*: a criterion for mate assessment? *Ibis* 137, no. 1 (1995): 35–43.

Greenwood, Paul J., Paul H. Harvey, and Christopher M. Perrins. Kin selection and territoriality in birds? A test. *Animal Behaviour* 27 (1979): 645–651.

Grellet-Tinner, Gerald, Xabier Murelaga, Juan C. Larrasoaña, et al. The first occurrence in the fossil record of an aquatic avian twig-nest with Phoenicopteriformes eggs: Evolutionary implications. *PLoS ONE* 7, no. 10 (2012): 1–14.

Griggio, Matteo, Francisco Valera, Alejandro Casas, et al. Males prefer ornamented females: a field experiment of male choice in the rock sparrow. *Animal Behaviour* 69, no. 6 (2005): 1243–1250.

Grüebler, Martin U., and Beat Naef-Daenzer. Survival benefits of post-fledging care: experimental approach to a critical part of avian reproductive strategies. *Journal of Animal Ecology* 79, no. 2 (2010): 334–341.

Guillette, Lauren M., and Susan D. Healy. Nest building, the forgotten behaviour. *Current Opinion in Behavioral Sciences* 6 (2015): 90–96.

Gulson-Castillo, Eric R., Harold F. Greeney, and Benjamin G. Freeman. Coordinated misdirection: A probable anti-nest predation behavior widespread in Neotropical birds. *Wilson Journal of Ornithology* 130, no 3 (2018): 583–590.

Hahn, Thomas P. Reproductive seasonality in an opportunistic breeder, the red crossbill, *Loxia curvirostra*. *Ecology* 79, no. 7 (1998): 2365–2375.

Halkin, Sylvia L. Nest-vicinity song exchanges may coordinate biparental care of northern cardinals. *Animal Behaviour* 54, no. 1 (1997): 189–198.

Hansell, M. H. The ecological impact of animal nests and burrows. *Functional Ecology* 7, no. 1 (1993): 5–12.

Hansell, M. H. The demand for feathers as building material by woodland nesting birds. *Bird Study* 42, no. 3 (1995): 240–245.

Hansell, Mike. *Built by Animals: The Natural History of Animal Architecture*. Oxford: Oxford Univ. Press, 2007.

Hansell, Mike. *Animal Architecture*. Oxford: Oxford Univ. Press, 2005.

Harrison, Hal H. *Wood Warblers' World*. New York: Simon & Schuster, 1984.

Hartman, C. Alex, and Lewis W. Oring. Orientation and microclimate of Horned Lark nests: the importance of shade. *Condor* 105, no. 1 (2003): 158–163.

Hauber, Mark E. *The Book of Eggs: A Life-Size Guide to the Eggs of Six Hundred of the World's Bird Species.* Chicago: Univ. of Chicago Press, 2014.

Headstrom, Richard. *A Complete Field Guide to Nests in the United States.* New York: Ives Washburn, 1970.

Healy, Sue, Patrick Walsh, and Mike Hansell. Nest building by birds. *Current Biology* 18, no. 7 (2008): R271–R273.

Heath, Mykela, and Mike Hansell. Weaving techniques in two species of Icteridae, the Yellow Oriole (*Icterus nigrogularis*) and Crested Oropendola (*Psarocolius decumanus*). Pp. 144–154 in *Studies in Trinidad and Tobago Ornithology Honouring Richard French,* Floyd E. Hayes and Stanley A. Temples, eds. St. Augustine, Trinidad: Univ. of the West Indies, 2002.

Heckscher, Christopher M., Syrena M. Taylor, and Catherine C. Sun. Veery (*Catharus fuscescens*) nest architecture and the use of alien plant parts. *American Midland Naturalist* 171, no. 1 (2014): 157–164.

Hill, Geoffrey E. Plumage coloration is a sexually selected indicator of male quality. *Nature* 350, no. 6316 (1991): 337–339.

Horvath, Otto. Seasonal differences in Rufous Hummingbird nest height and their relation to nest climate. *Ecology* 45, no. 2 (1964): 235–241.

Howe, Henry F. Sex-ratio adjustment in the Common Grackle. *Science* 198, no. 4318 (1977): 744–746.

Jackson, Jerome A., and Phillip G. Burchfield. Nest-site selection of Barn Swallows in east-central Mississippi. *American Midland Naturalist* 94, no. 2 (1975): 503–509.

Janicke, Tim, Steffen Hahn, Markus S. Ritz, et al. Vocal performance reflects individual quality in a nonpasserine. *Animal Behaviour* 75, no. 1 (2008): 91–98.

Jetz, Walter, Cagan H. Sekercioglu, and Katrin Böhning-Gaese. The worldwide variation in avian clutch size across species and space. *PLoS Biology* 6, no. 12 (2008): e303.

Johnston, John P., Will J. Peach, Richard D. Gregory, et al. Survival rates of tropical and temperate passerines: a Trinidadian perspective. *American Naturalist* 150, no. 6 (1997): 771–789.

Jones, Adam G., and Nicholas L. Ratterman. Mate choice and sexual selection: what have we learned since Darwin? *Proceedings of the National Academy of Sciences* 106, supp. 1 (2009): 10001–10008.

Joyce, Frank J. Nesting success of Rufous-naped Wrens (*Campylorhynchus rufinucha*) is greater near wasp nests. *Behavioral Ecology and Sociobiology* 32, no. 2 (1993): 71–77.

Kendeigh, Samuel Charles. Parental care and its evolution in birds. *Illinois Biological Monographs* vol. 22, nos. 1–3 (1952).

Kennedy, E. Dale. Determinate and indeterminate egg-laying patterns: A review. *Condor* 93, no. 1 (1991): 106–124.

Kerlinger, Paul. *How Birds Migrate.* Mechanicsburg, PA: Stackpole, 2008.

Klett, Albert T., Harold F. Duebbert, Craig A. Faanes, et al. Techniques for Studying Nest Success of Ducks in Upland Habitats in the Prairie Pothole Region. Resource Publication 158. Washington, D.C.: U.S. Fish and Wildlife Service, 1986.

Koenig, Walter D., and Joseph Haydock. Oaks, acorns, and the geographical ecology of acorn woodpeckers. *Journal of Biogeography* 26, no. 1 (1999): 159–165.

Koenig, Walter D., and Ronald L. Mumme. *Population Ecology of the Cooperatively Breeding Acorn Woodpecker.* Princeton, NJ: Princeton Univ. Press, 1987.

Koenig, Walter D., and Justyn T. Stahl. Late summer and fall nesting in the Acorn Woodpecker and other North American terrestrial birds. *Condor* 109, no. 2 (2007): 334–350.

Kulesza, George. An analysis of clutch-size in New World passerine birds. *Ibis* 132, no. 3 (1990): 407–422.

Lack, David. The significance of clutch-size. *Ibis* 89, no. 2 (1947): 302–352.

Lack, David Lambert. *Ecological Adaptations for Breeding in Birds.* London: Methuen, 1968.

Latif, Quresh S., Sacha K. Heath, and John T. Rotenberry. How avian nest site selection responds to predation risk: testing an "adaptive peak hypothesis." *Journal of Animal Ecology* 81, no. 1 (2012): 127–138.

Lederer, Roger J. Facultative territorialty in Townsend's Solitaire (*Myadestes townsendi*). *Southwestern Naturalist* 25, no. 4 (1981): 461–467.

Lima, Steven L. Predators and the breeding bird: behavioral and reproductive flexibility under the risk of predation. *Biological Reviews* 84, no. 3 (2009): 485–513.

Lomáscolo, Silvia B., A. Carolina Monmany, Agustina Malizia, et al. Flexibility in nest-site choice and nesting success of *Turdus rufiventris* (Turdidae) in a montane forest in northwestern Argentina. *Wilson Journal of Ornithology* 122, no. 4 (2010): 674–680.

Loye, J. E., and S. P. Carroll. Ectoparasite behavior and its effects on avian nest site selection. *Annals of the Entomological Society of America* 91, no. 2 (1998): 159–163.

Lyon, Bruce E. Egg recognition and counting reduce costs of avian conspecific brood parasitism. *Nature* 422, no. 6931 (2003): 495–499.

Mainwaring, Mark C., and Ian R. Hartley. Seasonal adjustments in nest cup lining in Blue Tits *Cyanistes caeruleus*. *Ardea* 96, no. 2 (2008): 278–282.

Mainwaring, Mark C., and Ian R. Hartley. The energetic costs of nest building in birds. *Avian Biology Research* 6, no. 1 (2013): 12–17.

Mainwaring, Mark C., Ian R. Hartley, Marcel M. Lambrechts, et al. The design and function of birds' nests. *Ecology and Evolution* 4, no. 20 (2014): 3909–3928.

Marsh, Regan H., Scott A. MacDougall-Shackleton, and Thomas P. Hahn. Photorefractoriness and seasonal changes in the brain in response to changes in day length in American goldfinches (*Carduelis tristis*). *Canadian Journal of Zoology* 80, no. 12 (2002): 2100–2107.

Martin, Thomas E. Habitat and area effects on forest bird assemblages: Is nest predation an influence? *Ecology* 69, no. 1 (1988): 74–84.

Martin, Thomas E. On the advantage of being different: nest predation and the coexistence of bird species. *Proceedings of the National Academy of Sciences* 85, no. 7 (1988): 2196–2199.

Martin, Thomas E. Avian life history evolution in relation to nest sites, nest predation, and food. *Ecological Monographs* 65, no. 1 (1995): 101–127.

Martin, Thomas E., Andy J. Boyce, Karolina Fierro-Calderón, et al. Enclosed nests may provide greater thermal than nest predation benefits compared with open nests across latitudes. *Functional Ecology* 31, no. 6 (2017): 1231–1240.

Mayntz, Melissa. How Birds Claim Territory. Thespruce.com. www.thespruce.com/how-birds-claim-territory-386444; accessed February 2019.

Mayr, Gerald. *Avian Evolution: The Fossil Record of Birds and Its Paleobiological Significance*. Chichester, West Sussex: John Wiley & Sons, 2016.

McFarland, K. P., and C. C. Rimmer. Horsehair fungus, *Marasmius androsaceus*, used as nest lining by birds of subalpine spruce-fir community in the northeastern United States. *Canadian Field-Naturalist* 110, no. 3 (1996): 541.

McKinney, Frank, Scott R. Derrickson, and Pierre Mineau. Forced copulation in waterfowl. *Behaviour* 86, no. 3 (1983): 250–293.

Milam, Erika L. *Looking for a Few Good Males: Female Choice in Evolutionary Biology.* Baltimore, MD: Johns Hopkins Univ. Press, 2010.

Møller, Anders Pape. Clutch size in relation to nest size in the swallow *Hirundo rustica. Ibis* 124, no. 3 (1982): 339–343.

Moreno, Juan. Avian nests and nest-building as signals. *Avian Biology Research* 5, no. 4 (2012): 238–251.

Muth, Felicity, and Susan D. Healy. Zebra finches select nest material appropriate for a building task. *Animal Behaviour* 90 (2014): 237–244.

Nager, Ruedi G., Pat Monaghan, and David C. Houston. The cost of egg production: Increased egg production reduces future fitness in gulls. *Journal of Avian Biology* 32, no. 2 (2001): 159–166.

Naish, D. The fossil record of bird behaviour. *Journal of Zoology* 292, no. 4 (2014): 268–280.

Nice, Margaret Morse. Nesting success in altricial birds. *Auk* 74, no. 3 (1957): 305–321.

Nice, Margaret Morse. *Studies in the Life History of the Song Sparrow.* Reprint. New York: Dover Publications, 1964.

Norell, M. Origins of the feathered nest. *Natural History* 104, no. 6 (1995): 58–61.

North American Bird Conservation Initiative. *The State of North America's Birds 2016.* Ottawa, ON: *Environment and Climate Change Canada*, 2016. www.stateofthebirds.org.

Nur, Nadav. The cost of reproduction in birds: an examination of the evidence. *Ardea* 55, no. 1–2 (1988): 155–168.

Olson, Storrs L. Why so many kinds of passerine birds? *BioScience* 51, no. 4 (2001): 268–269.

Orians, Gordon H., and Mary F. Willson. Interspecific territories of birds. *Ecology* 45, no. 4 (1964): 736–745.

Perkins, Sid. Missing Link between Dinosaur Nests and Bird Nests. Sciencemag.org. www.sciencemag.org/news/2015/11/missing-link-between-dinosaur-nests-and-bird-nests; accessed January 3, 2016.

Petersen, Wayne R., and W. Roger Meservey, eds. *Massachusetts Breeding Bird Atlas*. Amherst: Univ. of Massachusetts Press, 2003.

Pettifor, R. A., C. M. Perrins, and R. H. McCleery. Individual optimization of clutch size in great tits. *Nature* 336, no. 6195 (1988): 160–162.

Price, J. Jordan, and Simon C. Griffith. Open cup nests evolved from roofed nests in the early passerines. *Proceedings of the Royal Society of London B: Biological Sciences* 284, no. 1848 (2017): 1–8.

Proctor, Noble S., and Patrick J. Lynch. *Manual of Ornithology: Avian Structure and Function*. New Haven, CT: Yale Univ. Press, 1993.

Reudink, Matthew W., Peter P. Marra, Peter T. Boag, et al. Plumage coloration predicts paternity and polygyny in the American redstart. *Animal Behaviour* 77, no. 2 (2009): 495–501.

Richardson, D. S., and T. Burke. Extra-pair paternity in relation to male age in Bullock's orioles. *Molecular Ecology* 8, no. 12 (1999): 2115–2126.

Richardson, David S., and Ginger M. Bolen. A nesting association between semi-colonial Bullock's orioles and yellow-billed magpies: evidence for the predator protection hypothesis. *Behavioral Ecology and Sociobiology* 46, no. 6 (1999): 373–380.

Ricklefs, Robert E. Density dependence, evolutionary optimization, and the diversification of avian life histories. *Condor* 102, no. 1 (2000): 9–22.

Robertson, Gregory J., Evan G. Cooch, David B. Lank, et al. Female age and egg size in the Lesser Snow Goose. *Journal of Avian Biology* (1994): 149–155.

Robinson, Scott K., and David S. Wilcove. Forest fragmentation in the temperate zone and its effects on migratory songbirds. *Bird Conservation International* 4, no. 2–3 (1994): 233–249.

Roper, James J., and Rachel R. Goldstein. A test of the Skutch hypothesis: Does activity at nests increase nest predation risk? *Journal of Avian Biology* (1997): 111–116.

Rosenberg, Kenneth V., Adriaan M. Dokter, Peter J. Blancher, et al. Decline of the North American avifauna. *Science* 366, no. 6461 (2019): 120–124.

Rowley, Ian. The use of mud in nest-building—a review of the incidence and taxonomic importance. *Ostrich* 40, no. S1 (1969): 139–148.

Ruxton, Graeme D., and Mark Broom. Intraspecific brood parasitism can increase the number of eggs that an individual lays in its own nest.

Proceedings of the Royal Society of London B: Biological Sciences 269, no. 1504 (2002): 1989–1992.

Sanders, Laura. Life: Dinosaur dads as caretakers: Fossilized bones near nests probably came from males. *Science News* 175, no. 2 (2009): 14–14.

Schaefer, V. H. Geographic variation in the placement and structure of oriole nests. *Condor* 78, no. 4 (1976): 443–448.

Schmidt, Kenneth A., Richard S. Ostfeld, and Kristina N. Smyth. Spatial heterogeneity in predator activity, nest survivorship, and nest-site selection in two forest thrushes. *Oecologia* 148, no. 1 (2006): 22–29.

Schneider, Todd M. *Breeding Bird Atlas of Georgia.* Athens: Univ. of Georgia Press, 2010.

Scott, S. David, and Casey McFarland. *Bird Feathers: A Guide to North American Species.* Mechanicsburg, PA: Stackpole, 2010.

Scott, Virgil E., Keith E. Evans, David R. Patton, et al. Cavity-Nesting Birds of North American Forests. Agriculture Handbook 511. Washington, D.C.: U.S. Dept. of Agriculture, 1977.

Semel, Brad, and Paul W. Sherman. Intraspecific parasitism and nest-site competition in Wood Ducks. *Animal Behaviour* 61, no. 4 (2001): 787–803.

Shizuka, Daizaburo, and Bruce E. Lyon. Coots use hatch order to learn to recognize and reject conspecific brood parasitic chicks. *Nature* 463, no. 7278 (2010): 223–226.

Sibley, David. *The Sibley Guide to Bird Life and Behavior.* New York: Knopf, 2009.

Skagen, Susan Knight. Asynchronous hatching and food limitation: a test of Lack's hypothesis. *Auk* 105, no. 1 (1988): 78–88.

Skutch, Alexander F. The life history of Rieffer's Hummingbird (*Amazilia tzacatl tzacatl*) in Panamá and Honduras. *Auk* 48, no. 4 (1931): 481–500.

Skutch, Alexander F. Life histories of Central American birds: families Fringillidae, Thraupidae, Icteridae, Parulidae and Coerebidae. *Pacific Coast Avifauna* no. 31. Berkeley, CA: Cooper Ornithological Society, 1954.

Skutch, Alexander F. The nest as a dormitory. *Ibis* 103, no. 1 (1961): 50–70.

Skutch, Alexander F. Clutch size, nesting success, and predation on nests of Neotropical birds, reviewed. *Ornithological Monographs* no. 36 (1985): 575–594.

Slagsvold, Tore. Clutch size variation in passerine birds: the nest predation hypothesis. *Oecologia* 54, no. 2 (1982): 159–169.

Slagsvold, Tore. Clutch size variation of birds in relation to nest predation: on the cost of reproduction. *Journal of Animal Ecology* 53, no. 3 (1984): 945–953.

Smith, N. G. Some evolutionary, ecological, and behavioural correlates of communal nesting by birds with wasps or bees. *Proceedings of the International Ornithological Congress* vol. 17, (1980): 1199–1205.

Smith, Susan M. The "underworld" in a territorial sparrow: adaptive strategy for floaters. *American Naturalist* 112, no. 985 (1978): 571–582.

Soler, Juan José, Manuel Martín-Vivaldi, Claudy Haussy, et al. Intra-and interspecific relationships between nest size and immunity. *Behavioral Ecology* 18, no. 4 (2007): 781–791.

Soler, Juan José, Anders Pape Møller, and Manuel Soler. Nest building, sexual selection and parental investment. *Evolutionary Ecology* 12, no. 4 (1998): 427–441.

Soukup, Sheryl Swartz, and Charles F. Thompson. Social mating system and reproductive success in house wrens. *Behavioral Ecology* 9, no. 1 (1998): 43–48.

Starck, J. M. Evolution of avian ontogenies. *Current Ornithology* 10 (1993): 275–366.

Starck, J. M., and R. E. Ricklefs. Patterns of development in birds: the altricial-precocial spectrum. *Journal of Ornithology* 135, no. 3 (1994): 326.

Starck, J. Matthias, and Robert E. Ricklefs, eds. *Avian Growth and Development: Evolution within the Altricial-Precocial Spectrum.* New York: Oxford Univ. Press, 1998.

Stoddard, Mary Caswell, Ee Hou Yong, Derya Akkaynak, et al. Avian egg shape: Form, function, and evolution. *Science* 356, no. 6344 (2017): 1249–1254.

Suzuki, Mamoru. *Birds' Nests of the World.* Translation ed. Camarillo, CA: Western Foundation of Vertebrate Zoology, 2010.

Tieleman, B. Irene, Hendrika J. Van Noordwijk, and Joseph B. Williams. Nest site selection in a hot desert: trade-off between microclimate and predation risk. *Condor* 110, no. 1 (2008): 116–124.

Traylor, Jr., M. A., and J. W. Fitzpatrick. A survey of the tyrant flycatchers. *Living Bird* 19 (1982): 7–50.

Trevelyan, R., P. H. Harvey, and M. D. Pagel. Metabolic rates and life histories in birds. *Functional* Ecology 4, no. 2 (1990): 135–141.

Trivers, Robert L., and Dan E. Willard. Natural selection of parental ability to vary the sex ratio of offspring. *Science* 179, no. 4068 (1973): 90–92.

Veiga, José P., and Vicente Polo. Feathers at nests are potential female signals in the spotless starling. *Biology Letters* 1, no. 3 (2005): 334–337.

Warning, Nathanial, and Lauryn Benedict. Paving the way: Multifunctional nest architecture of the Rock Wren. *Auk* 132, no. 1 (2015): 288–299.

Wesolowski, Tomasz. On the origin of parental care and the early evolution of male and female parental roles in birds. *American Naturalist* 143, no. 1 (1994): 39–58.

Wickersham, Lynn E., ed. *The Second Colorado Breeding Bird Atlas.* Denver: Colorado Bird Atlas Partnership, 2016.

Wimberger, Peter H. The use of green plant material in bird nests to avoid ectoparasites. *Auk* 101, no. 3 (1984): 615–618.

Yom-Tov, Yoram. Intraspecific nest parasitism in birds. *Biological Reviews* 55, no. 1 (1980): 93–108.

Yom-Tov, Yoram. An updated list of some comments on the occurrence of intraspecific nest parasitism in birds. *Ibis* 143, no. 1 (2001): 133–143.

Young, Bruce E., Michael Kaspari, and Thomas E. Martin. Species-specific nest selection by birds in ant-acacia trees. *Biotropica* (1990): 310–315.

Young, Bruce E. Geographic and seasonal patterns of clutch-size variation in house wrens. *Auk* 111, no. 3 (1994): 545–555.

ADDITIONAL
PHOTO CAPTIONS
AND CREDITS

PAGE i: *Two Osprey building their nest on the top of a snag.* David Moskowitz, WA

PAGES ii–iii: *A Bushtit with material for nest construction.* David Moskowitz, WA

PAGE vi: *Cedar Waxwings feeding nestlings.* David Moskowitz, WA

PAGE vii: *An Inca Dove nestling peers out of its nest.* Casey McFarland, TX

PAGE x: *Veery eggs and nest.* Matt Monjello, CT

PAGE 1: *Cedar Waxwing nestlings.* David Moskowitz, WA

PAGE 5: *A Pileated Woodpecker at its nest cavity in a snag.* David Moskowitz, WA

PAGE 6: *A Red-tailed Hawk at its nest high in the crotch of a cottonwood tree.* David Moskowitz, WA

PAGE 7: *Royal Terns in a nest colony.* Matt Monjello, NC

PAGE 16: *Costa's Hummingbird nestlings.* Scott Olmstead, AZ

PAGE 17: *A Western Kingbird bringing food for nestlings.* David Moskowitz, WA

PAGE 38: *A Bullock's Oriole constructs a nest in an aspen tree.* David Moskowitz, WA

PAGE 39: *Mourning Doves on their nest.* Matthew Monjello, AZ

PAGES 56–57: *Cliff Swallows at their nests.* David Moskowitz, ID

PAGES 448–449: *A Cooper's Hawk feeding nestlings.* Jack Bartholmai, WI

PAGE 450: *Varied thrush nestlings.* David Moskowitz, WA

PAGE 451: *Common Loons on their nest.* Matthew Monjello, Maine

PAGE 456: *A Pelagic Cormorant on its nest.* Alan J. Vernon, CA

PAGE 457: *A Common Raven calling out by its cliffside nest.* David Moskowitz, WA

PAGE 458: *A female Yellow-headed Blackbird with insects in her bill for nestlings.* David Moskowitz, WA

PAGE 459: *A Pacific Wren with insects destined for nestlings.* David Moskowitz, WA

PAGE 470: *An abandoned nest in early winter.* Casey McFarland, CO

PAGE 471: *A Pygmy Owl feeding its fledgling.* David Moskowitz, WA

PAGE 472: *Cedar Waxwing nestlings.* David Moskowitz, WA

PAGE 473: *A Pygmy Owl in nest cavity.* David Moskowitz, WA

INDEX

Page references in *italics* refer to text graphics.

A Black-capped Chickadee excavating a nest in a rotten snag. David Moskowitz, WA

An Eastern Kingbird nest with a recently laid egg. David Moskowitz, WA

A Western Meadowlark sings in the evening light. David Moskowitz, WA

A Pacific Wren peers out from its nest. David Moskowitz, WA

A pair of Harlequin Ducks on a freshwater mountain stream during breeding season.
David Moskowitz, WA

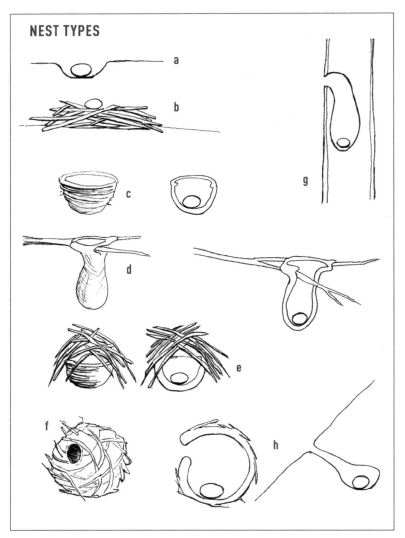

NEST TYPES

The basic nest types, each of which may have numerous variations: (a) scrape nest (p. 49); (b) platform nest (p. 49); (c) cup nest (p. 51); (d) pendulous or pensile cup nest (p. 51); (e) domed nest (p. 51); (f) globular nest (p. 51); (g) cavity nest (p. 51); (h) burrow nest (p. 52).

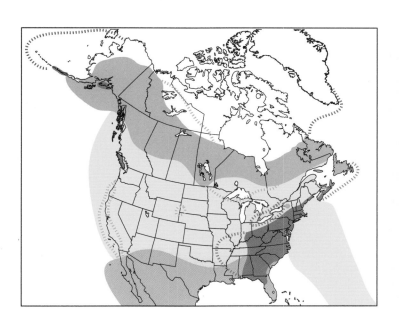

LEGEND TO RANGE MAPS

	RED: summer range
	BLUE: winter range
	PURPLE: year-round range
	YELLOW: traditional migration range

RED DASH LINE: approximate limits of summer range and/or postbreeding range

BLUE DASH LINE: approximate limits of irregular winter range

PURPLE DASH LINE: approximate limits of year-round range

YELLOW DASH LINE: approximate limits of migration range